Springer Complexity

Springer Complexity is an interdisciplinary program publishing the best research and academic-level teaching on both fundamental and applied aspects of complex systems—cutting across all traditional disciplines of the natural and life sciences, engineering, economics, medicine, neuroscience, social and computer science.

Complex Systems are systems that comprise many interacting parts with the ability to generate a new quality of macroscopic collective behavior the manifestations of which are the spontaneous formation of distinctive temporal, spatial or functional structures. Models of such systems can be successfully mapped onto quite diverse "real-life" situations like the climate, the coherent emission of light from lasers, chemical reaction-diffusion systems, biological cellular networks, the dynamics of stock markets and of the internet, earthquake statistics and prediction, freeway traffic, the human brain, or the formation of opinions in social systems, to name just some of the popular applications.

Although their scope and methodologies overlap somewhat, one can distinguish the following main concepts and tools: self-organization, nonlinear dynamics, synergetics, turbulence, dynamical systems, catastrophes, instabilities, stochastic processes, chaos, graphs and networks, cellular automata, adaptive systems, genetic algorithms and computational intelligence.

The three major book publication platforms of the Springer Complexity program are the monograph series "Understanding Complex Systems" focusing on the various applications of complexity, the "Springer Series in Synergetics", which is devoted to the quantitative theoretical and methodological foundations, and the "Springer Briefs in Complexity" which are concise and topical working reports, case studies, surveys, essays and lecture notes of relevance to the field. In addition to the books in these two core series, the program also incorporates individual titles ranging from textbooks to major reference works.

Indexed by SCOPUS, INSPEC, zbMATH, SCImago.

Series Editors

Understanding Complex Systems

Founding Editor: Scott Kelso

More information about this series at https://link.springer.com/bookseries/5394

Foreword

Synergetics is a rather strange scientific endeavour. It doesn't belong to the traditional scientific disciplines, but has ties to many of them. It doesn't deal with the large-scale phenomena of cosmic dimensions nor with the micro world of elementary particles/quantum fields. Rather, Synergetics is interested in phenomena that are mostly accessible to our senses, phenomena that we experience in our daily life that—quite often—we take for granted.

The world surrounding us abounds of all sorts of structures. A number of them are man-made, but many have come into existence without human help—they have come into existence by self-organization. Here many parts, elements,... cooperate to form highly ordered structures or to perform highly ordered actions. Synergetics aims at unearthing general principles/mechanisms that underlie self-organisation irrespective of the nature of the individual parts. An attempt to deal with this problem may have seemed absurd in view of the different nature of parts. But such principles have been found as substantiated by numerous examples.

One result is an important insight into the "mechanisms" of self-organization: It doesn't work by direct interventions (e.g., by putting atoms together "by hand", as it is done by force electron microscopy) but by suitable external conditions.

Actually, this book by Peter J. Plath, a pioneer of Synergetics since its early days, gives us a deep insight into a variety of processes. To unearth general principles across different disciplines requires, of course, a "look over the fence". At the early days of the Synergetics' enterprise, the situation in science was well characterised by my late friend, the eminent Russian scientist Juri Klimontovich who compared scientists with miners working in different mines, digging deeper and deeper, and having no contact among each other. He was a strong supporter of the idea to bring the different scientific disciplines together. Synergetics provides a common ground for such a joint enterprise—the deep problem of self-organization.

This book presents an excellent selection of characteristic phenomena of the formation of spatial, temporal and spatio-temporal patterns mainly in chemical systems, and sheds light on the history of their generation and discovery. The phenomena dealt with are Liesegang rings, Runge pictures, different kinds of dissipative structures/Turing patterns and fractal structures, to mention a few. Also, structure formation in social systems is considered. The authors include phenomena that are rarely studied such as the dynamics of beer foam.

I am intrigued by the discussion on “creativity”. The authors don’t share the general belief that only creative acts by individuals play the decisive role nor the joint action of a group, but what counts is the amplification of the creative idea of an individual by a group.

Nevertheless, the creative act itself remains an enigma: who would think of a snake biting into its own tail (Kekulé). And who would have thought that the central concept of thermodynamics, namely entropy, is no more essential for self-organization processes? Actually, in physical, chemical and biological systems, it is the power/matter input into a system.

I am sure this book will find a broad audience of researchers and students in synergetics.

Sindelfingen, Germany
November 2021

Hermann Haken

Peter J. Plath · Ernst-Christoph Haß ·
Hartmut Linde

Imagery Synergetics

Science of Cooperation

Springer

Peter J. Plath
Lychen, Germany

Ernst-Christoph Haß
Seddiner See, Germany

Hartmut Linde
Berlin, Germany

ISSN 1860-0832 ISSN 1860-0840 (electronic)
Understanding Complex Systems
ISBN 978-3-030-95605-9 ISBN 978-3-030-95607-3 (eBook)
https://doi.org/10.1007/978-3-030-95607-3

This Springer imprint is published by the registered company Springer Nature Switzerland AG
The registered company address is: Gewerbestrasse 11, 6330 Cham, Switzerland

Dedicated to the authors' wives
Elisabeth Swart
Claudia Haß
Gudrun Linde

Foreword

Synergetics is a rather strange scientific endeavour. It doesn't belong to the traditional scientific disciplines, but has ties to many of them. It doesn't deal with the large-scale phenomena of cosmic dimensions nor with the micro world of elementary particles/quantum fields. Rather, Synergetics is interested in phenomena that are mostly accessible to our senses, phenomena that we experience in our daily life that—quite often—we take for granted.

The world surrounding us abounds of all sorts of structures. A number of them are man-made, but many have come into existence without human help—they have come into existence by self-organization. Here many parts, elements,... cooperate to form highly ordered structures or to perform highly ordered actions. Synergetics aims at unearthing general principles/mechanisms that underlie self-organisation irrespective of the nature of the individual parts. An attempt to deal with this problem may have seemed absurd in view of the different nature of parts. But such principles have been found as substantiated by numerous examples.

One result is an important insight into the "mechanisms" of self-organization: It doesn't work by direct interventions (e.g., by putting atoms together "by hand", as it is done by force electron microscopy) but by suitable external conditions.

Actually, this book by Peter J. Plath, a pioneer of Synergetics since its early days, gives us a deep insight into a variety of processes. To unearth general principles across different disciplines requires, of course, a "look over the fence". At the early days of the Synergetics' enterprise, the situation in science was well characterised by my late friend, the eminent Russian scientist Juri Klimontovich who compared scientists with miners working in different mines, digging deeper and deeper, and having no contact among each other. He was a strong supporter of the idea to bring the different scientific disciplines together. Synergetics provides a common ground for such a joint enterprise—the deep problem of self-organization.

This book presents an excellent selection of characteristic phenomena of the formation of spatial, temporal and spatio-temporal patterns mainly in chemical systems, and sheds light on the history of their generation and discovery. The phenomena dealt with are Liesegang rings, Runge pictures, different kinds of dissipative structures/Turing patterns and fractal structures, to mention a few. Also, structure formation in social systems is considered. The authors include phenomena that are rarely studied such as the dynamics of beer foam.

I am intrigued by the discussion on "creativity". The authors don't share the general belief that only creative acts by individuals play the decisive role nor the joint action of a group, but what counts is the amplification of the creative idea of an individual by a group.

Nevertheless, the creative act itself remains an enigma: who would think of a snake biting into its own tail (Kekulé). And who would have thought that the central concept of thermodynamics, namely entropy, is no more essential for self-organization processes? Actually, in physical, chemical and biological systems, it is the power/matter input into a system.

I am sure this book will find a broad audience of researchers and students in synergetics.

Sindelfingen, Germany
November 2021

Hermann Haken

Preface

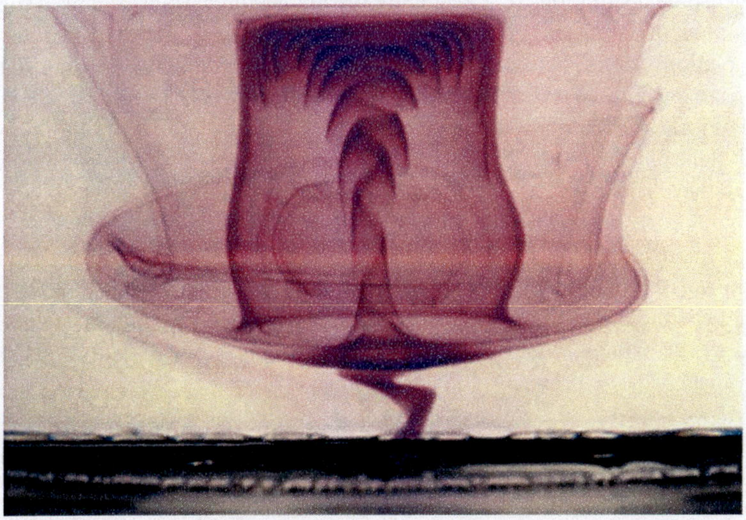

Flow behavior below a rotating disk without chemicals being involved (*Photo* P. J. Plath and P. Dembowski)

When I started in Bremen in 1973, the new university was renowned for its interdisciplinary project-based courses called "project studying". All my colleagues accepted these constraints, and we developed a corresponding curriculum for studying chemistry. My first research topics in this new position were the logic of chemical formulae and zeolite chemistry. During this time, we have organized project-orientated student research courses for both undergraduate and graduate students. This was a very new way of studying chemistry. The Roskilde University in Denmark was one model for us. Together with my colleague Nils I. Jaeger, we started our first student research project in synergetics with electrochemical investigations of the dissolution of iron and with the catalytic oxidation of methanol using Pd-supported catalysts. Especially this part of our student project became 1977 the starting point of a long and fruitful research period for both of us.

This work, in conjunction with very engaged students, convinced me that it became necessary to develop a special lecture course for chemists on synergetics in order to continue research and teaching on the high level which we had now achieved. I have been teaching mathematics for chemists, mathematical and physical chemistry for 10 years, but I had never taught synergetics before. This opportunity opened up to me for the first time in the winter semester of 1981/1982.

In the fall of 1981, I was invited to Yale University by the theoretical chemist Oktay Sinanoğlu and by Erwin Hiebert, professor of the History of Science at Harvard University. I was very much surprised to receive these invitations since I had never met these colleagues before. I remember very well the famous guest house of Harvard University where I stayed for the time of our discussions. Before returning to Bremen, I spent some days in Cambridge, Massachusetts, sitting in a coffee shop preparing my first lectures on chemical synergetics.

To get some inspirations for developing my first lectures in synergetics, I took the famous book of Werner Ebeling and Rainer Feistel "Physik der Selbstorganisation und Evolution" [1]. They described the development of the universe after the "big bang" from the point of self-organisation. For the creation of the geological as well as the biological structure of the earth, they used successfully their idea of the "photonic mill". But the physical way in which they argued was hard to understand for chemists, although they did not use too many mathematical formulae in the beginning of their book.

I was impressed by this way of thinking. However, working with my students, I recognized very soon that they could not simply translate the physical language into their chemical way of thinking. How could they understand that the atoms were created by the expansion of the universe by self-organisation? I had to learn very quickly that I had to express myself in the chemical language when talking to students of chemistry. I had to use the ideas they were familiar with, so to speak, like chemical formulae and chemical kinetics. Moreover, I had to illustrate all my statements with meaningful pictures, graphs and photos if I wanted to gain their attention. It was a long and asymptotic way to understand these problems and to create suitable pictures.

I have to thank all my students and co-workers for their patience and their efforts to understand my attempts to explain to them the new and amazing world of synergetics. Now, that I have not been teaching at the university for many years, the distance to the daily problems that were current at that time has grown and all what remains are the many inspiring ideas which we pursued in the projects or later research internships and one huge abundance of beautiful photographs and project reports. They testify to the enthusiasm and wealth of ideas with which student research is directly connected. It is important to me to revive this with a few selected contributions in this book.

In Ernst-Christoph Haß, I was lucky enough to find a congenial partner for the publication of this book, with whom I have been scientifically and amicably connected for more than 50 years since his diploma thesis.

From time to time, I heard something from W. Ebeling, H. Engel and L. Kuhnert about the very interesting and idiosyncratic work—e.g., about the light-sensitive

Belousov-Zhabotinsky reaction—by H. Linde from the Central Institute for Physical Chemistry of the Academy of Sciences of the GDR. When the opportunity arose, I invited him to speak about his recent work at one of my winter seminars on the Zeinisjoch. At that time, he was talking about his trip to Egyptian mountain desert, where he met Bedouins who showed remarkable sculptured stones. He was already retired. In his old age, he agreed with me that he would write nicely illustrated articles for this book from his large pool of unconventional works. His material given to me was processed by us for this purpose.

While writing the last chapters of this book, the question arose again, "What is synergetics?", to which Hermann Haken answered so precisely in the foreword of his book "Synergetics—an Introduction"[2]:

> What we investigate is the joint action of many subsystems (mostly of the same or of few different kinds) so as to produce structure and functioning on a macroscopic scale. On the other hand, many different disciplines cooperate here to find general principles governing self-organizing systems.

That is the guiding principle under which this book is written as well.

The idea for this book came up shortly after my retirement from the university. But it took me a long time to get around to implementing it. The book "Contributions to the History of Synergetics" [3, 4], which I wrote together with Hermann Haken, Werner Ebeling and Yuri M. Romanovsky, initially required my full attention. And as time went on, it became more and more difficult for me to immerse myself in the old subjects and experiments in such a way that I could think and write about them in new ways. The original texts had to be rethought and revised, formulae recalculated and numerous images recreated.

To my great pleasure, I was supported by former companions from long ago, such as Prof. Wolfgang Lefèvre with valuable hints on Jean.-B. Lamarck and the antiquarian bookseller Cornelia Albrecht from Lychen, who provided me with valuable old books for many chapters. Professor Stefan Müller helped me with critical hints to the chapter on Runge images and Dr. Jan K. Plath provided us with the program MeVisLab 3.4 for test purposes and used it to develop cellular automata to simulate the patterns on sea shells. We are indebted to all of them!

I would like to thank Springer Verlag, especially Dr. Thomas Ditzinger and the staff members Holger Schaepe and Rajangam Ramamoorthy, also on behalf of Dr. Ernst-Christoph Haß, many times for their understanding and almost unlimited patience and help with the completion of this book.

Flow behavior just beneath a rotating disc electrode while chemical reactions are involved (left: flow pattern; right: resulting etching pattern of the electrode disc): dissolution of iron in concentrated $FeCl_3$ solution [5]. (*Photo* P. J. Plath and M. Baune)

Lychen, Germany
September 2021

Peter J. Plath

References

1. Ebeling, W., Feistel, R.: Physik der Selbstorganisation und Evolution, Akademie-Verlag Berlin (1982)
2. Haken, H.: Synergetics – An Introduction, Second Enlarged Edition, Springer-Verlag Berlin, Heidelberg New York, (1978) p. VIII, Preface to the first edition (1976)
3. Haken, H., Plath, P.J., Ebeling, W., Romanovsky, Y.M.: Beiträge zur Geschichte der Synergetik – Allgemeine Prinzipien der Selbstorganisation in Natur und Gesellschaft, Springer Spektrum, Springer Fachmedien Wiesbaden (2016)
4. Haken, H., Plath, P.J., Ebeling, W., Romanovsky, Y.M.: Beiträge zur Geschichte der Synergetik – Allgemeine Prinzipien der Selbstorganisation in Natur und Gesellschaft, translated into Russian, Moscow (2017)
5. Baune, M., Plath, P.J.: Chemically induced Hydrodynamic Pattern Formation: Slowly Rotating Disk Electrode under Dissolving Conditions and Genesis of Spatial Bifurcation, Bifurcation and Chaos, Vol. 12, 10 (2002) 2209–2217

Contents

Chapter 1
Images from the History of Synergetics

Peter J. Plath

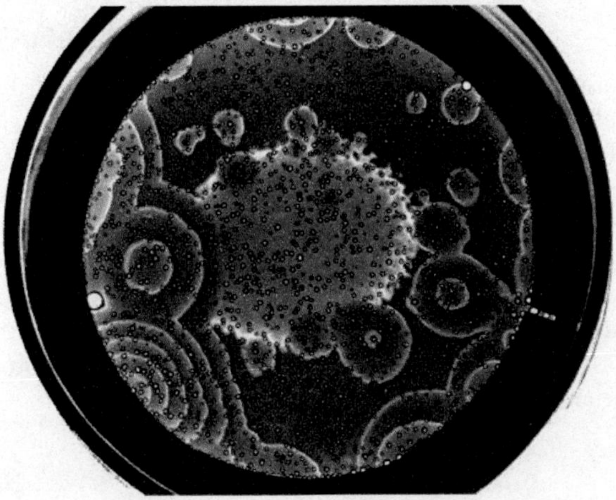

Fig. 1.1 Waves and phase waves at the Belousov–Shabotinsky reaction (BZR) in a Petri dish. Both are purely chemical waves. They move with quite different velocities and shape of their fronts (photo: P. J. Plath). With the discovery of oscillations and waves in the BZR, the ideas of synergetics by Hermann Haken became popular in chemistry too

1.1 Introduction

The number of publications dealing with the basic idea of synergetics and its application is enormous. There are also a large number of publications devoted to the work of Hermann Haken, the father of synergetics, as well as the history of synergetics [1]. In this post, I will not deal with the well-known and very illustrative examples of synergetics, such as limit cycle behaviour, BZ-reaction, scroll waves, Feigenbaum

P. J. Plath et al., *Imagery Synergetics*, Understanding Complex Systems,
https://doi.org/10.1007/978-3-030-95607-3_1

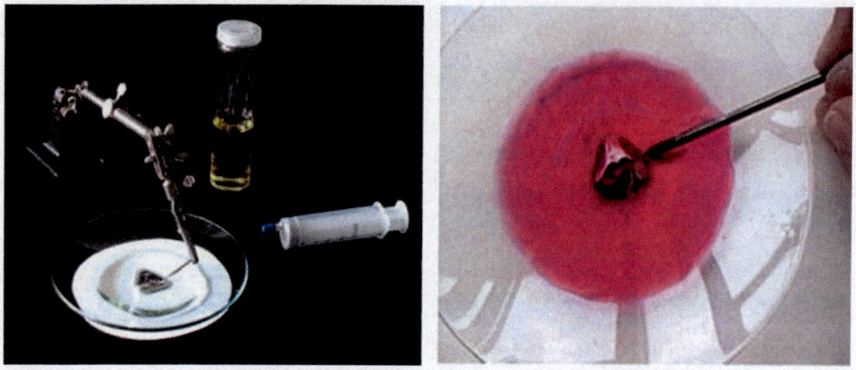

Fig. 1.2 **Left** Experimental setup for executing a pulsating mercury heart in nitric acid solution. The mercury drop starts to beat if it is touched by an iron wire ([3] with friendly permission); **right** Excerpt from the film "Das Quecksilberherz" by "Prof. Blume's Tip of the Month March (2007)" (with friendly permission of R. Blume) [4]

scenario, reconstruction of α-waves, Rayleigh-Bénard convection, Rössler's chaos attractor, Mandelbrot set, etc. (see for illustration Fig. 1.1).

I will primarily deal hopefully with such phenomena of synergetics which, at least today, should not always be familiar to those interested in synergetics.

For example, let me refer to the remark of Ferdinand Runge in 1829 on the "strange behavior of mercury in contact with nitric acid and iron" [2]. Going back to the prehistory of synergetics, A. Volta (1800), W. Henry (1800) and R. Ritter (1802) had investigated the beating mercury heart [3, 4] first (Fig. 1.2). At this time the pattern formation of the mechanically pulsating mercury drops in nitric acid created disturbing images indeed [2].

1.2 Historical Experiments

For very long time nobody had been interested in these experiments. However when synergetics saw the light of day one needed illustrative experiments to explain the new scientific vision pupils and students. It was the merit of Möllenkamp et al. [5] to rediscover the beating mercury heart and to repeat the historical experiments using modern equipment. J. Berkemeier und H. G. Purwins have made this experiment on the internet in 2011 [3].

To my knowledge the beating mercury heart was the first known system in which a chemical reaction creates not only an electrical oscillation but a mechanically oscillation with different geometric patterns of the mercury drop as well.

Later on, when F. Runge worked as chemist and entrepreneur near Oranienburg near Berlin he was interested in dyeing of fabrics and paper with the new Alizarin staining and tar colors. Correlated to this work he investigated the spreading of

Fig. 1.3 Self organized pattern formation of Berliner Blau (Turnbulls Blue) and insoluble Turnbulls Blue in filter paper (originally produced by Lothar Kuhnert (1985); a present to the author 1988 at the occasion of the Wartburg conference "Dynamical Networks"). F. R. Runge selected a similar picture as his personal "chemical coat of arms"

chemically reacting droplets in filter paper. In our days this is well known by the term "Runge Pictures" (Fig. 1.3). He published two very exiting books, full of a variety of examples of such reactions [6, 7].

F. Runge was very much fascinated by the self organization of these reactions:

> "This picture above is my chemical coat of arms (Wappen). It is not artificially produced but it emerged naturally". … "I name the creative force of these pictures the 'Bildungstrieb der Stoffe' (impulsion of formation of substances)." [6]

Lothar Kuhnert and U. Niedersen from the Academy of Science GDR brought back this long-lost knowledge into consciousness again with the rise of synergetics in 1987 [8].

$$\begin{aligned}
&\text{Gelbes Blutlaugensalz yields Lösliches Berliner Blau (Turnbulls Blau)}\\
&Fe^{3+} + K_4\left[Fe^{II}(CN)_6\right] \rightarrow 3K^+ + K\left[Fe^{III}Fe^{II}(CN)_6\right]\\
&\text{Rotes Blutlaugensalz yields Lösliches Berliner Blau (Turnbulls Blau)}\\
&Fe^{2+} + K_3\left[Fe^{III}(CN)_6\right] \rightarrow 2K^+ + K\left[Fe^{III}Fe^{II}(CN)_6\right]\\
&\text{Excess of } Fe^{3+} \text{ or } Fe^{2+} \text{ yields Unlösliches Berliner Blau (insoluble Turnbulls Blue)}\\
&Fe^{III}\left[Fe^{III}Fe^{II}(CN)_6\right]_3 \cdot 14 - 16H_2O
\end{aligned} \tag{1.1}$$

The early work of Wetzlar [9] and Fechner [10] on the oscillations of discharging electrochemical cells or accumulators respectively have been often cited in electrochemical review articles. At those times chemistry and physics were not so strongly divided as in our days. So, it is not so surprising that investigations on oscillating discharge phenomena were not restricted to chemical batteries of electrochemical cells but extended on general phenomena of electrical discharge processes. In the second part of the nineteenth century physicists like Helmholtz, Thomson, Kirchhoff, Hertz, and A. J. von Oettingen (1836–1920) were very much interested in the oscillating discharges of electric capacitors like batteries of Leyden bottles (Fig. 1.4). In his magister thesis (1862) A. J. von Oettingen from Dorpat (Dorpat: in our days Tartu in Estland) investigated very carefully the oscillating behaviour of these discharge processes [11]. Referring to his thesis he published his investigations on the interferences of oscillating electric discharges of Leidens batteries in 1888 and 1890 [12, 13]:

> Schon seit 27 Jahren kennt man den oscillatorischen Charakter der Batterieentladung und niemand zweifelt daran, dass auch zwei solcher oscillatorischer Entladungen in ein und demselben Drahte sich superponieren können….

Chemical, electrical and thermodynamic phenomena were fascinating research topics and they formed the basis for the technical revolution in the nineteenth century. The dynamics of those processes, especially oscillating phenomena were widely discussed in the society. So, it is not surprising that the economists K. Marx and F. Engels looked for such phenomena in industrial and finance processes too [14].

> Es verhält sich mit den industriellen Zyklen so, daß derselbe Kreislauf, nachdem der erste Anstoß einmal gegeben, sich periodisch fortsetzen muß."… "Die akute Form des periodischen Prozesses mit ihrem bisherigen zehnjährigen Zyklus scheint in eine mehr chronische, länger gezogene, sich auf die verschiedenen Industrieländer verschiedenzeitig verteilende Abwechslung von relativ kurzer, matter Geschäftsbesserung mit relativ langen entscheidungslosem Druck gewichen zu sein.

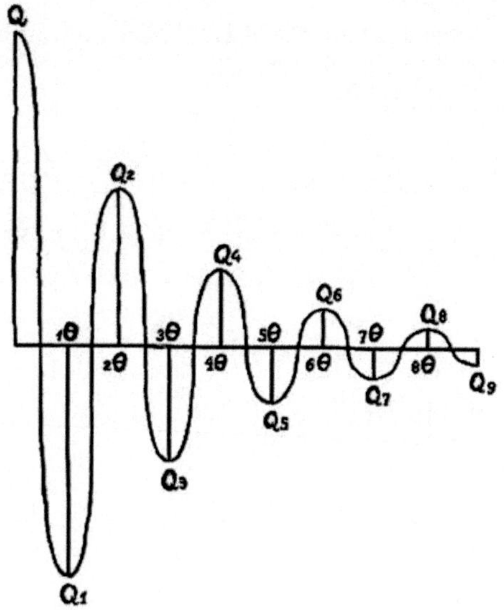

Fig. 1.4 A. J. von Oettingen: Doctor Thesis 1862: Oscillatory discharge of Leiden bottles 1862 [11]

Vielleicht aber handelt es sich nur um eine Ausdehnung der Dauer des Zyklus des Welthandels, 1815–1847 lassen sich annähernd fünfjährige Zyklen nachweisen; von 1847 – 67 ist der Zyklus entschieden zehnjährig; sollten wir uns in der Vorbereitungsperiode eines neuen Weltkrachs von unerhörter Vehemenz befinden?...

Translation:

With industrial cycles, the same cycle has to continue periodically after the first impetus has been given. ... The acute form of the periodic process with its previous ten-year cycle seems to be a more chronic, longer drawn period to have given way to the different industrialized countries, alternating between a relatively short, weak business improvement with a relatively long, undecided pressure. ...

But perhaps it is only an extension of the duration of the cycle of world trade, 1815 - 1847 almost five-year cycles can be demonstrated; from 1847-67 the cycle was decidedly ten years long; should we be in the preparatory period of a new world crash of unheard of vehemence?...

Economists are engaged in the study of these cyclic processes until today. C. Suter investigated the indebtedness of the third world for example [15] (see Fig. 1.5 left).

For given actual reason, i.e. the indebtedness of the third world [15, 16], it should be remembered that the idea of social classes has been developed first by Karl Marx and Friedrich Engels. In his book "Die Abstiegsgesellschaft" ("The Relegation Society") O. Nachtwey [16] poses the politically incorrect question in 2016:

Kehren mit der Abstiegsgesellschaft die *soziale* Frage und der damit verbundene (Klassen-) Konflikt zurück?

Do the relegated society return the social question and the (class) conflict associated with it?

And he answered:

Um einen Teil der Antwort vorwegzunehmen: Ja, die soziale Frage kehrt zurück, aber nicht in der Form, in der man sie früher gekannt hatte. ... Neue kollektive Akteure entstehen erst

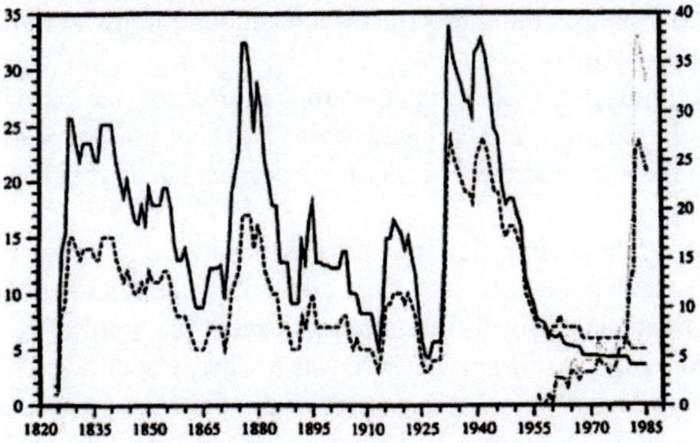

Fig. 1.5 **Left** Suter Christian, (Neuchâtel), "Indebtedness of the third world". Frankfurt am Main, Anton Hain [15] (with friendly permission of Ch. Suter); **right** Friedrich Engels (1820–1895) http://gutenberg.spiegel.de/autor/friedrich-engels-147

> in größeren, über mehrere Episoden verlaufenden Konflikten, in denen sich gemeinsame Praktiken und Deutungen entwickeln.

> To anticipate part of the answer: yes, the social question is returning, but not in the form in which it was known before. … New collective actors only emerge in larger, multi-episode conflicts in which common practices and interpretations develop.

However, the politically incorrect nature of the question does not mean that the question is scientifically not correct.

The idea of K. Marx and F. Engels involves that the class formation as well as their decay depends on the size of the inequality of income or wealth respectively. Classes have been defined in their book "Die Deutsche Ideologie" (written: 1845–1846) by the cooperative actions of individuals [17] which indeed means social pattern formation:

> Die einzelnen Individuen bilden nur insofern eine Klasse, als sie einen gemeinsamen Kampf gegen eine andere Klasse zu führen haben; im übrigen stehen sie einander selbst in der Konkurrenz wieder feindlich gegenüber

> Individuals only form a class insofar as they have to fight together against another class; otherwise they are hostile to each other even in the competition. (translated by the author)

They also formulate in the next sentence for the first time the principle of cyclic causality where the classes act as folders.

> Auf der anderen Seite verselbständigt sich die Klasse wieder gegen die Individuen, so daß diese ihre Lebensbedingungen prädestiniert vorfinden, von der Klasse ihrer Lebensstellung und damit ihre persönliche Entwicklung angewiesen bekommen, unter sie subsumiert werden.

> On the other hand, the class becomes independent again against the individuals, so that they find their living conditions predestined, i.e. they are instructed by the class of their position in life and thus their personal development, is subsumed under them. (translated by the author)

At the end of the nineteenth century two chemists R. Liesegang (1869–1946) and Wi. Ostwald (1853–1932) emerged with quite fascinating works. R. Liesegang discovered periodic patterns during executing precipitation reactions with inorganic substances in gels 1896 [18] (Fig. 1.6).

Ostwald [20] investigated carefully the oscillating gas production during the dissolution of the new produced chromium in mineral acids (Fig. 1.7). For this purpose he developed the first analog recorder and the first thermostat which have ever been produced.

Let me emphasize that none of these fantastic discoveries in the nineteenth century resulted from main stream research at the time they were created. In this sense these discoveries were industrially and scientifically meaningless except the work of K. Marx and F. Engels which got political relevance. There was no theory at all, which could explain these surprising observations. Moreover, there did not exist practical applications at those times.

This situation changed at first in 1910 by the work of A. Lotka (1880–1949) [21], since A. Lotka in 1925 and one year later V. Volterra (1860–1940) applied these differential equations to biological problem of the robber booty type:

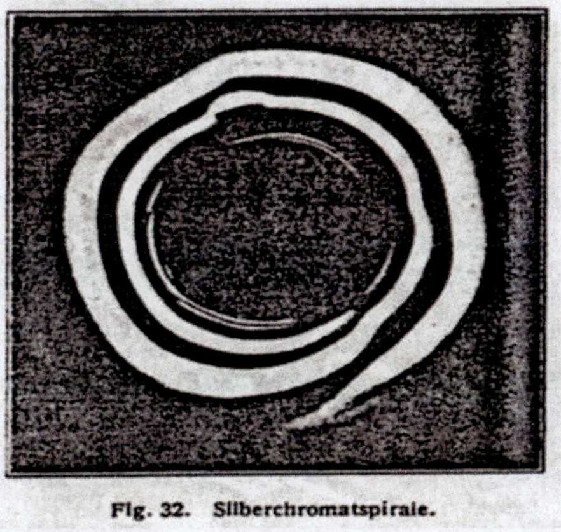

Fig. 1.6 Rhythmic precipitation reactions in gels: a spiral of silver chromate [19]

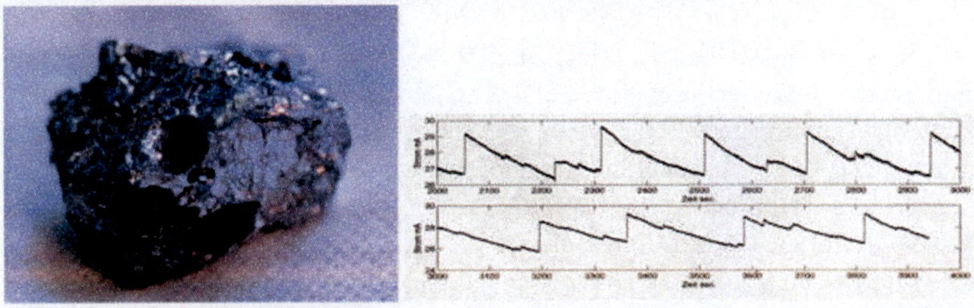

Fig. 1.7 **a** Piece of the original chromium which has been used by Wi. Ostwald. **b** Galvanostatic current oscillation with just this piece of chromium during its dissolution in mineral acids (executed by Th. Rabbow and P. Plath in 2002)

$$
\begin{aligned}
&\text{chemical equations}\\
&\mathrm{A} + \mathrm{X_1} \xrightarrow{\mathrm{k_1}} 2\mathrm{X_1}\\
&\mathrm{X_1} + \mathrm{X_2} \xrightarrow{\mathrm{k_2}} 2\mathrm{X_2}\\
&\mathrm{X_2} \xrightarrow{\mathrm{k_3}} \mathrm{F}\\
&\mathrm{A} \rightarrow \mathrm{F}
\end{aligned}
\tag{1.2}
$$

$$
\begin{aligned}
&\text{kinetic equations}\\
&\frac{dc_1}{dt} = k_1 c_a c_1 - k_2 c_1 c_2\\
&\frac{dc_2}{dt} = k_2 c_1 c_2 - k_3 c_2
\end{aligned}
\tag{1.3}
$$

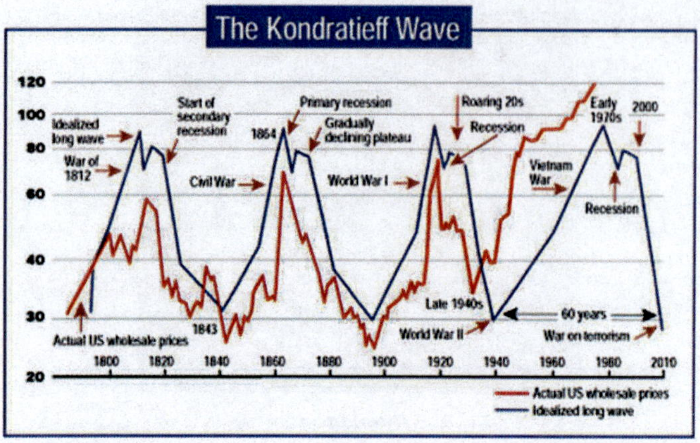

Fig. 1.8 The Kondratieff wave: wholesale prices as a function of time/years. Peaks and troughs are associated with major political or cultural events (quantumpranx.wordpress.com) [23]

At the same time when A. Lotka and V. Volterra worked on the population dynamics, the Soviet economist Nicolai Kondratieff (Kondratjew) (*1892–†1938 (executed), rehabilitated 1987) developed his idea of the cyclic development of the economy [22]. After empirical investigations of two long waves he predicted right the third wave at the end of the twenties of the twentieth century (stock market crash, Black Friday, Great Depression). His ideas were taken up by J. Schumpeter (1883–1950) when he took the lecture on mathematical economic theory at Harvard in 1932. Schumpeter called the long waves of economic life: Kondratieff cycles [23] (see Fig. 1.8).

In the fortieth and early fiftieth of the twentieth century two major steps in the prehistory of synergetics have been made. Both of them resulted from pure basic research:

(a) The discovery of electro-chemical waves 1941 by the famous electrochemist K. F. Bonhoeffer (1899–1957) [24]. Bonhoeffer was familiar with the work of Lotka which is documented in a series of his publications on the oscillations of the dissolution of iron in nitric acid. He described his observed electrochemical oscillations by the zero isoclines of the well known system of differential equations (1.4) (Bonhoeffer K. F.). This way he used the idea of limit cycle behaviour to describe periodic chemical reactions.

$$\frac{dx}{dt} = -ax^3 + bx + c$$
$$\frac{dy}{dt} = dx^2 + ex - f \qquad (1.4)$$

(b) The discovery of the oscillations of the Cer-catalysed Bromation of citric acid (1950/1959) [25] by B. P. Belousov (1893–1970); (Fig. 1.9)

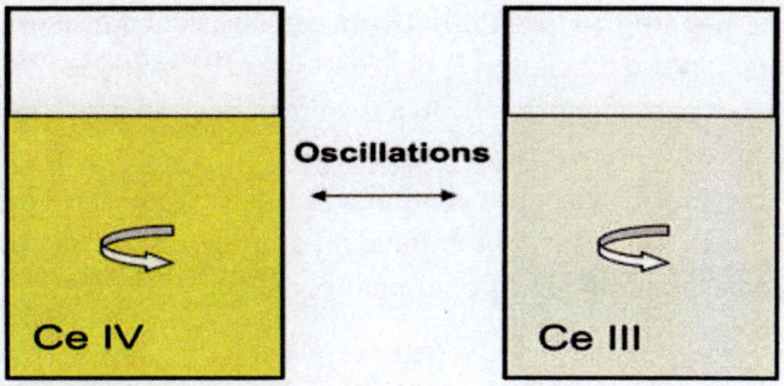

Fig. 1.9 Scheme of the oscillations Belousov reaction, which have been observed in a closed system (batch reactor) and could be observed visually by the transition between the yellow Ce^{IV} and the colorless Ce^{III} ions

A. M. Zhabotinsky (1938–2008) [26], animated by S. E. Scholl (1961), followed up the work of Belousov. Zhabotinsky replaced the original system of Belousov by the well known bromation of malonic acid catalysed by ferroine indicator. In his first works he let the reactants flow into the stirred tank reactor realizing a quasi open system (Fig. 1.10), which enables him to observe oscillation with almost constant amplitude for a while. A detailed description of the history of this discovery is given by A. T. Winfree [27], L. Kuhnert and U. Niedersen [8] and the author [1].

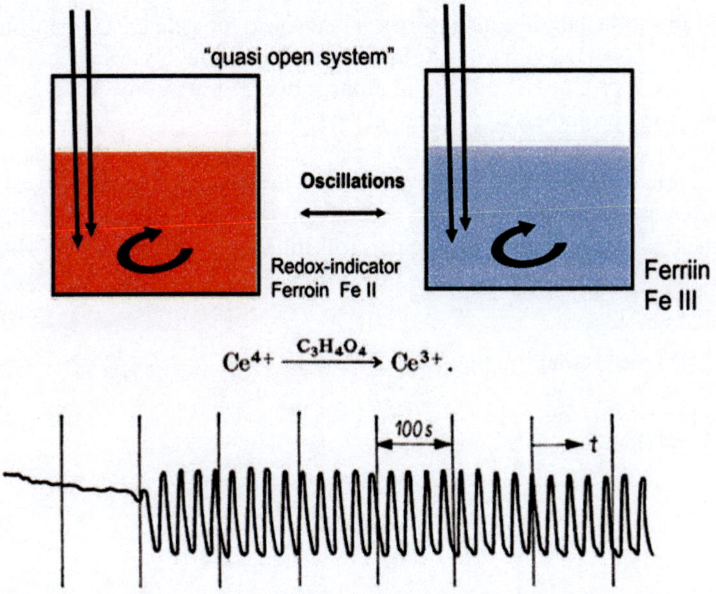

Abb. 7-2. Aufzeichnung der Anfangsphase der Reaktion (über die Lichtabsorption). Konzentrationen der Ausgangssubstanzen (in M): $KBrO_3$ = 0,067; $C_3H_4O_4$ = 1; Ce = 0,001; t = 18 °C. Lösungsmittel 3 N H_2SO_4

Fig. 1.10 **a** Scheme of the ferroine catalysed bromation of malonic acid which has been developed by A. M. Zhabotinsky. **b** Oscillations occurred in a 3 ml cuvette recorded by light transmission; [26] and [8, pp. 83–89]

Working on the conflict between classical thermodynamics and real life evolution and pattern formation Ilya Prigogine (1917–2003) [28, 29] developed the concept of the export of entropy from a system under consideration to its surrounding (1967/68). This way, regarding open systems (1.5) and (1.6) (where A is an equal constant flow and E can grow without limitation) I. Prigogine developed in 1977 a set of kinetic equations which have an auto-catalytic reaction step: $2X + Y \rightarrow 3X$ in its corresponding set of chemical equations [30], which is well known as *Brüsselator model* (Eqs. 1.5 and 1.6).

$$
\begin{aligned}
&\mathrm{A} \rightarrow \mathrm{X} \\
&2\mathrm{X} + \mathrm{Y} \rightarrow 3\mathrm{X} \\
&\mathrm{B} + \mathrm{X} \rightarrow \mathrm{Y} + \mathrm{D} \\
&\mathrm{X} \rightarrow \mathrm{E}
\end{aligned}
\tag{1.5}
$$

$$
\begin{aligned}
\frac{dX}{dt} &= A + X^2Y - BX - X \\
\frac{dY}{dt} &= BX - X^2Y
\end{aligned}
\tag{1.6}
$$

Such an auto-catalytic equation was very unusual for a chemist, who would never write: $2X + Y \rightarrow 3X$, but $Y \rightarrow X$. Let me emphasise that his system of differential equations is very similar to those of Bonhoeffer. The new true was his chemical interpretation: the autocatalytic chemical system:

So it was very hard to understand the chemical background of this model at those times. However, he introduced this description for constructing the non-linear ordinary differential equation, which he needed for creation of a limit-cycle (Fig. 1.11). Nevertheless, in the eighties, Prigogine's fruitful Ansatz inspired a lot of chemists to look for similar auto-catalytic reactions.

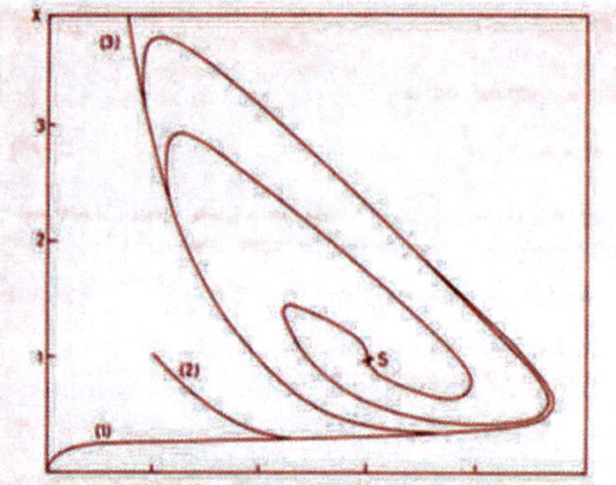

Fig. 1.11 Original drawing of the limit cycle behaviour of the Brüsselator [31] (see Eqs. (1.5) and (1.6))

1.3 The Early Days of Synergetics [32]

Remembering all this convincing findings it is astonishing that it took about hundred years from its beginning that these ideas became united under one roof by Hermann Haken when he created "Synergetics" in the early seventies.

In April 1972 Hermann Haken organized the first international "Symposium on Synergetics" in Schloß Elmau, which is documented in the proceedings "Synergetics—Cooperative Phenomena in Multi-Component Systems" which appeared 1973 by B. G. Teubner, Stuttgart [33]. It needed some years more that this idea maturated in the scientific community. In May 1977 Hermann Haken organized a second international workshop on synergetics at Schloß Elmau. There he stated "Synergetics is a rather new field of interdisciplinary research which studies the self-organized behaviour of systems leading to the formation of structures functioning's" [34].

Summer 1977 was just the date, when N. Jaeger and me at the physical chemistry department of Bremen University got acquaintance with the ideas of structure formation in irreversible processes by the small book of Werner Ebeling which just appeared [35].

Two years later U. F. Franck and E. Wicke organized the first meeting of the Bunsen-Gesellschaft für Physikalische Chemie on chemical oscillations in Aachen [36] (Fig. 1.12). Meanwhile a lot of physical-chemists came up with ideas like chemical oscillations, structure formation in irreversible processes, etc. and they presented their first results on these topics to a wide spread auditorium. Among them there were P. Gray (England), A. Zhabotinsky (USSR), and from Germany U. F. Franck, E. Wicke and O. E. Rössler, who brought the term chaos to the chemists mind at first. He formed the powerful idea: "Chaos … can be realized in simple chemical systems" [37] (see Fig. 1.13). This was very stimulating and inspiring to look for experimentally too.

$$
\begin{aligned}
\dot{x} &= x(a_1 - k_{-1}x - z - y) + k_{-2}y^2 + a_3 \\
\dot{y} &= y(x - k_{-2}y - a_5) + a_2 \\
\dot{z} &= z(a_4 - x - k_5 z) + a_3
\end{aligned}
\tag{1.7}
$$

For this set of three non-linear kinetic equations he offered a set of five chemical equations which can be interpreted conventionally by bi-molecular reactions (Eq. 1.8). This was a very important step for chemists to better understand the coupled differential equations of the type of Eq. (1.7), as it was in line with their ideas about the molecular process.

A few months earlier, O. E. Rössler published in Springer Series in Synergetics "Synergetics—Far from Equilibrium" his ideas on chaos and strange attractors in chemical kinetics [38]. With this book of A. Pacault and C. Vidal the term "*far from equilibrium*" became a winged word characterizing selforganized structure formation.

Fig. 1.12 U. F. Frank (very left) and O. E. Rössler (left on top) at the Discussion Meeting "Kinetics of Physicochemical Oscillations", Aachen 1979

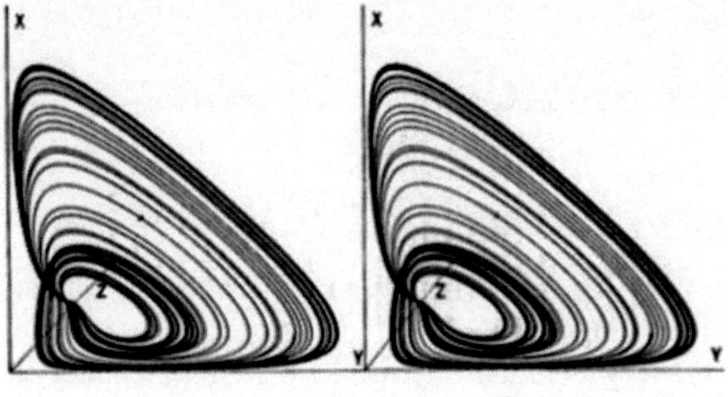

Fig. 1.13 Chaotic oscillations of Eq. (1.7). Stereoscopic display (two parallel projections). This is a copy of the original figure published by O. E. Rössler et al. [37]

$$
\begin{aligned}
A_1 + X &\underset{k_{-1}}{\overset{k_1}{\rightleftarrows}} 2X \\
Y + X &\underset{k_{-2}}{\overset{k_2}{\rightleftarrows}} 2Y \\
A_5 + Y &\underset{k_{-3}}{\overset{k_3}{\rightleftarrows}} A_2 \\
Z + X &\underset{k_{-4}}{\overset{k_4}{\rightleftarrows}} A_3 \\
A_4 + Z &\underset{k_{-5}}{\overset{k_5}{\rightleftarrows}} 2Z
\end{aligned}
\tag{1.8}
$$

As a result, rapid development of Synergetics continued in all scientific disciplines. Working on the BZ reaction, Richard Noyes 1919–1997 and Kenneth Showalter observed truly sustained chemical oscillations in continuously stirred tank reactors (CSTR). They postulated [39] and observed (Fig. 1.14):

> Most of the initial studies were done in closed batch reactors. ... Truly sustained oscillations are possible in principle with - an open – continuously stirred tank reactor (CSTR)

1.3.1 Chemical Waves in the BZ Reaction

We remember the great potential of Fick's second law for structure formation if some critical values get overstepped. Now, combining Fick's second law with reaction equations a great variety of new structure formations should arise. Such reaction–diffusion equations (1.9) played a very import role in development of synergetics, since a local oscillation in an excitable chemical medium becomes a chemical wave by diffusion. The term $f_1(c_1, c_2, \cdots, c_n)$ stands for any kind on non diffusion processes like classical chemical kinetics. The most exciting example for this kind of structure formation was the Belousov–Zhabotinsky reaction (see Fig. 1.1).

$$
\frac{\partial \mathrm{c_i}}{\partial \mathrm{t}} = \mathrm{f_i}(\mathrm{c_1}, \mathrm{c_2}, \cdots, \mathrm{c_n}) + \mathrm{D_i}\nabla^2 \mathrm{c_i} \tag{1.9}
$$

Haken used a picture of this kind of pattern formation in BZR for the cover of his famous and wide spread book "Erfolgsgeheimnisse der Natur" [42]. Self exciting cyclic waves, so-called target patterns (see Fig. 1.1), cyclic waves in excited media and spiral waves can be observed, if one execute this reaction in a thin layer for example in a Petri-dish. With enormous effort this reaction has been studied in all details theoretically as well as experimentally since decades. The waves in BZR

Fig. 1.14 Original drawings of the oscillating BZ reaction under CSTR conditions [39–41] taken from R. A. Schmitz, K. R. Graziani and J. L. Hudson and reported by Showalter et al. as well as Noyes

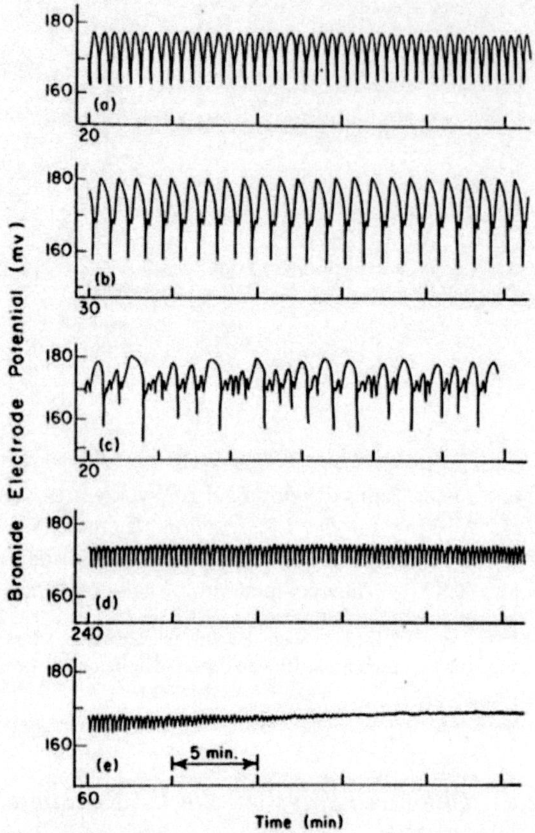

FIG. 1. Potentials of a bromide ion specific electrode observed experimentally for a 25.1 cm^3 reactor with $[CH_2(COOH)_2]_0 = 0.30M$, $[NaBrO_3]_0 = 0.14M$, $[NaBr]_0 = 3.5\times10^{-6}\,M$, $[H_2SO_4]_0 = 0.20M$, $[Fe(phen)_3^{2+}]_0 = 1.25\times10^{-4}M$. Flow rates corresponded to k_0 values in sec^{-1} of (a) 6.64×10^{-4}, (b) 1.328×10^{-3}, (c) 2.656×10^{-3}, (d) 2.988×10^{-3}, (e) 4.648×10^{-3}. Reproduced with permission from R. A. Schmitz, K. R. Graziani, and J. L. Hudson, J. Chem. Phys. 67, 3040 (1977).

became a key phenomenon for a huge variety of similar pattern formation processes [43, 44].

All these different patterns can be summarized by the idea of a solitary chemical wave. In earlier times they have been named by V. I. Krinsky as auto-waves [43]. Such waves annihilate if they collide and they cannot be reflected if they touch a wall. Hans Meinhardt demonstrated the wide occurrence of chemical waves in his very illustrative book "The Algorithmic Beauty of Sea Shells" [45].

These waves have been simulated successfully by reaction diffusion equations of type of the activator–inhibitor model for example, but also by cellular automata

Fig. 1.15 Spirals during single crystal catalysis of CO-oxidation on Pt(110) [49]

models [46–48]. However, the question remains what are the cooperative subsystems which lead to the observed patterns.

G. Ertl and his group [49] observed spiral waves on the surfaces of Pd single crystals during the catalytic oxidation of CO (Fig. 1.15). Although only a few atoms and molecules respectively are the acting subsystems, these processes could be modelled successfully by the reaction diffusion systems which are based on the Langmuir–Hinshelwood mechanism expanded to surface sub-layers acting as storages for oxygen.

There is a huge number of very careful studies on the formation of chemical spirals and their movement especially in BZ-like systems [50]. However, it should be mentioned that there is a very important detection of C-Amp spirals in the communication of Dictyostelium dicoideum cells by C. J. Weijer et al. [51].

According to my opening paragraph I shall restrict myself on those discoveries of synergetics which developed slightly beside the well known mainstream.

For example there is the detection of ripples occuring during abrasive water-jet-cutting by G. Radons and R. Friedrich [52] (Fig. 1.16 left). Cutting processes of quite different nature are accomplished very often by such pattern formation in the cutting face. Cutting processes are of enormous importance in production techniques. This work of Radons and Friedrich encouraged us to look for similar pattern formation in micro-etching processes. We found formation of ripples during laser-jet etching as well [53] (Fig. 1.16 right).

Moreover in the last two decades the investigation of turbulence made great progress. The new results in this field are strongly correlated to the ideas of synergetics [54]. Again the results are of importance for technical processes especially in wind turbines [55, 56] and in electro-polishing [57].

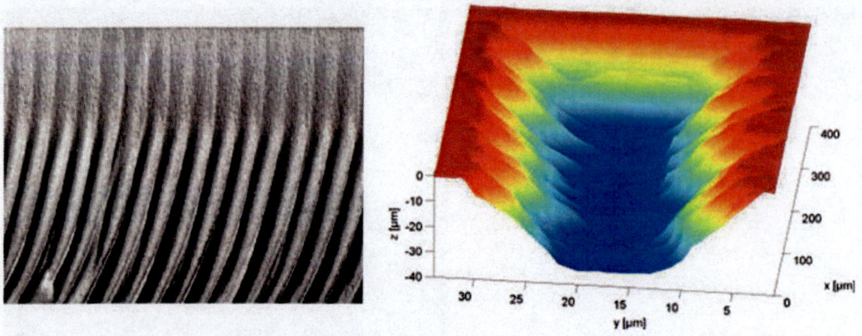

Fig. 1.16 Ripple formation in cutting processes; **left** water jet cutting [52]; **right** laser jet etching [53]

1.3.2 *Outlook—Ongoing Problems Which Could Be Solved by Synergetics*

During the symposium *Complexity and Synergetics* 2015 [58] I have been asked what are the future problems on which synergetics could be applied successfully.

Two actual problems on which I am working personally are reported in special contributions of the proceedings of "Complexity and Synergetics": aging of accumulator batteries [59] and oxidation of CO on supported catalysts [60, 61].

Very briefly I shall mention two further problems only:

Going back in the past F. Runge worked on chemical reactions occurring in filter paper. The dynamics of these wonderful pictures (Figs. 1.3 and 1.17) are not well understood until today.

Let me give you a small example of a common ongoing investigation in cooperation with St. C. Müller. Taking a sheet of filter paper and let a potassium dichromate solution diffuse continuously into the paper, a more or less cyclic, dark yellow fleck occurs. After drying, water diffuses into the fleck from a centered spot forming a cyclic front again, which is followed by an instable "washing out front" of the potassium salt crystals. It looks like a displacement pattern in a Hele-Shaw arrangement [62]. For sure this kind of research topic belongs to the very fundamental topics of basic research in synergetics.

Another very actual task of research in synergetics could be to investigate the movement of large groups of peoples. Regarding the ancient transmigration into the Roman Empire (Fig. 1.18) there are indications that it could be understandable in the framework of the ideas of an economic basin of attraction. Tribes surrounding the Roman Empire were attracted by the strength of its economy, since their own economy was very pure in comparison. The decay of the social and still matriarchal structures of their original societies created the well known varying warring groups which fought against the Roman Empire. Following the synergetic principle of cyclic causality they did not only destroy the Empire but their own original societies creating

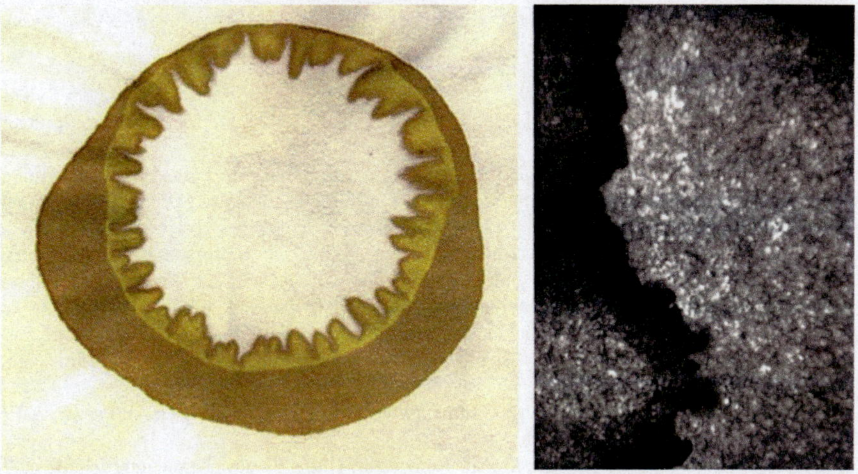

Fig. 1.17 Front instability during displacement of potassium chromate in soaked and dried filter paper. **Left**: (photo: P. J. Plath); **right**: the water is coming from the right side of the photo; a very sharp black front on the windward side (German: Luv Seite) is created whereas the leeward side (German; Lee) goes slower into the original yellow area of the again wetted paper; (photo: St. C. Müller)

more and more fighters. At least they settled being integrated in the Roman Empire which dissolved itself.

Today's transmigration of refugees into the European Union seems to be very similar to the migration of the ancient tribes into the Roman Empire. Again Europe acts as economic basin of attraction for peoples of more or less the surrounding countries. Again their emigration destroys their own countries. Again we fight against the "rebels" so-called terrorists, again we let them settle in Europe and we try to integrate them. I shall stop here to continue to list the similarities. We should try to understand the socio-dynamics of these processes!

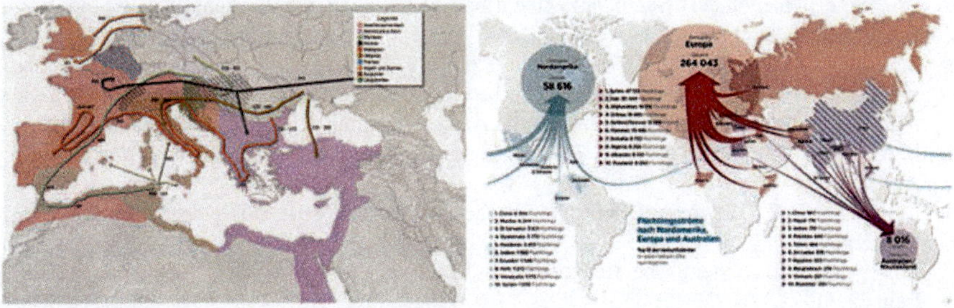

Fig. 1.18 **Left** Transmigration to the Roman Empire [63]; **right** Transmigration of refugees to Europe 2015 [64]

References

1. Haken, H., Plath, P.J., Ebeling, W., Romanowsky, Y.M.: Beiträge zur Geschichte der Synergetik – Allgemeine Prinzipien der Selbstorganisation in Natur und Gesellschaft. Springer Spektrum, Wiesbaden (2016)
2. Haken, H., Obituaries, in: "Selforganisation in Complex Systems: The Past, Present, and Future of Synergetics", (Pelster, A., Wunner, G.; editors) Springer:Complexity, Springer International Publishing Switzerland (2016) 283–285
3. Kröger, B.; "Hermann Haken – His Roadmap to Synergetics", in: "Selforganisation in Complex Systems: The Past, Present, and Future of Synergetics", (Pelster, A., Wunner, G.; editors) Springer: Complexity, Springer International Publishing Switzerland (2016) 265–281
4. Runge, F.F.; „Merkwürdiges Verhalten des Quecksilbers in Berührung mit Salpetersäure und Eisen", Annalen der Physik und Chemie (Hrsg. Poggendorff) 15. Band der ganzen Folge, Leipzig (1829) pp. 95–98
5. Berkemeier, J., Purwins, H.-G, (2011) https://www.uni-muenster.de/Physik.AP/Purwins/DE/Quecksilberherz-de.html
6. Blume, R.; Prof. Blumes Tipp des Monats März (2007) https://www.chemieunterricht.de/dc2/tip/03_07.htm
7. Möllenkamp, H., B. Flintjer, B., Jansen, W., CHEMIKON (1994) 117–125
8. Runge, F.F., in "Der Bildungstrieb der Stoffe", Oranienburg (1855) Selbstverlag
9. Runge, F.F, Zur Farbenchemie, Berlin (1850), Verlag E.S. Mittler Sohn
10. Runge, F.F., Liesegang, R.E., Belousov, B.P., Zhabotinsky, A.M., Selbstorganisation chemischer Strukturen, eingeleitet von Kuhnert,U.L. und Niedersen, U., Ostwalds Klassiker der exakten Wissenschaften Vol. 272 (1987) Akademische Verlagsgesellschaft Geest & Portig K.-G. Leipzig
11. Wetzlar, G.: Ueber die Reduction der Metalle durch einander auf nassem Wege", in: Journal für Chemie und Physik (Schweigg. Journal) 49 (1827), S. 88–109, 128–144, 470–489
12. Fechner, M.G.T., Über Die Umkehrung der Polarität in der Einfachen Kette. In: Jahrb. d. Ch. u. Ph. 23 (1828), Nr. 2, S. 129–151
13. Oettingen, A.J. von: „Der Rückstand der Leidener Batterie als Prüfungsmittel für die Art der Entladung", Abhandlung zur Erlangung der Magisterwürde der Physico-mathem. Facultät der Kaiserl. Universität zu Dorpat (1862) p. 4
14. Oettingen, A. J. von, Annalen der Physik 270, 7 (1888) 570–575
15. Oettingen, A.J. von, Annalen der Physik 276, 5 (1890) 83–92 (Ueber die oscillatorische Entladung metallischer Conductoren)
16. Marx, K., Das Kapital Bd. III,(Dietz Verlag Berlin (1949) pp.. 533.
17. (published for the 1st time London, 4 Oktober 1894 by F. Engels)
18. Ch., Suter: (Neuchâtel) Schuldenzyklen (indebtedness) in der Dritten Welt. Anton Hain, Frankfurt am Main (1990)
19. http://www.3sat.de/mediathek/?mode=play&obj=60325, Sendung Kulturzeit vom 11. 7. 2016 mit Ernst Granditz; O. Nachtway, „Die Abstiegsgesellschaft –Über das Aufbegehren in der regressiven Moderne", edition suhrkamp 2. Aufl. Suhrkamp Verlag Berlin (2016) p.183
20. Marx, K., Engels, F.: Feuerbach. In: Die Deutsche Ideologie, 4, p. 53. Dietz Verlag Berlin, Aufl. (1960)
21. Liesegang, R.E., Liesegangs Photographisches Archiv, Nr. 801; Düsseldorf, 1.Nov. (1896); Heft XXI
22. Liesegang, R.E., Chemische Reaktionen in Gallerten, Verl. Theodor Steinkopf, Dresden, Leipzig, (1924) S. 67
23. Ostwald, Wi., "Periodische Erscheinungen bei der Auflösung des Chroms in Säuren" Z. phys. Chemie 35 (1900) 33–76 and Z. phys. Chemie 35 (1900) 204–256
24. Lotka, A.J.: Zur Theorie der periodischen Reaktionen. Z. phys. Chemie **72**, 508–511 (1910)
25. Kondratjew, N.D.: Die langen Wellen der Konjunktur. Archiv für Sozialwissenschaften und Sozialpolitik, Berlin **56**, 573–609 (1926)

26. Schumpeter, J., Business Cycles. A Theoretical, Historical and Statistical Analysis of the Capitalist Process, New York/London 1939. (deutsch: Konjunkturzyklen. 1961)
27. https://quantumpranx.wordpress.com/the-kondratieff-theory/
28. Bonhoeffer, K.F., Renneberg; W., Z. f. Physik **118**, 1/2 (1941) 389–400
29. Belousov, B.P., Sb. Ref. Radiats Med. za,: Medgiz. Moscow I **1959**, 145 (1958). (Russian)
30. Zhabotinsky, A.M.: Doklady Akademija Nauk SSSR **157**, 392–395 (1964)
31. Winfree, A.T.; J. Chem. Educ.(1984) 661–663
32. Prigogine, I., Nicolis, G.: On symmetry-breaking instabilities in dissipative systems. J. Chem. Phys. **46**, 3542–3550 (1967)
33. Prigogine, I., Lefever, R.: On symmetry-breaking instabilities in dissipative systems II. J. Chem. Phys. **48**, 1695–1700 (1968)
34. Prigogine, I.: "Zeit, Struktur und Fluktuationen (Nobel-Lecture). Angew. Chem. **90**, 704–715 (1978)
35. Prigogine, I.: Vom Sein zum Werden, p. 112. Verlag, München, Zürich, R. Piper Co. (1979)
36. Plath, P.J., "Synergetic Impact of Synergetics – Remembrances of a Chemist", in: "Selforganization in Complex Systems: The Past, Present, and Future of Synergetics", (A. Pelster, G. Wunner; editors) Springer : Complexity, Springer International Publishing Switzerland (2016) 59 – 74
37. Haken, H., "Synergetics – Cooperative Phenomena in Muli-Component Systems". B.G. Teubner Stuttgart (1973)
38. Haken, H.: Synergetics - A Workshop. Springer Verlag Berlin, Heidelberg, New York, Preface (1977)
39. Ebeling, W., „Strukturbildung bei irreversiblen Prozessen", Mathem. Naturwiss. Bibliothek Bd. 60, BSB B.G. Teubner, Leipzig (1976)
40. Franck, U.F., Wicke, E., "Kinetic of Physicochemical Oscillations", Vol. II Reprints of submitted Papers , Fotodruck J. Mainz, Aachen (1976)
41. Willamowski, K.-D., Rössler, O.E., "Irregular oscillations in a realistic abstract quadratic mass action system", in : Proceedings of the Discussion Meeting – Kinetics of Physicochemical Oscillations in Aachen pp. 505–513 (1979)
42. Rössler, O.E., "Chaos and Strange Attractors in Chemical Kinetics"; in: "Synergetics – Far from Equilibrium" (A. Pacault, C. Vidal, editors) Springer-Verlag Berlin, Heidelberg, New York (1979) 107–113
43. Showalter, K., Noyes, R., Bar-Eli, K.: A modified Oregonator model exhibiting complicated limit cycle behavior in a flow system. J. Chem. Phys. **69**(6), 2514–2524 (1978)
44. Schmitz, R.A., Graziani, K.R., Hudson, J.L.: J. Chem. Phys. **67**, 3040 (1977)
45. Noyes, R.M.: Oscillations in Homogenous Systems. Ber. Bunsenges. Phys. chem. **84**, 295–303 (1980)
46. Haken, H.: Erfolgsgeheimnisse der Natur - Synergetik: die Lehre vom Zusammenwirken. Deutsche Verlags-Anstalt, Stuttgart (1981)
47. Krinsky, V.I.: Autowaves and Structures Far from Equilibrium, Springer Series in Synergetics vol. 28 pp. 9 – 19 Springer-Verlag Heidelberg (1984)
48. Winfree, A.T.: When Time Breaks Down,- the Three Dimensional Dynamics of Electrochemical Waves and Cardiac Arrhythmias. Princeton University Press (1987)
49. Meinhardt, H.: The Algorithmic Beauty of Sea Shells, Springer-Verlag (1998)
50. Schwietering, J., Plath, P.J.: „Wachsende Muster", Wissenschaft und Fortschritt - Zeitschrift für interdisziplinäres Denken; 42, 2 (1992) pp.73 - 75 and pp. 41, 12 (1991) US.III
51. Gerhardt, M., Schuster, H., T, J.J., A Cellular Automaton Model of Excitable Media Including the Effect of Curvature and Dispersion, Science **247** 1563 – 1466 (1990)
52. Markus, M., Krafczyk, M., Hess, B.: Randomized automata for isotropic modeling of two and three-dimensional waves and spatiotemporal chaos in excitable media, in: Holden, V.A., Markus, M., Othmer, H.G. (eds.)Nonlinear Waves Processes in Excitable Media. Plenum Press, New York (1991)
53. Nettesheim, S., von Oertzen, A., Rotermund, H.H., Ertl, G.: Reaction diffusion patterns in the catalytic CO-oxidation on Pt(110): Front propagation and spiral waves. J. Chem. Phys. **98**, 9977–9985 (1993)

54. Brandtstädter, H., Braune, M., Schebesch, I., Engel, H.: Experimental study of the dynamics of spiral pairs in light-sensitive Belousov-Zhabotinskii media using an open-gel reactor. Chem. Phys. Lett. **323**(1–2), 145–154 (2000)
55. Dormann D., Vasiev B., Weijer C.J.; Propagating waves control Dictyostelium discoideum morphogenesis; Biophys. Chem. (1998) May 5;72(1–2):21–35.
56. Radons, G., Ditzinger. T., Friedrich. R., Henning, A., Kouzmichev, A., Westkämper, E., Nonlinear Dynamics and Control of Ripple Formation in Abrasive Water-Jet Cutting, in: Radons, G., Neubauer, R. (eds.) Nonlinear Dynamics of Production Systems, pp. 391 – 410. Wiley-VCH Verlag GmbH & Co KGaA, Weinheim (2004)
57. Rabbow, T.J., A. Mora, A., Haase, M., Plath, P.J., „Self organised Structure formation in Organised Microstructuring by Laser-Jet-Etching", International Journal of Bifurcation and Chaos Vol. 27, No. 01, 1750001 (2017)
58. Friedrich, R., Peinke. J.; Description of a Turbulent Cascade by a Fokker-Planck Equation, Phys. Rev. Lett. **78**, 863 (1997)
59. Peinke, J., Barth, St., Böttcher, F., Heinemann, D., Lange. B.; Turbulence, a challenging problem for wind energy, Physica **A 338** (2004) 187 – 193
60. Anvari, M., Rahini Tabar, R., Peinke, J., Wächter, M., Dynamics and Synchronisation in Wind Farms, in: a Tube, in: Complexity and Synergetics, Müller, St.C., Plath, P.J., Radons, G., Fuchs, A., (eds.)Springer International Publishing AG (2018) pp. 383–388
61. Sydow, U., Buhlert, M., Plath. P.J., Fractal Characteristics of Electropolished Metal Surfaces, in: Vernetzte Wissenschaften – Crosslinks in Natural and Social Sciences, , Plath, P.J., Haß, E.-C., (eds.), Logos Verlag Berlin (2008) pp. 87–97
62. Müller, St.C., Plath, P.J., Radons, G., Fuchs, A., (eds.) Symposium "Complexity and Synergtics" in Hannover July 2015 , Springer International Publishing AG (2018)
63. Haß, E.-C., Knicker, K., Sydow, U., Schulz, M., Plath, P.J., Battery-Determination and Forecast via Synergetics, in: Complexity and Synergetics, Müller, St.C., Plath, P.J., Radons, G., Fuchs, A., (eds.)Springer International Publishing AG (2018) pp. 139–153
64. Ballandis, C., Plath, P.J.: A New Discrete Model for the Non-Isothermic Dynamics of the Exothermic CO-oxidation an Palladium supported catalyst. J. Non-Equilib. Thermodyn. **25**, 301–324 (2000)
65. Plath, P.J., Ballandis , C., Catalytic Oxidation of CO – A Striking Example of Synergetics – Experimental Studies of Coupled Zeolite Catalyst Wafers, in a Tube, in: Complexity and Synergetics, Müller, St.C., Plath, P.J., Radons, G., Fuchs, A., (eds.)Springer International Publishing AG (2018) pp. 87–100
66. Damme van , H.; Flow and Interfacial Instabilities in Newtonian and Colloidal Fluids (or the birth, Life and Death of a Fractal) in: The Fractal Approach to Heterogeneous Chemistry – Surface, Colloids, Polymers (Avnir, D., ed.), John Wiley & Sons, Chichester, New York, Brisbane, Toronto, Singapore (1989) 199–226
67. Vökerwanderungkarte". Lizenziert unter Gemeinfrei über Wikimedia Commons https://commons.wikimedia.org/wiki/File:Voelkerwanderungkarte.png#/media/File:Voelkerwanderungkarte.png
68. Handelsblatt; Autor: Reuters; Date: 24.04.2015 18:29 Uhr; http://www.handelsblatt.com / BAMF/UNHCR and: https://www.roternagel.de/2016/02/26/warum-kommen-fl%C3%BCchtlinge-nach-europa/

Part I
Synergetic View on Historic Experiments

Chapter 2
The Swinging Chromium

Oscillations During Dissolution of Metals with Local Cells

Thomas Rabbow and Peter J. Plath

Fig. 2.1 Photo of an original piece of chromium that Wilhelm Ostwald used for his experiments on the swinging chromium and that also served us today to repeat his experiments (photo: P. J. Plath, T. Rabbow)

Dedicated to Margarete (Gretel) Brauer, Eberhard and Grete Brauer, and Wilhelm Ostwald.

P. J. Plath et al., *Imagery Synergetics*, Understanding Complex Systems,
https://doi.org/10.1007/978-3-030-95607-3_2

2.1 Introduction

Hundred years ago, in 1900, Wilhelm Ostwald published his two guiding articles on "periodic appearances during the dissolution of chromium in acids" (Periodische Erscheinungen bei der Auflösung des Chroms in Säuren) in the "Zeitschrift für physikalische Chemie" [1, 2]. This extraordinary work belongs to the first comprehensive scientific articles on the complex dynamics of chemical processes.

Within the last few decades a great variety of pattern forming processes in electrochemical systems have been described scientifically by use of synergetics the theory of dynamic systems. By chance, one of us met the very honest person G. Brauer who takes as a granddaughter of Wilhelm Ostwald great care on her grandparent's heritage and the impressive small exhibition appertaining to their house. She especially loves her grandfather's [1, 2] and her father's [3] experiments on the oscillating dissolution of chromium in hydrochloric acid. Next to the worldwide first thermostat, a small but heavy carton with the handwritten inscription "Schwingendes Chrom" is the heart of this originally exhibition. And indeed, in this carton there were pieces of the first chromium ever been produced on which Wilhelm Ostwald and his former co-worker E. Brauer had studied the oscillatory behaviour of the acidic dissolution of chromium. It is worthwhile visiting the Wilhelm Ostwald museum in his "Haus Energie" in Großbothen and to talk to all the nice and helpful people who take care on Wilhelm Ostwald's life work.

It was very fascinating to repeat Wilhelm Ostwald's experiments using his original chromium and our day's equipment. Mrs. G. Brauer (Fig. 2.1) agreed in this idea and gave us one piece of this historical chromium (Fig. 2.2).

Fig. 2.2 Margarete (Gretel) Brauer (Photo: *Beate Bahnert*), "Gretel Brauer" Mitteilungen der Wilhelm-Ostwald-Gesellschaft zu Großbothen e.V. 13. Jg. 2008, Heft 2; ISSN 1433–3910

2.2 Historical Remarks

Since the work of G. Th. Fechner in 1828 [4] one can find reports on periodic chemical reactions. Wilhelm Ostwald mentioned all these phenomena in his textbook Lehrbuch zur "Elektrochemie—Ihre Geschichte und Lehre" in 1896 [5] as well as in his famous articles on the oscillating dissolution of chromium in 1900 [1, 2].

Based on the work of J. Keir (1790; 1828) [6], and G. Wetzlar (1827) [7] on the passivity of iron and CuO in HNO_3 respectively, G. Th. Fechner used a galvanometer for his investigations on the system iron/silver in a solution of $AgNO_3$ in nitric acid. He was the first scientist who recognized electrochemical oscillations.

> Ich habe jedoch bei wiederholten Versuchen bemerkt, daß die Erscheinung hierbei gewöhnlich noch nicht stehenblieb, vielmehr das Auflösen des Eisens und Wiederblankwerden nebst Auflösung des gefällten Silbers wohl vier bis sechsmal, oft sehr schnell hintereinander, abwechselten, wobei jedesmal die Ablenkung der Magnetnadel auf das entgegengesetzte übersprang, bis das Eisenstäbchen zuletzt unwirksam liegenblieb, cited concerning Ref [4].

In 1833 J. F. W. Herschel [8] observed electrochemical oscillations during the dissolution of iron in nitric acid, the concentration of which was in the range that neither passivation nor activation of iron would be a stable process. He is the first who observed passivation waves on the iron. During this time many scientists were fascinated by electrochemistry which appeared as a new and exciting branch of physics. The chemist C. F. Schönbein [9] was interested in the oscillations during corrosion of iron in nitric acid, wherefore he coupled directly several iron wires getting synchronous oscillations

> Bringt man aber die Drähte entweder innerhalb oder außerhalb der Säure in leitende Verbindung, so findet an dem ganzen Drahtsystem die Pulsation haarscharf gleichzeitig statt, und tritt dauernde Indifferenz an einem Drahte ein, so erfolgt die nämliche in demselben Augenblicke an allen übrigen Drähten, cited concerning Ref [9].

The physicist J. P. Joule investigated coupling phenomena of oscillating electrochemical systems as well [10]. This first period of interest in the dynamic behaviour of electrochemical systems is characterized by the collection of new and fascinating phenomena but not by systematic physico-chemical investigations. More than half a century later Wilhelm Ostwald introduced quite another view to these new phenomena. As an excellent experimentalist he developed all the tools to understand the oscillating electrochemical systems based on the fundamental ideas of the new interdisciplinary science: physical chemistry.

He got one of the first pieces of metallic chromium (Fig. 2.1), produced in 1894 by H. Goldschmidt by aluminothermics. In order to record the oscillations in the hydrogen production during the dissolution of chromium in acids, he constructed the "Chemograph" [1, p. 36, 2, p.212], which was the first x-t-recorder. Moreover, he developed the first thermostat [2, p. 216] for his measurements.

In order to study the acidic dissolution he collected some pieces of chromium in a bag of net-veil putting it in an aqueous 2 N HCl solution. He recognized that oxidising agents like HNO_3 or $KBrO_3$ act as activators which are shortening the period length of the oscillations [1, p. 71], whereas reducing agents like formaldehyde H_2CO, rhodanide SCN^-, or iodide J^- work as inhibitors, enlarging the period length. Summarizing, he observed periods between one minute and one hour in the cyclic behaviour. Increasing temperature the frequency of the oscillations increases as well, and for strong cooling all oscillations vanish.

When his chromium had used up, because of the extensive investigations, he got a second contingent of new and very pure chromium by Goldschmidt (1894?). But with these new samples he could not achieve oscillating behaviour again wherefore he concluded

> … daß die auffallende Eigentümlichkeit meines Metalls von irgend einer Beimischung herrührt, welche sich in den älteren Proben befand und in der neuen nicht mehr vorkam, cited concerning ref. [1, p. 52].

Some additive substances in the first pieces of chromium might be responsible for the oscillatory behaviour of this first sample, although he could not support this assumption by documentary evidence.

This was the starting point for our own investigations. On the one hand, one can get very pure chromium in our days and on the other hand we got pieces of chromium of the very first sample of Wilhelm Ostwald. Moreover, well established electrochemical methods are of our disposal.

2.3 Experiments

2.3.1 *Pure Chromium*

The metal chromium is well known with respect to the formation of a stable oxide surface layer. This layer is protecting the less noble pure metal from dissolution. Concerning the fundamental work of K. F. Bonhoeffer [11–14] and Franck [15–17] the current voltage function reflects this typical behaviour of electrochemical systems. In the region of the Flade-potential [18] which is the turning point in the function $I = f(E)$ between the region of active dissolution and passivation, oscillations can be observed usually.

For this reason we recorded these $I = f(E)$ function carrying out a cyclic voltamogram (Fig. 2.3) with our days chromium of high purity (99.99%, Heraeus). The chromium anode was coated with epoxy resins. Free metal surfaces of 13 mm^2 and 30 mm^2 respectively remain for being dissolved in 2 N hydrochloric acid HCl. Wilhelm Ostwald used a 2 N HCl solution as well. We used a silver sheet (23 mm × 43 mm) as counter electrode. The distance between both electrodes had been fixed to 7 cm.

The scanning rate of the potential sweep (potentiostat WENKING HP 72) was 11 mV/s. An Ag/AgCl electrode (E_0 = 222 mV) with a Haber Luggin capillary had been used as reference electrode in our three electrode arrangement. To compare different chromium samples the surface had always been grinded before the experiment.

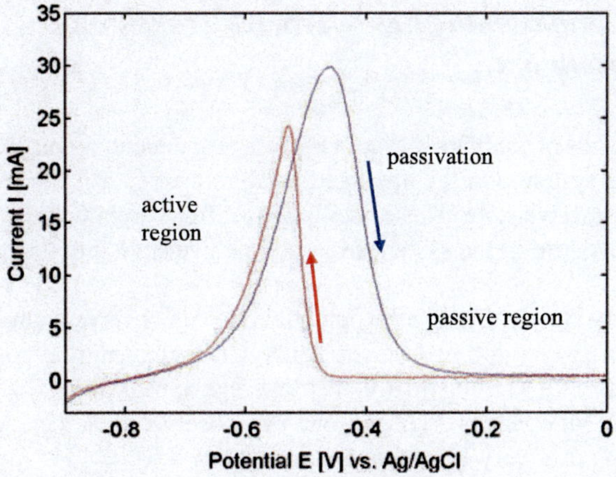

Fig. 2.3 Hysteresis in the cyclic voltamogram $I = f(E)$ of pure chromium in 2 N HCl (scan rate 11 mV/s)

Figure 2.3 shows the typical hysteresis behaviour of a metal which can be passivated at higher potential. Such kind of current voltage diagrams strongly indicate the possibility to achieve electrochemical oscillations close to the Flade potential. For potentials less than –0.8 V there is an cathodic polarization of the chromium electrode where hydrogen is formed. In the "active" range from approx. -0.75 V to approx. -0.46 V no formation of hydrogen is observed and the strong increase in the current is correlated with the dissolution of the chromium.

$$Cr \rightarrow Cr^{3+} + 3\,e^{-}$$

Which is correlated to the formation of the green complex.

$$\left[Cr^{III}Cl_3(H_2O)_3\right] * 3H_2O$$

At E = -460 mV versus Ag/AgCl (or E = -238 mV vs. NHE) the maximum is reached and corresponds well with the Flade-potential which sometimes is defined by the maximum of the current–voltage relation. The dependence of the Flade-potential from pH of the electrolyte was experimentally determined by Kolotyrkin [19] as:

$$E_{Flade} = -250 - 58 \cdot pH$$

Passing this maximum, the current reaches a very low level again, indicating the passive behaviour which is caused by the formation of oxidic layers on the metal surface. Turning back the scan a hysteresis behaviour can be observed [20, 21] as it has been explained in detail by R. Otterstedt [22]. Repeating this cyclic voltamogram several times, no strong quantitative reproducibility could be achieved. Especially between the first and the second cycle strong deviations can be observed.

2.3.2 *Oscillations at the Active Passive Transition of Chromium*

Two different kinds of oscillations could be observed depending on the history of the electrochemical system. On the one hand oscillations occur if the active system is shifted into the passive state. These oscillations differ very much from those which are generated by shifting the system from the passive state into the direction to the active state.

Scanning very slowly through the potential one can observe damped oscillations already at about E = −440 mV versus Ag/AgCl. Increasing the potential step by step, oscillations with a strong drift in their average appear.

At E = −428 mV versus Ag/AgCl one can observe stable characteristic oscillations (Fig. 2.4) of a very low main frequency $f = 4.5*10^{-3}$ Hz (period length T = 222 s). The system oscillates at a level of $I \approx 30$ mA with an amplitude of about 3 mA. Within one large saw-tooth current oscillation cycle the system is often interrupted by sharp increases of the current with small amplitudes.

Especially in his first article [1] on the periodic dissolution of chromium in acids, Wi. Ostwald reported on quite similar patterns of the oscillations (Fig. 2.5). For sure, he carried out his experiments in another way. He could not use a potenstiostatic equipment. First he passivated totally the chromium metal in an acid (HCl) solution of $K_2Cr_2O_7$ for several days. During this time chromium did not loose his metallic brightness. Before his experiments he mostly activated the sample by touching it with cadmium metal in the 2 N HCl solution. Then he recorded the hydrogen development. The main difference of his procedure to the potentiostatic treatment is that his system could freely choose potential and current by itself, whereas in our system the voltage was fixed or varied in a controlled way. So, we have really two different experimental conditions, wherefore one cannot expect similar time series. However, both experiments have in common strong activation within a very short time. The deactivation takes a much longer time.

Using potentiostatic conditions this deactivation process is often interrupted by smaller activation peaks (Fig. 2.5). Therefore it is very interesting to compare these results with the experiment in which Wi. Ostwald investigated the coupling/decoupling [1] of two pieces of chromium in just the same HCl solution (Fig. 2.6). Figure 2.6 shows the superposition of the hydrogen production of two decoupled pieces of chromium. In that case he also observed interruptions of the

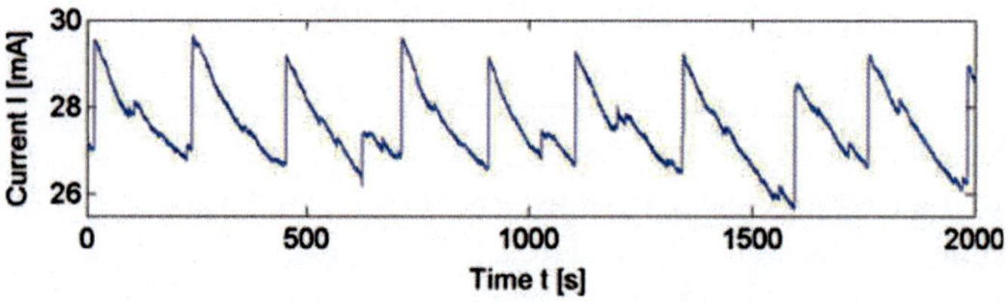

Fig. 2.4 Oscillations at the transition from active to passive state at E = -428 mV versus Ag/AgCl

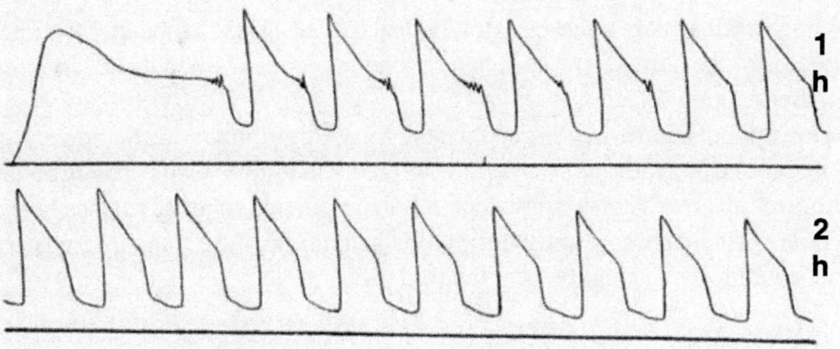

Fig. 2.5 Original time series for the dissolution of chromium in 2 N HCl measured by Wilhelm Ostwald [1]. The ordinate represents the pressure of the gas which was developed during the dissolution

oscillations cycle by small peaks that are caused by the smaller piece of chromium. It seems that the smaller activation peaks in our experiments are caused by small areas of the electrode surface which are somehow decoupled from the other parts of the surface.

Indeed, deactivation means that the metallic surface of the electrode will be covered with an oxidic layer protecting the metal from being solved. Obviously, this layer is not formed in a homogeneous way, covering the whole surface at the same time. Small parts of this layer can suddenly be destroyed. It seems, as if they split of the surface. Immediately this freshly created metallic area contributes to the current and one can observe small but sharp increases of the current in the time series

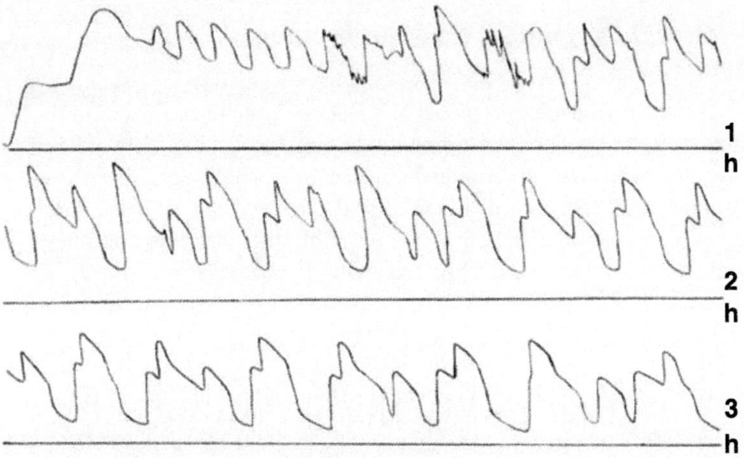

Fig. 2.6 Original time series for the dissolution of chromium in 2 N HCl measured by Wilhelm Ostwald [1]. Two pieces of chromium have been placed in the same solution. The ordinate represents the pressure of the gas which was developed during the dissolution

while the overall system becomes slowly deactivated. This means that the deactivation process of the surface causes a heterogeneous process since it is decomposed in several areas acting differently.

In correspondence to this procedure the very large jumps in the current can be correlated to the activation of the whole surface at one time. This activation process synchronizes all area. This is equivalent to a phase-transition in the surface behaviour. The whole system acts as a unity during this activating phase transition whereas the system decomposes during the deactivation process.

2.3.3 *Oscillations at the Passive Active Phase Transition*

Scanning the potential stepwise from the passive state at−200 mV (vs. Ag/AgCl) with 10 mV/min between each step into the cathodic direction, the system gets unstable at about E = −350 mV. The system shows very fast chaotic oscillations with the main frequencies of about 0.088 Hz and 0.097 Hz (period lengths T = 10 s–11 s) at E = −370 mV and low current (6–9 mA) instead of frequencies in the range of 0.02 to 0,0045 Hz (period lengths T = 50 s–222 s) during the active passive transition discussed above (Fig. 2.7).

Especially the shapes of the oscillations differ strongly during both scanning directions. In case of the passive active scan one can always observe a strong drift in the time series towards higher current i.e. towards the active state of the electrode. The oscillating system cannot be stabilized in this region.

2.3.4 *Ostwald's Original Chromium*

In order to compare the measurements of Wilhelm Ostwald and Eberhard Brauer, we have carried out cyclic voltamograms with the original piece of Ostwald's chromium which we got from Mrs. G. Brauer in 1 N and 2 N HCl solutions. These cyclic voltamograms (Fig. 2.8a, d) were recorded with the raw piece of Ostwald´s chromium without epoxy resin and also with a raw piece of pure chromium of undefined size (blue indicates a forward scan, while red shows the backward scan).

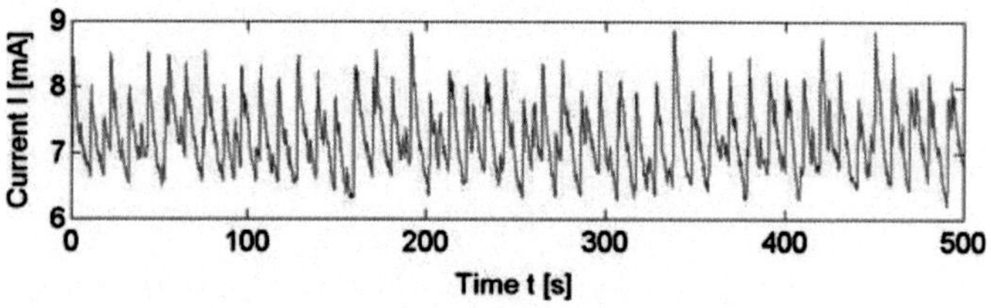

Fig. 2.7 Oscillations at the passive active transition at the potential E = −370 mV versus Ag/AgCl

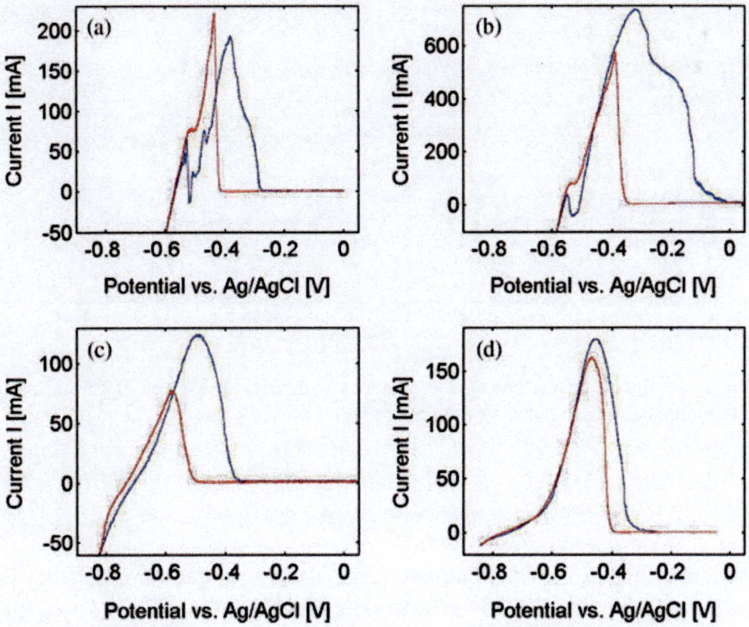

Fig. 2.8 Cyclic voltamograms of chromium taken with a scanning velocity of 1 mv/s: **a** Ostwald's chromium in 1 N HCl; **b** Ostwald's chromium in 2 N HCl; **c** pure chromium in 1 N HCl; **d** pure chromium in 2 N HCl

Surprisingly the cyclic voltamograms of Ostwald's chromium (Fig. 2.8a, b) obey some remarkable features in the active region at about-0,5 V versus Ag/AgCl compared to the cyclic voltamograms of the very pure chromium showed in (Figs. 2.3 and 2.8c, d).

Pregnant maxima in the active region can be detected in case of Ostwald´s chromium which might be responsible for the oscillations which he observed. These maxima are missing in the cyclic voltamograms of pure chromium (Fig. 2.8c, d). In addition there is a strong development of hydrogen in the first active region of Ostwald´s chromium whereas no hydrogen production in this region could be observed with pure chromium.

Furthermore, potential oscillations which are accompanied with the oscillation of the hydrogen production occur by its own if we put Ostwald´s piece of chromium in 2 N HCl solution. Figure 2.9 shows the time series of the potential versus Ag/AgCl.

In the minima of the potential oscillations (Fig. 2.9b) at about $E = -0.58$ V (versus Ag/AgCl) a strong hydrogen production takes place. In order to compare the original time series of W. Ostwald (e.g. Fig. 2.5) with the potential oscillations which we got from his piece of chromium (Fig. 2.10b) shows the negative potential as a function of time.

Thus there is a strong correlation between the hydrogen production and the maxima of the time series in (Fig. 2.9b). Only the shape of the oscillations is not

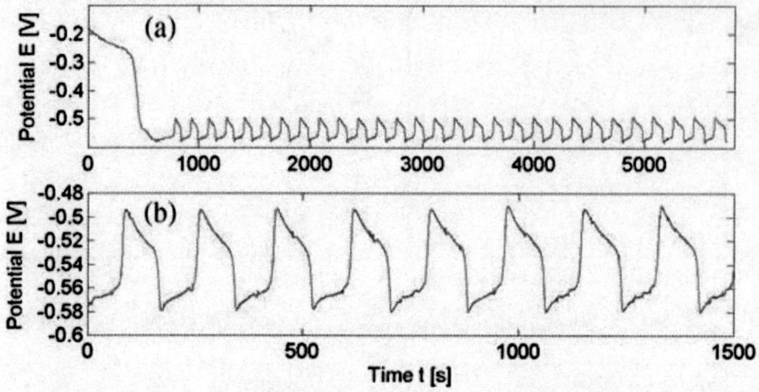

Fig. 2.9 **a** Development of the potential oscillations with Ostwald´s piece of chromium in 2 N HCl. **b** Magnified part of the potential oscillations with Ostwald´s piece of chromium in 2 N HCl

quite the same since he measured directly the hydrogen production. Also they have a different period length of T = 354 s (Ostwald´s experiment) and T = 175 s (potential oscillations).

Figure 2.10a shows the time series of W. Ostwald´s experiment. By digitising a scanned image (Fig. 2.5) it was possible to analyse the data and reconstruct the attractor of his original hydrogen oscillations.

Figure 2.11 shows both the attractor of Ostwald´s experiment and the attractor of the potential oscillations. Both have in common the typical structure of relaxation oscillations.

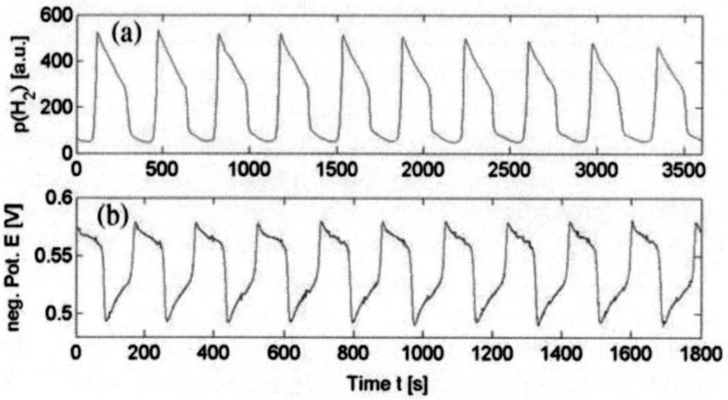

Fig. 2.10 **a** Digitised and rectified time series of hydrogen pressure oscillations of Ostwald´s experiments (compare Fig. 2.5). **b** Inverse representation of the magnified part of the potential oscillations (Fig. 2.9b) with Ostwald´s piece of chromium in 2 N HCl, whereby the maxima are correlated directly with the hydrogen production

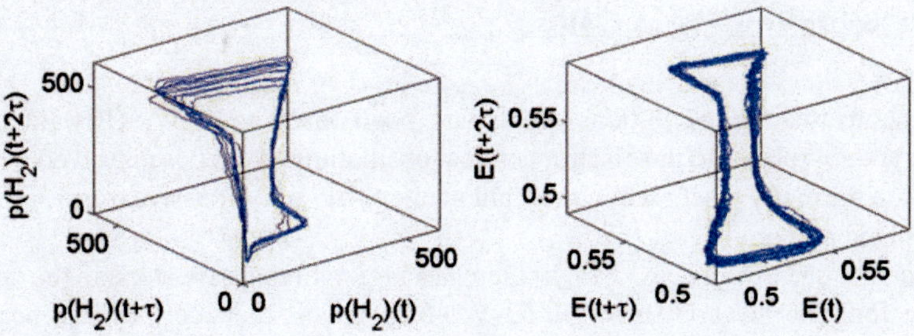

Fig. 2.11 Left: Attractor of Ostwald´s experiment of oscillating hydrogen pressure (compare Fig. 2.10a, Fig. 2.5) with $\tau = 21$ s. Right: Attractor in the delay phase space with $\tau = 8$ s of the potential oscillation of Ostwald´s original piece of chromium the time series of which is shown in (Fig. 2.10b)

The time series of the pure chromium in 2 N HCl differs strongly from the time series of Ostwald´s original piece of chromium (see Fig. 2.12). For example, the potential for which the current is zero of the intermittent oscillations is placed at about $E = -0.71$ V instead of $E = -0.57$ V in case of Ostwald´s chromium. Nevertheless, oscillations in the active region of the current—potential function occur also in case of the pure chromium. But they differ a lot in their position, shape, and character (Fig. 2.12) from those of Ostwald's original piece of chromium.

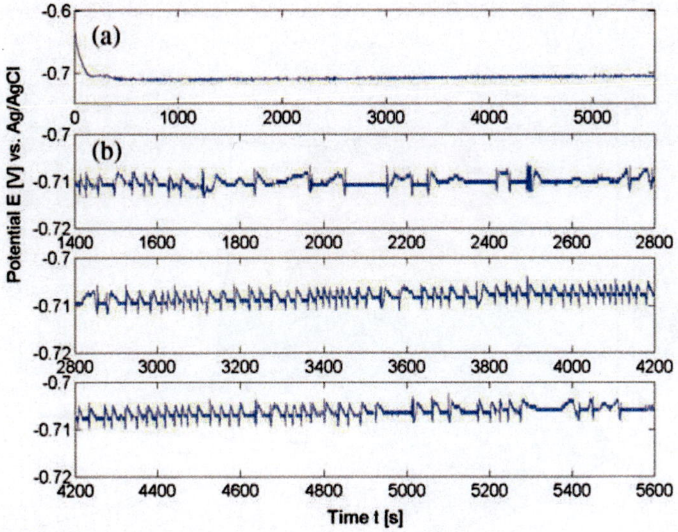

Fig. 2.12 Time series of pure chromium in 2 N HCl solution **a** An overview of the very small oscillation, and **b** an amplified representation of this oscillations

2.4 Oscillating Local Cells

It is obvious that the oscillations which have been observed by Wi. Ostwald and E. Brauer are not related to the electrochemical oscillations of metals dissolved in acids which are normally discussed in the framework of the active passive transition close to the Flade potential.

As can be seen from (Fig. 2.8a, b) there is a region of negative slope in the current voltage function at about E = −0.57 V which means that there exist a negative differential resistance which is responsible for the occurrence of the oscillation in this region [23, 24]. But there is no region like this in case of the pure chromium metal if one scans this metal with a velocity of 1 mV/s (Fig. 2.8 c and Fig. 2.8d).

Since it is known that the scanning velocity will influence the shape of the current voltage function we determined experimentally the two variable function $\frac{\partial I}{\partial E}$ with v equals the scanning velocity for the three situations: **a** the pure chromium in 1 N HCl, and Ostwald´s chromium in **b** 1 N HCl and **c** in 2 N HCl solution. Fig. 2.13 shows the surface $I = f(E, v)$ in the range from v = 10mV/sec to v = 1mV/s for the pure chromium in 1 N HCl.

The partial derivative $\frac{\partial I}{\partial E}$ of function $I = f(E, v)$ is positive semi-definite in the whole range up to the maximum but a shoulder is developing if one goes to smaller scanning velocities v. This shoulder might develop to a saddle point or even a maximum for the for the limit point $v \rightarrow 0$. This can be assumed because of the occurrence of intermittent oscillations on the pure chromium in 2 N HCl (see Fig. 2.12a, b).

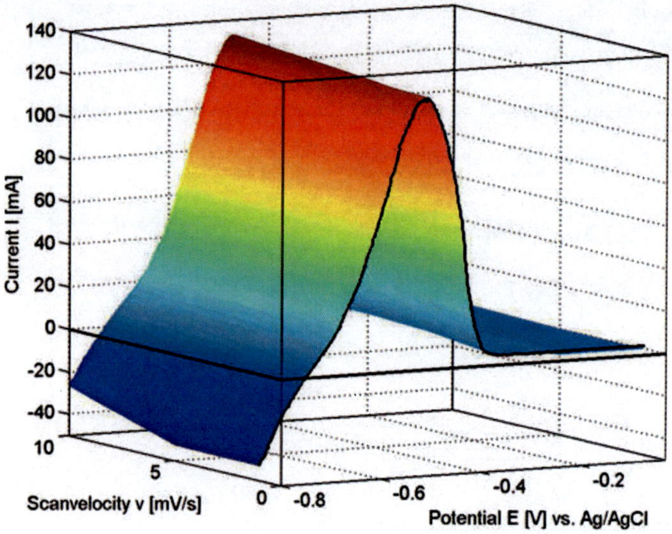

Fig. 2.13 The function $I = f(E, v)$ for the dissolution of pure chromium in 1 N HCl. v is the scanning velocity

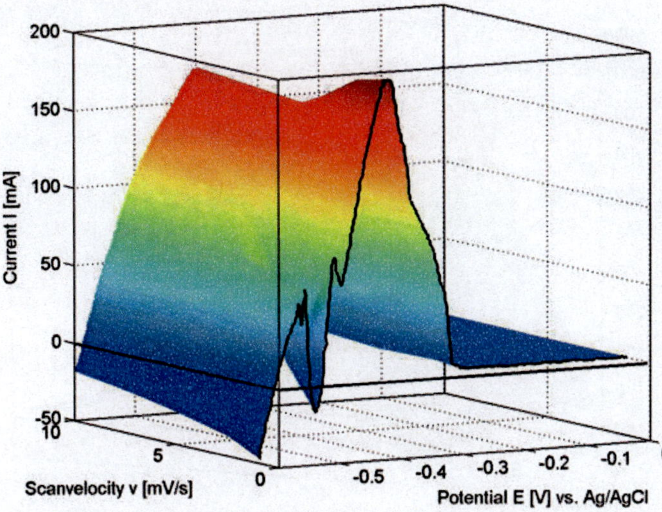

Fig. 2.14 The function $I = f(E, v)$ for the dissolution of Ostwald´s piece of chromium in 1 N HCl. v is the scanning velocity

In case of Ostwald´s piece of chromium one can observe a cusp catastrophe in the surface $I = f(E, v)$ for the 1 N HCl solution. A strong maximum arises if one diminishes the scanning velocity even only down to $v = 1mv/s$ (Fig. 2.14). Moreover a very complicated structure developes in the surface $I = f(E, v)$ with a cascade of cusp catastrophes.

For higher concentrations of the acid the main cusp catastrophe is not so well developed but it still remains and we can take it for sure that for the limit point $v \to 0$ the system posses a strong negative resistance $\frac{\partial I}{\partial E}$ (see Fig. 2.15).

In any case the potential oscillations as well as the oscillations in the hydrogen production are due to the occurrence of the negative slope in the derivation $\frac{\partial I}{\partial E}$ for very slow scanning velocities $v = 0$. It is obvious that Ostwald and Brauer could not choose any other scanning velocity than $v = 0$ since they put their pieces of chromium simply in the acid solution. Repeating their experiment we could proof that their hydrogen oscillations are strongly correlated with the potential oscillations in the region of the negative slope of the derivation $\frac{\partial I}{\partial E}$ in the range of the potential E = −0.57 V up to E = −0.49 V.

From the synergetic point of view the question is not essential what is the chemical mechanism which might cause the occurrence of the cusp folding in the surface $I = f(E, v)$. An explanation for this behaviour could be the assumption that in the chromium metal there are involved for example some metallic impurities forming local cells at the surface of the chemically inhomogeneous surface parts of the chromium alloys [25]. Another possibility could be that the surface is physically inhomogeneous because of different crystallographic planes which are exposed to the electrolyte [Vetter p. 599]. In any case the local excess voltage of hydrogen will

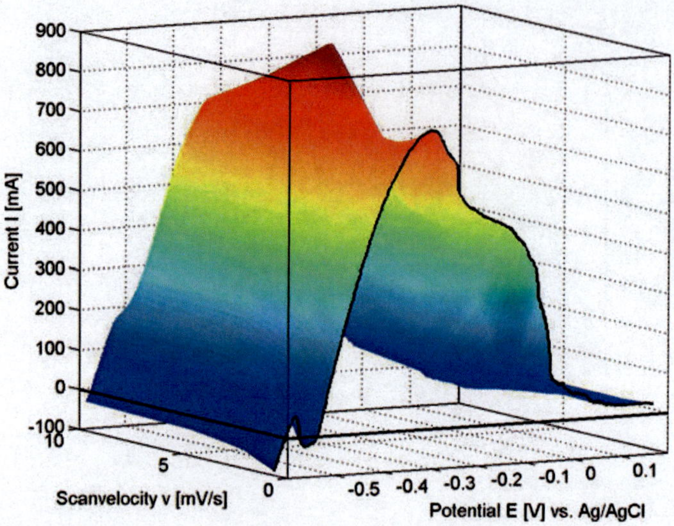

Fig. 2.15 The function $I = f(E, v)$ for the dissolution of Ostwald´s piece of chromium in 2 N HCl. v is the scanning velocity

be influenced differently compared to the homogeneously acting metallic surfaces. This scenario becomes visible if the local systems have time enough to develop itself, i.e. for very slow scanning velocities.

Let us assume that the surface is divided into two different areas which are distributed among each other randomly. The behaviour of the local cells can be described schematically by the addition of the current voltage functions of the processes which are involved. [25, 26]. Furthermore, the total current should be zero, as in case of Ostwald´s experiments, i.e. the mixing potential E_m. E_m can be estimated as the potential for which the sum of all partial currents becomes zero.

The dissolution of the chromium is accompanied by the production of hydrogen; i.e. there are at least two independent electrode processes for example:

$$\mathrm{Me} \rightarrow \mathrm{Me}^{z} + +z * e^{-}$$

and

$$\mathrm{H}^{+} + 2 * e^{-} \rightarrow \mathrm{H}_2$$

contributing to the total current which can be observed in the cyclic voltamogram. The potential for which the total current is zero is called the mixing potential E_m [25, 26]. For this potential both current flows are equal in their absolute values.

Figure 2.16 shows a sketch of the overvoltage of hydrogen during the dissolution of a less noble metal in an acid. The distance between the zero points of the metal curve and the total current curve is called the anodic overvoltage $\eta_1$of the hydrogen

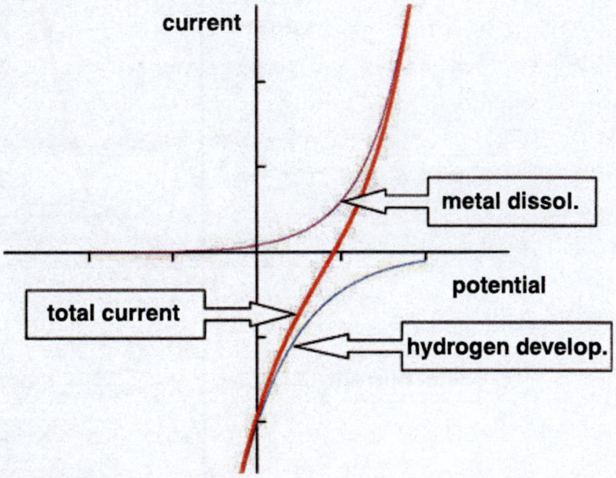

Fig. 2.16 Sketch for the construction of the mixing potential E_m for a less noble metal accompanied by the production of hydrogen [Vetter P. 589]

and η_2 the cathodic overvoltage is the distance between the corresponding zero points of the hydrogen curve and the total current curve.

Such a total current function I = f(E) could be observed in principle for the dissolution of the pure chromium in the cyclic voltamograms (Fig. 2.8c, d).

In this framework one may understand now the behaviour of the pure chromium as well as Ostwald's piece of chromium in hydrochloric acid.

Figure 2.8c shows a typical total current function I = f(E) for the less noble metal of the pure chromium in 1 N HCl with the development of hydrogen. For higher concentrations of the acid (Fig. 2.8d), the partial function $I_{hydr.} = f(E)$ is shifted to more positive values whether the total current function gets a shoulder and is slightly shifted to more positive potentials as well.

However, this description does not correlate totally with the observations for Ostwald´s pieces of chromium. There we can find a non monotonous behaviour of the total current function in the active region of the cyclic voltamogram (Fig. 2.8a, b).

Assuming, that the surface of the very polycrystalline chromium (see Fig. 2.2) is composed at least of two different kind of surfaces—never mind what is their chemical composition—one can describe approximately (Fig. 2.17) the experimental curves (Fig. 2.14). The partial surface—of the less noble metal 1—is dissolving very fast at a lower potential and gets passivated very soon (metal 1 curve in Fig. 2.17) with increasing potential.

The metal 2 surface is dissolved remarkably only for higher potential. Therefore there exists a range in the potential for which the anodic current is converted into a cathodic one because of the excess in the hydrogen production. In the region for which the potential causes a stronger dissolution of the metal 2 surface the hydrogen production is strongly diminished. This is just in correspondence with our observations in case of Ostwald´s piece of chromium.

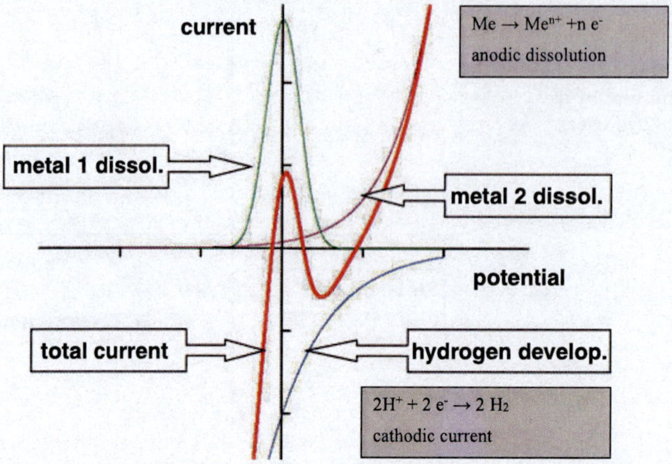

Fig. 2.17 Sketch of the total current function I = f(E) (red) during corrosion of a chemical or physical inhomogeneous surface assuming equilibrium for each involved process

This statement can only be taken as a hypothesis but not as a proofed explanation. Indeed we observed not only one maximum in the total current function but several one.

This could mean that maybe several different surfaces participate to the total current. On the other hand slow scanning velocities enable Ostwald´s system to develop transient states of self organised oscillations which can also be observed in the cyclic voltamogram. The strong first maximum in the surface $I = f(v,E)$ See Figs. 2.14 and 2.15 indicates that our hypothesis is valid at least for this area of the scanning velocity v and the potential E.

However, all these models do not explain sufficiently the oscillations. The assumption of Traud and Wagner [26] on the balanced currents is based on the equilibrium of the system, but not on a situation far from equilibrium as in case of oscillations.

Although we observed potential oscillations during the dissolution of Ostwald's chromium the overall current is zero since there does not exist an electronic circuit beyond the electrolyte in this experiment. So, we are forced to assume local currents based on the existence of local cells as in case of corrosion. We can compare this situation with the galvanostatic potential oscillation—in case of $I = 0$—of a system with a N-shaped current voltage function.

Up to now, we dicussed only the situation at very slow scanning velocities. Now, regarding the (Figs. 2.14 and 2.15), one can observe a folding in the surface $I = f(E, v)$. For higher scanning velocities v no negative differential resistance NDR [27] is observable in the surface whereas a NDR occurs if the scanning velocity becomes small enough. Such a NDR might be caused by the development of hydrogen bubbles. The velocity of this bubble creation process is obviously in the range of the slow scanning velocities used in the our experiments. For higher scanning velocities the system gets out of the range producing enough hydrogen to form hydrogen gas bubbles whether the NDR becomes hidden [27].

The produced hydrogen creates very small hydrogen bubbles at the metal surface. The bubbles will grow, while hydrogen is produced and they will undergo coalescence if they touch each other. If the bubbles reach a critical size they will suddenly leave the surface and they will rise up in the electrolyte because of buoyancy. This is a collective phenomenon and will create a big jump in the H_2 pressure as observed by Wi. Ostwald. Afterwards the surface gets almost cleaned of hydrogen and the potential is raising again. Suddenly a big jump in the potential can be observed caused by the active/passive phase transition of the less noble metal 1 and the visible hydrogen production stops totally.

During creation and coalescence of bubbles the hydrogen pressure and the potential decrease slowly according to an increase of the anodic overvoltage on the noble metal 2. Meanwhile metal 1 becomes activated from its passive state.

In the framework of U.W. Franck's model for the oscillating anodic dissolution of metals in acid electrolytes the pH—dependency of the Flade potential is responsible for the oscillations. The hydrogen development diminishes strongly the acidity in front of the metal 1, i.e. the pH-value is increasing while hydrogen is produced. Assuming that the growing pH-value will shift the Flade potential E_{Fl} of the metal 1 to lower values the metal 1 gets passivated [20] (Fig. 2.18).

After reaching its minimum, the. acidity will increase again in the vicinity of the metal surface because of diffusion of hydrogen ions from the bulk electrolyte towards the less acid region close to the metal surface.

Following the arguments given by R. Otterstedt [20] the H^+ concentration h_0 close to the metal surface can be identified with the constants of the current voltage curve for the metal dissolution

$$f = p_1 \varphi_0^3 \exp\left(-p_2 \varphi_0^2\right) \tag{2.1}$$

Current passage overvoltage = Faraday current

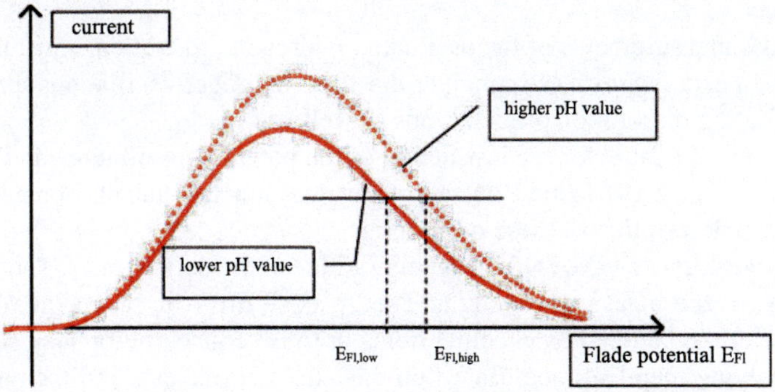

Fig. 2.18 Sketch of the pH dependency of the current on the Flade potential E_{Fl} concerning $g_A(\varphi, h_0)$[20]

$$g_1(\varphi, h_0) = p_1\varphi_0^3 \exp\left(-p_2\left(\varphi_0 + p_3 h_0^2\right)^2\right) \tag{2.2}$$

$$g_2 = -p_5 \exp(-b\varphi + c - d\ pH) \tag{2.3}$$

$$C_d \frac{\partial \varphi_0}{\partial t} = I_{tot} - g_1(\varphi_0, h_0) + g_2(\varphi_0, h_0) \tag{2.4}$$

$$\frac{\partial h_0}{\partial t} = \frac{D}{\delta}(H_{bulk} - h_0) + \mu h_0 \kappa \frac{\partial \varphi_0}{\partial \vec{n}} \tag{2.5}$$

2.5 Phenomenological Two Variable Model

But up to now, this hypothesis does not at all describe the oscillations in the hydrogen production which have been observed by Wi. Ostwald [1, 2] and E. Brauer [3]. As we have shown (Fig. 2.10a, b) these oscillations are strongly correlated with the oscillations in the potential. If we have a strong negative potential at about –0.58 V in 2 N HCl one can observe a strong hydrogen production. The oscillations in the potential form a limit cycle (Fig. 2.11) which strongly reminds to relaxation oscillations which can simply be described by the well known types of two variable models. Since the potential is one of the variables and the total current is fixed to zero we have to find a second variable of physical meaning which is not linearly dependent on the potential.

It is well known that local cells exist in the surfaces of alloys. Assuming that in the original piece of chromium such local cells occur as impurities. Local currents would be the consequence. Since potential oscillations could be observed (see Fig. 2.9) during the dissolution of Ostwald´s original piece of chromium but not at all in pure chromium we have to assume accompanying local current oscillations in local cells. But all these oscillating local cells have to be coupled because of the experimental observations.

Ostwald´s measurements of the oscillating hydrogen production is just the proof for this hypothesis. Now, we can conclude that these hydrogen oscillations correspond to the oscillating dissolution of metal ions as well.

One can use the attractor reconstruction of the potential oscillations in the delay phase space (Fig. 2.19) to find out a simple two dimensional phenomenological mathematical description of these oscillations.

For this purpose we have chosen a generic value of the delay time τ. The "x-axis" E(t) represents the measured potential. The "y-axis" E(t + τ) may symbolizes the mean value of the coupled local cell currents or th H^+ concentration h_0 just in front of the dissolving metal surface. Based on this attractor one can find the non-linear zero iso-clines of the system just by fitting a function to the run of the curve. In order

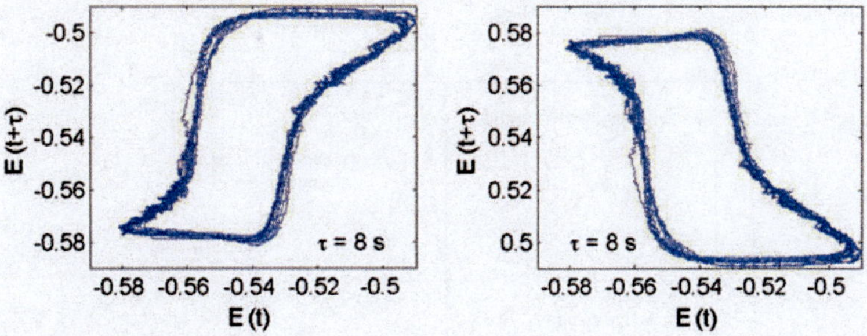

Fig. 2.19 **left**: Attractor of the potential oscillations (Ostwald´s original piece of chromium) in the two-dimensional delay phase space (delay-time $\tau = 8$ s). **right**: same attractor with possitive E(t + τ) axis

to obtain a zero iso-cline with a negative slope of the function E (t + τ) = F(E(t)) the attractor was reconstructed with positive values for E(t + τ) (Fig. 2.19 right).

The second zero iso-cline could just be derived, assuming a linear differential equation. A simple phenomenological two dimensional systems which enables to reconstruct qualitatively well the dynamics of the observed system by the following equations:

$$\varepsilon \frac{dx}{dt} = -2abx \exp(-bx^2) + c \exp(x) - kx - y \tag{2.6}$$

$$\frac{dy}{dt} = mx - y \tag{2.7}$$

Using this set of equations (Eqs. 2.6 and 2.7) one can try to find the best approximation of the reconstructed limit cycle (Fig. 2.20).The constants of this fit are:

a = 1b = 3 c = 0,1

k = 1m = 10ε = 1

For sure, these constants will change with the chosen value of the delay time constant τ, but nevertheless since we have chosen a generic value of τ, the slightly different values of the constants will correspond to topological equivalent representations of the limit cycle. Figure 2.20 shows the attractor of the dynamics of the two variable model concerning the chosen set of constants for these equations.

To compare the dynamics of the phenomenological two variable model and the potential oscillations (Fig. 2.21) shows both time series. They have in common thypical shape of relaxation oscillations, i.e. a fast activation and deactivation which is mainly determined by the slope of the second linear differential equation and two parts of a much slower development of the sysem. In this slower part the system runs along the non-linear zero iso-cline. According to the value of the time scale separation ε the system runs more (small ε) or less (bigger ε) close to the zero iso-cline (Eqs. 2.6 and 2.7).

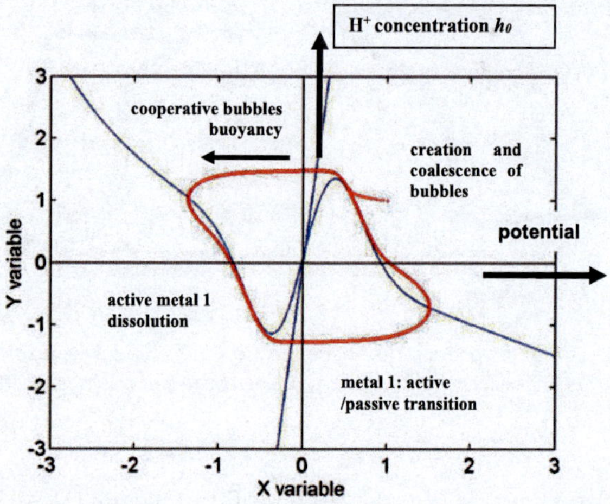

Fig. 2.20 Sketch of the limit cycle dynamics (red) which may describe the observed potential oscillations and zero-isoclines (blue) of the correlated attractor concerning (Eqs. 2.6 and 2.7).

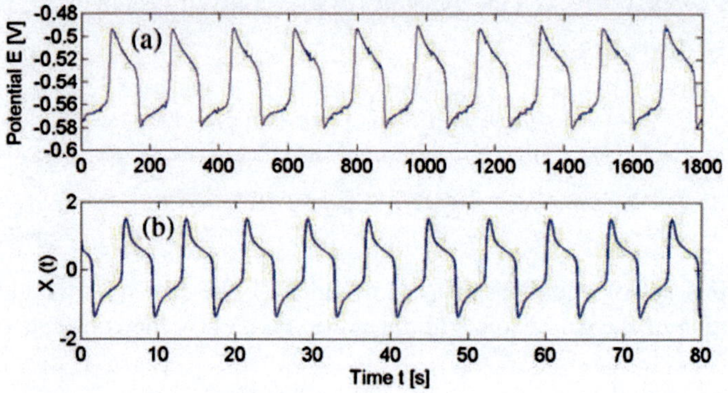

Fig. 2.21 **a** Time series of potential oscillations of Ostwald´s original piece of chromium. **b** Time series of x(t) from the two-variable model (Eqs. 2.6 and 2.7)

Although our phenomenological two variable model does not reproduce the original frequency of oscillations, which is a question of correct values of the parameter pair m and ε, it is qualitatively in accordance with the shape of potential oscillations.

2.6 Conclusions

It is a pity that these fundamental works of Wi. Ostwald [1, 2] and E. Brauer [3] on the great variety in the dynamics of electrochemical systems got almost lost in the actual scientific knowledge since about 100 years.

For a long time one could not really understand the very complicated dynamics of this electrochemical system. Especially because of the fact that Wi. Ostwald could not reproduce his experiments with more pure samples of chromium one put these excellent scientific results either into the box of failed experiments because of impurities or in the box of scientific curiosities which are of no generic meaning.

It was a great luck that some pieces of Ostwald´s original chromium survived for 100 years in the protection of Mrs G. Brauer. So we were able to repeat really Ostwald´s and Brauer´s original experiments. Using our days equipment we proofed that impurities were responsible for their results.

But in contrast to common meaning these "impurities" did not make worthless these historical experiments. Instead of being neglectable these "impurities" form local cells indeed which cause the observed oscillations.

So, what Wi. Ostwald and E. Brauer really detected that were the oscillations of local cells during dissolution of chromium. These detections open a new way in the understanding of corrosion, since they enable us to get much deeper knowledge on the dynamics of this complex phenomenon of great technical importance.

Acknowledgements In recognition of the untiring work of Mrs. Dr. Grete Brauer in the Wilhelm Ostwald Gesellschaft and Mrs Prof. Katharina Krischer (University of Munich) who rescued these works from oblivion we were enabled to bring back to our days scientific consciousness the fundamental meaning of these works of Wi. Ostwald and E. Brauer.

References

1. Ostwald, W.: Perodische Erscheinungen bei der Auflösung des Chroms in Säuren I. Z. Phys. Chem. **35**, 33–76 (1900)
2. Ostwald, W.: Perodische Erscheinungen bei der Auflösung des Chroms in Säuren II. Z. Phys. Chem. **35**, 204–256 (1900)
3. Brauer, E.: Über das Elektrische Verhalten des Chroms bei der Auflösung in Säuren. Inaugural-Dissertation, Universität Leipzig, Verlag Wilhelm Engelmann Leipzig (1901)
4. Fechner, M.G.Th.: Zur Elektrochemie – Ueber Umkehrungen der Polarität in der einfachen Kette. Schweigger J. Phys. Chem., Halle. **53**, 129–151 (1828)
5. Ostwald, W.: Elektrochemie – Ihre Geschichte und Lehre, Verlag Veith and Comp., Leipzig (1896)
6. Keir, J.: Philos. Trans. **80**, 359 (1790); and Von der Präcipitation des Silbers aus SalpeterSäure durch Eisen. Schweigger J. Phys. Chem., Halle. **53**, 151 (1828)
7. Wetzlar: Schweigger J. Phys. Chem., Halle. **49**, I. 470 II. 88 and 129 (1827)
8. Herschel, L.F.W.: Ann. Chim. Phys. **54**, 87 (1833) and Pogg. Ann. **32**, 211 (1834)
9. Schönbein, Ch.F.: Pogg. Ann. **38**, 444 (1836) and Arch. l'électicité. **2**, 269 (1812)
10. Joule, J.P.: Philos. Mag. **24**, 106 (1841)
11. Bonhoeffer, K.F.: Über periodische chemische Reaktionen I. Z. Elektrochem. **51**(1), 24–29 (1948)
12. Bonhoeffer, K.F., Brauer, E., Langhammer, G.: Über periodische Reaktionen II – die kathodische Polarisation von Eisen in Salpetersäure. Z. Elektrochem. **51**(1), 29–37 (1948)
13. Bonhoeffer, K.F., Haase, V., Langhammer, G.: Über periodische Reaktionen III – der Refraktärzustand frisch-passiven Eisens in konzentrierter Salpetersäure. Z. Elektrochem. **51**(2), 60–67 (1948)

14. Bonhoeffer, K.F., Langhammer, G.: Über periodische Reaktionen IV – Theorie der kathodischen Polarisation von Eisen in Salpetersäure. Z. Elektrochem. **51**(2), 67–72 (1948)
15. Franck, U.F.: Chemische Oszillationen. Angew. Chem. **90**(1), 1–16 (1978)
16. Franck, U.F., Fitz-Hugh, R.: Periodische Elektrodenprozesse und ihre Beschreibung durch ein mathematisches Modell. Z. Elektrochem. – Ber. Bunsenges. Phys. Chem. **65**(2), 156–168 (1961)
17. Franck, U.F., Meunier, L.: Gekoppelte periodische Elektrodenvorgänge. Z. Naturforsch. **8b**, 396–406 (1953)
18. Flade, F.: Z. Phys. Chem. **76**, 513 (1911)
19. Kolotyrkin, Y.M.: Electrochemical behaviour and anodic passivity mechanism of certain metals in electrolyte solutions. Z. Elektrochem. **62**(6/7), 664–669 (1958)
20. Otterstedt, R.: Raumzeitliche Strukturbildung bei der anodischen Auflösung von Kobalt in Phosphorsäure – Experimente und Modellierung. Universität Bremen, Doktorarbeit (1997)
21. Plath, P.J.: Jenseits des Moleküls – Raum und Zeit in der Chemie. Buch (222 Seiten) – Reihe Facetten. Vieweg & Sohn Verlagsgesellschaft mbH, Braunschweig, Wiesbaden (1997)
22. Otterstedt, R.D., Plath, P.J., Jaeger, N.I.: Phys. Rev. E. **54**, 3744–3751 (1996)
23. Hudson, J.L., Tsotsis, T.T.: Electrochemical reaction dynamics: a review. Chem. Eng. Sci. **49**, 1493–1572 (1994)
24. Mazous, N., Flätgen, G., Krischer, K.: Phys. Rev. E **55**, 2260–2266 (1997)
25. Vetter, K.J.: Elektrochemische Kinetik, p. 589. Springer, Berlin, Göttingen, Heidelberg (1961)
26. Wagner, C., Traud, W.: Z. Elektrochem. **44**, 391 (1938)
27. Strasser, P., Eiswirth, M., Koper, M.T.M.: J. Electroanal. Chem. **478**, 50–66 (1999)

Chapter 3
Liesegang Structures

Peter J. Plath

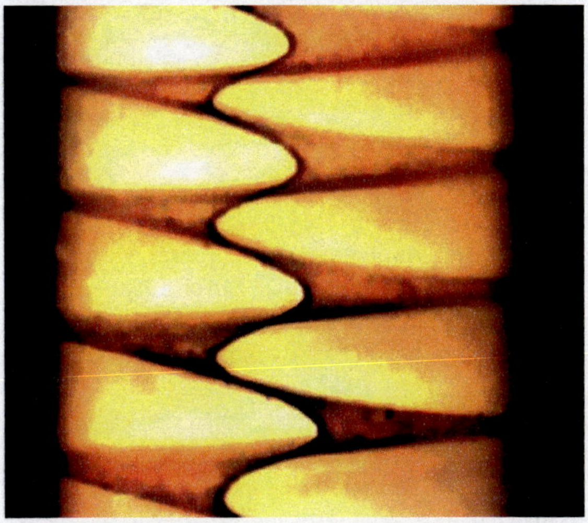

Fig. 3.1 Self-organized helicoid precipitation patterns of lead chromate in agar–agar gel in a test tube (size in diameter 1.56 cm). (*Photo* W. Jacobi, P.J. Plath)

This chapter on Liesegang structures [1] is largely based on the work of my coworkers Werner Jacobi [2], who prepared the famous Liesegang screw surfaces in the test tubes as shown in Fig. 3.1 Claudia Müller, Uwe Sydow [3], Robert Lipski, Bernd Görtler, and A. Fiedler whom I thank for their fruitful cooperation and the carefully executed experiments. Special thanks are due to S. Hollatz and T. Plikat, whom I owe to the experimental elucidation of the three-dimensional Liesegang structures, which are reported for the first time in detail in this book. A. Deutsch and K. Koblitz, I would like to thank for providing the photos on the Figs. 3.2 and 3.3 respectively.

Almost half a century later than F.F. Runge (1896), Raphael Eduard Liesegang (born November 1, 1869 in Düsseldorf, 13 November 1947 in Bad Homburg) worked

P. J. Plath et al., *Imagery Synergetics*, Understanding Complex Systems,
https://doi.org/10.1007/978-3-030-95607-3_3

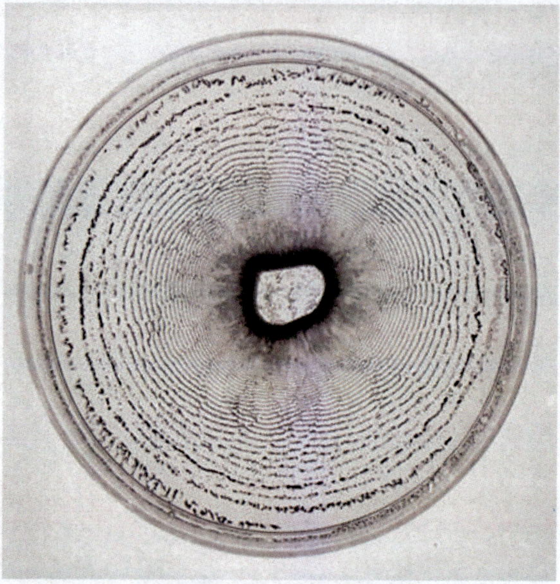

Fig. 3.2 Formation of Liesegang rings during the precipitation of KJ (potassium iodide) in a Petri dish. Inner Electrolyte: $Pb(NO_3)_2$; outer electrolyte: solid KJ (*Photo* W. Jacobi, P.J. Plath—in 1984). Liesegang named such ring structures A-lines

in the company of his father, which he took over together with his two brothers in the same year when his father died. They produced light-sensitive gel-like silver chlorate layers on glass plates for photographic purposes. He applied a drop of silver nitrate solution to a thin layer of gelatine which had already been solidified on a glass plate and soaked with potassium dichromate. After one day, he observed an annular periodic precipitation of slightly soluble silver dichromate with a diameter of 5 cm (Figs. 3.2 and 3.3).

In the light of the new structures found in physics executing gas evacuations in "evacuated" gas-discharge tubes, "in which often stratifications of light" occur, he named his discovery "A-lines" and published them with success.

The analogies that led Raphael Liesegang to call his discovery of the cyclic, periodic structures in precipitation reactions in gels "A-lines" is not at all as outlandish as they may seem to many readers today, and I present them here as his working hypotheses:

> In den Crookesschen Röhren, welche seit der Röntgenschen Entdeckung auch für die Photographie von Bedeutung erlangt haben, kommen häufige Schichtungen des Lichtes vor, welche wie stehende Wellen aussehen. … Bei meinen Gallerten-Versuchen habe ich einige Erscheinungen beobachtet, welche vielleicht zu einer Erklärung dieser Lichtschichtungen führen können.

> In the Crookes tubes, which have also become important for photography since Röntgen's discovery, frequent layers of light appear, which look like standing waves. ... In my jelly experiments I have observed some phenomena, which may perhaps lead to an explanation of these layers of light.

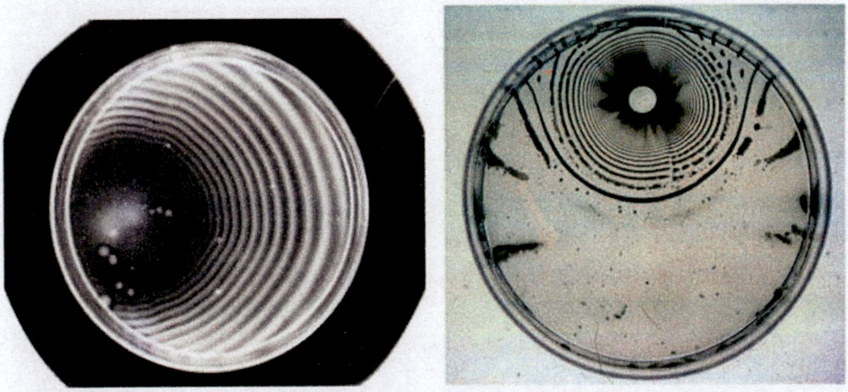

Fig. 3.3 No reflection of the precipitation waves at the vessel rim of the Petri dish (*Photo* W. Jacobi, P.J. Plath (1984)) **b** internal electrolyte: $5{\cdot}10^{-3}$ Mol/l $Pb(NO_3)_2$; external electrolyte: crystals of KJ (potassium iodide)

The precipitation patterns he discovered are indeed waves, so called “chemical waves” or as V.I. Krinsky calls tthem: “autowaves”. [4]

> However, “autowaves differ fundamentally from waves in traditional (conservative) media. They propagate at the expense of energy taken from the active media and therefore cannot be considered as conservative systems. The shape and amplitude of autowaves remain constant during propagation, whereas the amplitude of classical waves rapidly falls with the distance and the waveform is distorted by dispersion. In the case of autowaves, there is no reflection from either the medium boundary or inhomogeneities. Unlike waves in conservative media (solitons and soliton-like solutions) two colliding autowaves annihilate rather than penetrate one another, and, therefore no interference takes place.”

If you bring the outer electrolyte not in the center of the Petri dish, but near its edge, then you can see very well that the precipitation waves are not reflected on the edge of the Petri dish but extinguished as shown in Fig. 3.3.

If the experimental arrangement is changed in such a way that the outer electrolyte is applied to two opposite positions of the Petri dish, the propagating precipitation waves must inevitably strike. As shown in Fig. 3.4, the precipitation waves are not superimposing, but the precipitating waves which strike are annihilating themselves.

The precipitation waves are therefore attributable to the class of chemical waves or autowaves. This does not change the fact that in these experiments they are not stable in their shape, and indeed lose intensity even with the distance from their place of origin. That’s because in these closed systems the “chemical energy reserve” is used up by the precipitation reaction, namely the ions involved in this reaction were used up and cannot be replenished arbitrarily. The shape of the waves, however, depends on the particular local concentration conditions (Figs. 3.3 and 3.4).

Bringing the outer and inner electrolytes together, then strong concentration gradients can occur in the boundary region of the inner electrolyte, whereby structures are possible, leading in the further course of the diffusion and precipitation process, for example, to rings, discs or spirals.

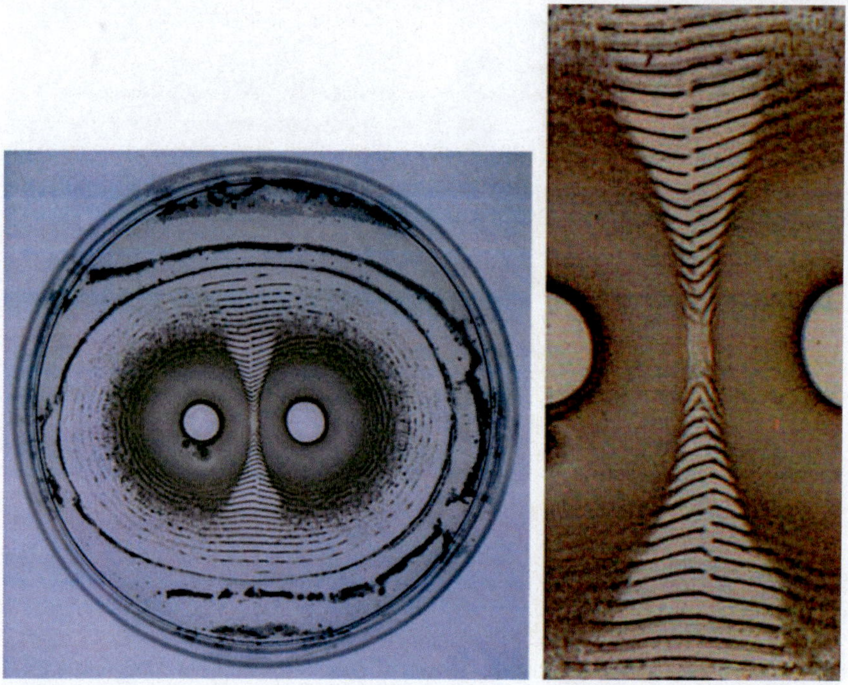

Fig. 3.4 **a** Annihilation of colliding precipitation waves (*Photo* W. Jacobi, P.J. Plath in 1984); **b** Enlarged detail of (**a**)

This can lead to very confusing precipitation patterns, which are strongly reminiscent of turbulent flows. Out of such turbulent precipitation patterns, classic Liesegang rings can develop in the further diffusion of the outer electrolyte into the inner electrolytes (see Fig. 3.5).

There are known a lot of sparingly soluble salts with which one can create similar precipitation patterns in gels. For example if one dilutes $Pb(NO_3)_2$ in 0.5 … 1.5% agar–agar in a Petri dish and places some crystals of potassium iodide in the center of the dish after the gel has been cooled, then concentric precipitations of lead iodide are obtained after a few hours.

He also carried out many experiments on these spatial patterns, which are now known as Liesegang structures, in the test tube and observed fantastic pattern formations such as periodic slices, helical surfaces and spirals. But his beautiful structures had no further significance for the development of photochemistry. These patterns seemed to be an exceptional case in the precipitation reactions, just curious curiosities occurring only in gels, and so they had no further influence on the development of physical chemistry except colloid-chemistry.

In his later work, Raphael Liesegang turned to the structures of the beautiful agates, where he could show that his "Liesegang structures" are at least of decisive

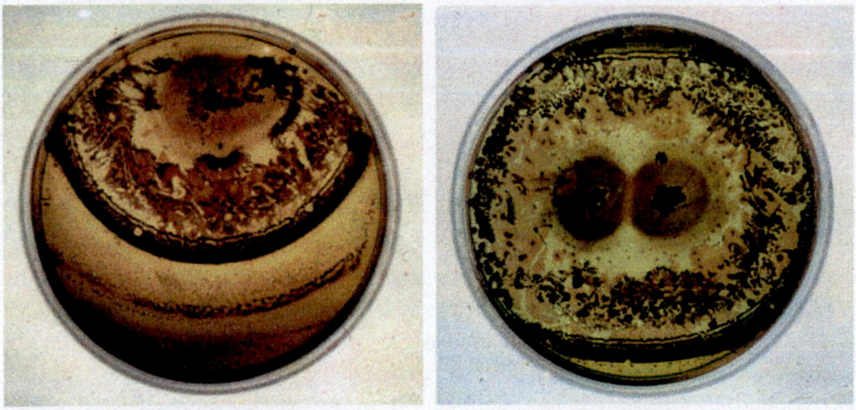

Fig. 3.5 Turbulent precipitation patterns. The inner electrolyte consists of crystals that dissolve in their immediate environment by partially destroying the gel there. Gel and the highly concentrated solution form a relatively sharp border. The precipitation then takes place in a turbulent manner only at some distance from this borderline in a wide range. (*Photos* W. Jacobi, P.J. Plath in 1984)

Fig. 3.6 Cut through a three-dimensional spiral occurring in a flint bulb. (*Photo* P.J. Plath)

importance in the formation of rocks from the viewpoint of mineralogists and geologists (Fig. 3.6) [5]. That they also occur in mountain formation, in a much larger range of scales, was only shown by Karl-Heinz Jacob (TU-Berlin) in the 1990's [6]. It should be noted at this point that his ideas on the role of Liesegang structures in electric fields have contributed to violent controversies in today's discussion on earth crust formation and continental shifts [7].

Both, Runge's and Liesegang's works as well as the publications of Beloussov and Zhabotinsky are portrayed quite carefully in their scientific-historical context, and described by Lothar Kuhnert and Uwe Niedersen in their book, "Selbstorganisation chemischer Strukturen", in the series "Ostwalds Klassiker der exakten Wissenschaften" [8].

Raphael E. Liesegang published his first papers on pattern formation in gelatine in 1896 [1–3]. In general, the experimental preparation runs as follows: a gel is prepared which contains an ionic species (inner electrolyte) that can later be precipitated by a suitable counter ion (outer electrolyte).

The experiment is usually carried out in a one- or two-dimensional arrangement, e.g. in a test tube or in a Petri dish; the precipitation reagent is then placed as a solid or in concentrated solution on to the gel column or in the middle of the gel layer respectively. During the diffusion of the counter ion into the gel, the precipitation reaction occurring sometimes produces concentric rings or sharp disks of precipitate. These structures are commonly referred to in the literature as Liesegang rings.

Moreover, it is a very simple chemical experiment, which is still full of surprises. A lead-salt solution is prepared in a 100 ml graduated flask (2–4 ml,10^{-1} Mol Pb $(N0_3)_2$), add 25 ml of a 3% hot agar-gar solution (filiform agar–agar DAB6 from Riedel de Haen) which is cooked until clear and then fill with distilled water until mark is reached. Now cook the resulting lead-containing sol again (one hour on a boiling water bath) so that all streaks disappear and fill some test tubes with the hot sol.

Now, leave it in the test tube sol is allowed to gel under room temperature. Overlaying carefully the solidified lead containing gel with a saturated solution of potassium iodide which still contains several grains of potassium iodide in order to obtain a saturated solution throughout the experiment.

If the test tube prepared in this way is left to stand for several hours, sometimes at least days, one observes slowly emerging gorgeous structures in the precipitation reaction which begins. This includes the formation of periodically occurring discs (rings) and also of helical surfaces (spirals).

3.1 Historical Notes

Since their first observation by R. E. Liesegang (1896), the occurring of these highly symmetric patterns during precipitation reactions in gels were a challenge for Chemists and physicists.

As mentioned above, in 1896 R. E. Liesegang coated a glass plate with gelatin, to which he had previously added to $K_2Cr_20_7$. After forming the gel, he added a few drops of $AgN0_3$ and after one day he observed the formation of circular patterns (compare Fig. 3.2).

In 1914, for the first time, Liesegang also reported on the development of helical surfaces in special test tube tests. Also Wo. Ostwald (1925) and R. Fricke (1926) found in their experiments, in addition to the periodic disks, the hard-to-understand helical surfaces (Fig. 3.7).

The first attempt to interpret Liesegang's structures—especially the periodic formation of the disks (rings), Ostwald (1897) made already a year after their discovery. According to his hypothesis, a supersaturated state occurs, from which the spontaneous precipitation takes place. The ions still remaining in the vicinity are

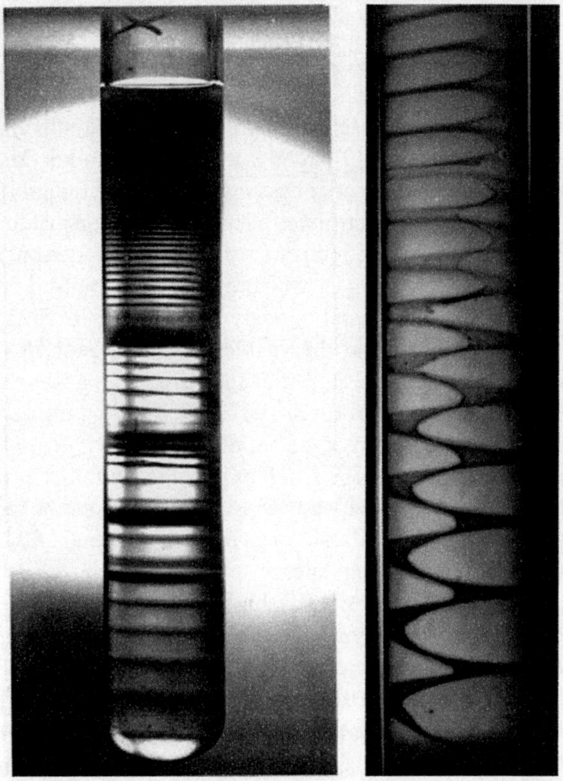

Fig. 3.7 Discs (often called rings) and helicoids (often called spirals) precipitation patterns of lead chromate in agar–agar gel. (*Photos* W. Jacobi, P.J. Plath)

then diffused in the direction of the condensation nuclei in order to be precipitated there as well. Thus, in the direct vicinity of the respective precipitation ring, depletion occurs at the ions of the inner electrolyte which is dissolved in the gel. Through this zone of low concentration of the inner electrolyte, the ions of the outer electrolyte have to first diffuse before reaching a zone of sufficient concentration of the inner electrolyte, so that super-saturation and then again precipitation may take place.

With this hypothesis one succeeded in understanding the steady increase of the distance between the rings qualitatively, but there are also systems from which this "distance law" is not followed. Moreover, extra condensation nuclei added had no significant influence on ring formation. Other observed phenomena could also be identified which cannot be satisfactorily described by this "super-saturation hypothesis":

- The presence of artificially supplied nuclei does not have a significant influence on ring formation
- There are a number of systems which do not obey the "ring distance law", but rather the distance between them decreases with increasing distance from the point of application of the external electrolyte.

- Ostwald's theory does not provide any explanation for the often very pronounced fine structure of the rings.

As a result, further attempts have been made to investigate the interesting phenomenon of Liesegang's structure-formation to understand better.

Thus, Bradford [9] stood out in 1916 with his "theory of adsorption" trying to explain the often clear zones between the precipitation rings by an adsorption of the ions of the inner electrolytes on the precipitation just formed. Only in 1954 Stern [10] contradicted to this attempt to explain, and assigned only a tendency for the adsorption of the ions to reinforce structure formation, but did not regard it as its cause. Dhar [11] on the other hand, already represented, in 1922, the very strong assumption that the formation of the visible precipitate precedes the formation of a colloidal dispersion. The appearance of such dispersions was experimentally confirmed by J. Ross and his co-workers in 1982 [12]. The ring formation occurs according to N. R. Dhar by coagulation of the colloidal particles.

Already Ostwald [13] referred to the solubility products in his "diffusion wave theory" of the participating ions. In his opinion, the structure formation is based on the interaction of diffusion waves of the internal and external electrolyte as well as of the colloidal precipitate.

All these hypotheses are based solely on equilibrium thermodynamics thus the Liesegang phenomena were not conceived as a consequence of self organization of the system far from equilibrium. This is achieved by J. Ross and his co-workers. They assumed that the formation of nuclei first begins homogeneously, and only later coagulation starts accompanied by the formation of the clear zones. In doing so, they are assuming "autocatalytic" growth of the nuclei beyond a certain particle size.

It was only by the work of Ortoleva [14] and Ross [15] that interest became apparent at the Liesegang structures again. It was recognized that these structures are particularly impressive examples of the pattern formation processes in reaction diffusion systems in addition to the well-known structures in the Beloussov-Zhabotinsky reaction.

The peculiarity of the Liesegang system is that it is the spatial one and temporally largely stationary formation of patterns in the precipitation of a sparingly soluble salt in the gel.

If the precipitation of lead iodide PbJ_2 is carried out in the usual, every beginner of chemistry trusted way in aqueous solution in the test tube, the familiar form of the cloudy, yellow precipitate is obtained, which is slow on the ground (Fig. 3.8).

So, for a well trained chemist it is really unbelievable that the presents of a gel in the water will change the situation so dramatically!

Is the test arrangement for the Liesegang experiments as is often customary, from a test tube in a vertical arrangement, filled with lead—containing agar–agar gel for example, as already mentioned, in addition to the "periodic" ring patterns also spiraling patterns, which Liesegang pointed out as early as 1914. Strictly speaking, these patterns are not rings and spirals, but disks and helical surfaces. How to understand this?

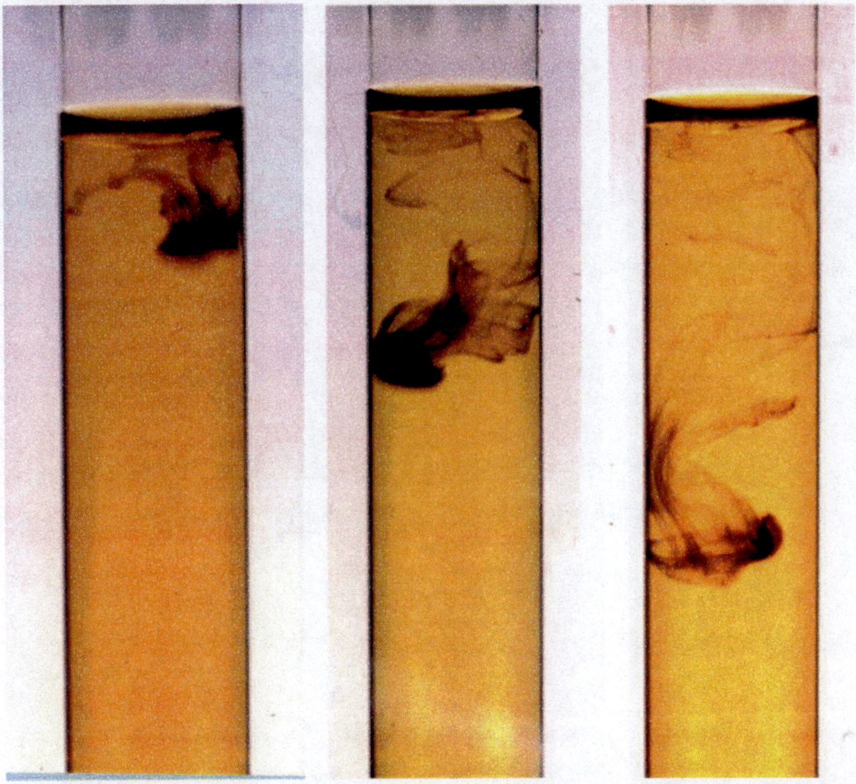

Fig. 3.8 Precipitation reaction of lead chromate in aqueous potassium chromate solution. (*Photo* P.J. Plath)

Among other things, it is just the occurrence of these helical surfaces—even Ross [16] regarded them as curiosities still in 1986—which questioned the earlier statements (Fig. 3.9).

But not only classical geometric forms occur in Liesegang's pattern formation, but also finely branching precipitation patterns can be seen, filling the test tube like the web of the roots of a tree.

Such patterns are a clear note that fractal geometries might be of importance in the formation of the precipitation patterns (Fig. 3.10).

Fig. 3.9 In addition to the simple screw surfaces, two of these elements can also be twisted—so called double helical surfaces. Even within one test tube, sometimes transitions can be observed from double helical surfaces to simple helical surfaces, to discs and even into regions of granular precipitation (*Photos*; W. Jacobi, P.J. Plath)

3.2 Walking on a Fractal Network?

In the case of the precipitation of the sparingly soluble lead or else silver salts, it is easy to follow the shift of the readily recognizable precipitation front as a function of time in the test tube experiments since the experiment lasts up to one week. Liesegang had already carried out such measurements.

During the precipitation of lead chromate the starting position for the diffusion of the chromate ions of the external electrolyte is the boundary between the solid lead gel and the supernatant saturated solution of the potassium dichromate. Over the test

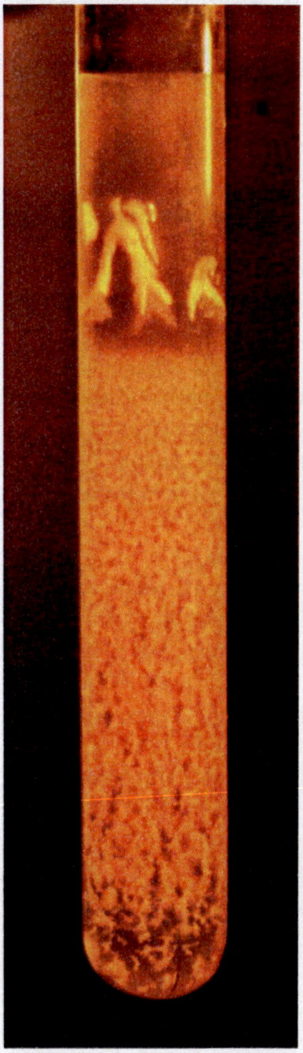

Fig. 3.10 Fractal, crazy root similar precipitation pattern of lead iodide in agar- agar gel at 5 °C (6 mM $PbNO_2$ in 0,75% agar–agar).; walking dimension: $d_w = 2,11 \pm 0,02$ (the fractal dimension d_w of a random walk on an aggregate). The finger-like structure visible in the upper part of the test tube was first formed bottom –top as a white, amorphous, finger-like precipitate after the formation of the yellow spin, probably consisting of $K(PbJ_3)$ which was stable at 5 °C but changed to yellow PbJ_2 at room temperature

time this boundary layer is not shifted for several days. However, this limit follows slowly the precipitation front after in case of the precipitation of lead iodide by using a saturated potassium iodide solution as the outer electrolyte. The reason for this shift of the phase boundary solution/gel is the hygroscopic character of potassium iodide (Fig. 3.10).

The mean displacement r of the front, which takes place by diffusion, thus results from the difference of the front displacement r_f and the displacement of the starting position of the diffusion r_l: $r = \left(r_f - r_0\right) - (r_l - r_0) = r_f - r_l$, where r_l is the distance of the phase boundary solution / Gel from the original starting position r_0 at time $t = 0$.

The precipitation tests extend over several days. If we determine the slope of the straight line (linear regression function) over these long test times from the double-logarithmic plot: $\log(r) = f(\log(t))$ of the system the lead iodide/agar–agar, the mean of the scaling exponent could be determined to $d_w = 2.25$ (Fig. 3.9 centered).

Now, plotting the logarithm of the mean shift r as a function of the logarithm of time, the slope of the resulting straight lines yields the inverse $1/\mathrm{d_w}$ of the exponent $\mathrm{d_w}$ of the displacement law: $\mathrm{x^{d_w}} \approx \mathrm{t}$. In the case of diffusion in a homogeneous medium or on a regular lattice, $\mathrm{d_w} = 2$, independent of the topological dimension d of the space in which the diffusion takes place; this can either be $\mathrm{d} = 1, 2$ or 3. It follows for the latter case the known formulation of the Einstein–Smoluchowski diffusion law: $r^2 = 2dDt = 6Dt$, where $d_w = 2$ is set. The value of r reaction is difficult to measure (at least at the beginning of the precipitation) so that when the hygroscopic potassium iodide is used, the function $r_l = r_l(t)$ has to be estimated using the Einstein–Smoluchowski relation: $r^2 = 6D_l t$. D_l is the estimated effective diffusion coefficient for those diffusion processes, which lead to the dissolution of the gel and thus to the displacement of the phase boundary solution/gel (Figs. 3.10 and 3.11).

This is a very surprising result! This is a value as it is typical for the impeded diffusion on fractal networks of the percolation cluster type. The exponent $d_w > 2$ is the dimension of the random migration on a fractal network embedded in three-dimensional space. With the help of the Alexander-Orbach conjecture that the ratio of the walking dimension d_w to the dimension of the network d_f on which the random migration takes place, is equal to 3/2, the dimension d_f of the underlying spatial fractal can be estimated: it is for the observed case $d_f = 1.51$. However, for an extended percolation cluster in the three-dimensional space, the following

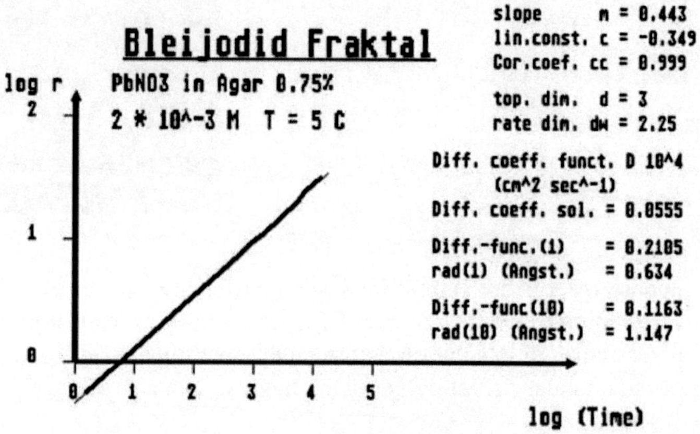

Fig. 3.11 Piecewise linear representation (thick line) of the function $\boldsymbol{log}(\boldsymbol{r}) = f(\boldsymbol{log}(\boldsymbol{t}))$ for the Liesegang experiment depicted in Fig. 3.8 with the patterning of a web in the precipitated lead iodide system at $\boldsymbol{T} = 5\,°\mathrm{C}$. The thin line is the linear regression function. The walking dimension of the migration is $\boldsymbol{d_w} = 2, 25$, the Stokes radius for $\boldsymbol{r} = 10$ has the value $\hat{\boldsymbol{r}} = 1, 14$ **Å**, and the diffusion coefficient for the redissolution of the gel was estimated to be $\boldsymbol{D_l} = 5.55{*}10^{-6}\mathbf{cm^2 s^{-1}}$

dimension should apply: $d_f = 2.5$. This apparent disproportion of the dimensions is to be understood if one considers that only those movements of the diffusing particles are of importance for the displacement of the precipitation front, in which the particle does not get lost in a dead end. But this is precisely the path of a single particle under Brownian motion. The dimension of such a path is just $d_f = 1.5$. On the other hand, this is also the dimension of the set of all paths in a percolation cluster, leading from one point of the clusters to another point of the same cluster, without taking notice of the dead-ends. The amount of all these paths is called the "backbone" of the percolation cluster.

A particle that falls into a dead end will be trapped in this for a long time, and with subsequent particles it will most likely form a no more movable crystal in such a dead end. According to this assumption a movement of the diffusing particles happens only on the backbone of the percolation cluster, provided they are important for the measurement of the front displacement.

This concept of the structure of the space in which the diffusion takes place, derived from the determination of the scaling exponent d_w, suggests inter alia that the role of agar–agar in the transition from the sol to the gel is the formation of a whole vessel-filling percolation cluster, still containing many sol particles.

Between this agar–agar percolation cluster and the agar–agar sol particles, a further percolation cluster extends to the aqueous phase solution in which the migration processes of the ions and nuclei's still to be discussed take place. This percolation cluster consists of the system of holes and pores of the gel material—in our case agar–agar. In this sense, agar–agar gel allows the formation of the aqueous percolation cluster on which the migration takes place, but it also forms the "solid" framework which keeps hold on largely the precipitation products at their sites of formation, for example in the "dead ends".

3.3 Diffusion of the Ions or Crystal Nuclei?

The question now arises as to which particles are actually involved in their diffusion is considered here? Let us estimate the order of magnitude of the diffusing particles starting from the modified Einstein–Smoluchowski relation and our experiments. Forming the logarithm of this relation yields the value for the axis section C of the straight line equation in the three-dimensional case with the topological dimension $d = 3$ of the embedding space:

$$r^{d_w} = 2dDt = 6Dt$$
$$\log r = 1/d_w \log t + 1/d_w(\log D + \log 6)$$
$$C = 1/d_w(\log d - \log(2d))$$

The experiments show that this constant C assumes values between $C = -1$ *to* $C = +1$ under our conditions i.e. using the natural logarithm and the time unit: hours. This gives values for the diffusion coefficient in the range between: $D = 5{\cdot}10^{-6}\text{cm}^2\text{s}^{-1}$ *to* $D = 5{\cdot}10^{-4}\text{cm}^2\text{s}^{-1}$.

$$\log D = d_w C - log(2d)$$
$$D = exp(d_w C - log(2d))$$

If $d_w > 2$, let us write $d_w = 2 + \vartheta$, then r^2/t is no longer a constant, but a function of r:

$$r^{2+\vartheta} = r^2 r^\vartheta = 2dDt$$
$$r^2/t = 2dD/r^\vartheta$$
$$\mathfrak{D} = D/r^\vartheta.$$

The further the diffusion front moves away from its starting point, the smaller becomes this diffusion function $\mathfrak{D}$, the more diffusion appears to be inhibited. If the migrating front has covered the distance of 1 cm, the classical diffusion coefficient is formally returned. Thus, at a greater distance, the value of the diffusion function $\mathfrak{D}$ is less than the diffusion coefficient D; At distances r, which are less than 1 cm, the value of $\mathfrak{D}$ is therefore greater than the diffusion coefficient D.

From the knowledge of the diffusion coefficient D or of the diffusion function $\mathfrak{D}$ the radius $\mathfrak{r}$ of the diffusing particles assumed to be ideal spheres can be estimated by use of Einstein's relation for diffusion in the three-dimensional space:

$$\mathrm{D} = \frac{\mathrm{kT}}{6\pi\eta\mathfrak{r}} \text{ or } \mathfrak{D} = \frac{\mathrm{kT}}{6\pi\eta\mathfrak{r}} \text{respectively.}$$

η denotes the dynamic viscosity of the medium through which the diffusion takes place. Assuming that the viscosity η of the water at 20° Celsius, is approximately $\eta = 1 \cdot 10^{-2}\text{gcm}^{-1}\text{s}^{-1}$ and using the equation given by Einstein (1905) for the radius $\mathfrak{r}$ of the particles

$$\mathfrak{r} = kT/6\pi\eta D \text{ and } \mathfrak{r} = kT/6\pi\eta\mathfrak{D}$$

respectively, the order of magnitude of 0.5 to $3 \cdot 10^{-8}$ cm can be estimated for the radius $\mathfrak{r}$ of the walking particles depending on the distance r traveled. For the average radius of the diffusing particles is, of course, a function of the distance traveled: the radius $\mathfrak{r}$ increases for $d_w > 2$ as the distance increases.

These values $\mathfrak{r}$ are typical for ionic radii in diluted aqueous solutions, if these are determined from measurements of the ion mobility or from diffusion measurements of the type described here with the aid of the Stokes law. At the temperature of 18 °C where the dynamic viscosity of water is $\eta = 1.041 \cdot 10^{-2}$Poise (g $\text{cm}^{-1}\text{s}^{-1}$), the following Stokes radii τ of the ions are obtained: J^-: 1, 16 Å, CrO_4^{2-}: 1.08 Å, and K^+: 1.21 Å [17]. These stokes radii are considerably smaller than the crystal radii of the ions, e.g. for J^{-1}: 2.16 Å, Pb^{2+}: 1.32 Å and for K^+: 1, 33 Å [18]. The Stokes radii are an expression of the mobility of the ions in the solvent. The smaller the Stokes radius, the more mobile is the ion. The above-mentioned estimation of the function $r_l = r_l(t)$ with $r^2 = 6Dt$ for the displacement of the solution / gel phase boundary in the re-dissolution of the gel due to the use of e.g. KJ is now carried out in such a way that at a distance of $r = 10$cm, the Stokes radius of the observed anion comes out,—which in this case is the iodide ion—which is known from the literature (cf. Fig. 3.11).

If $d_w = 2$, then for every ionic species in a particular solvent is obtained a specific radius. A remarkable fact of the Liesegang experiments is that for $d_w > 2$ a Stokes radius τ is obtained which increases steadily with time or with the distance from the starting point, whereas for $d_w < 2$ a constantly decreasing Stokes radius τ is obtained.

As a consequence, in the Liesegang experiments, the Stokes radius is thus no longer an ionic- and solvent-specific constant.

A growing Stokes radius r of the ions at $d_w > 2$ could be understood if it is assumed that in the course of the experiment more and more slowly growing nuclei of the precipitous sparingly soluble salt are involved in the diffusion. This would make it possible to describe the disabled diffusion which we previously tried to understand with the help of the idea of a hike on a fractal percolation cluster.

3.4 Accelerated Diffusion—Chemical Turbulence

However, those experiments generate problems for the interpretation for which $d_w < 2$—at least at the beginning of the experiment. At the beginning for $r < 1$cm, unusually large stokes radii τ are observed, which are rapidly becoming smaller as time passes. This could be explained by the fact that the diffusion here begins with many but relatively small instable nuclei, which then decay with time so that the mean Stokes radius decreases steadily. In this way, an accelerated diffusion is realized, since with any decay of a nucleus more and more diffusing particles are formed at the respective point of decay. The diffusing and subsequently dissolving or disintegrating nuclei lead in this way temporarily to a correlated movement of the later on still smaller nuclei or ions respectively resulting from the dissolution.

This correlated movement takes place only for short periods, and is caused by the chemical process of the resuscitation of even still very small nuclei of the sparingly soluble salt. With good reasons one can describe this temporarily correlated movement which is caused by the dissolution process as chemical turbulence. In this context, turbulence means nothing but the fact that particles for a certain time perform a correlated movement whereby they become smaller and smaller and thus accelerate the diffusion of the ions or molecules of the precipitate.

This explanation, however, appears to be at odds with the assumptions of nuclei growth in inhibited diffusion. How can one assume at one time of a growing nucleus radius τ, but the other time from a shrinking nucleus radius τ? The experiment might help us out of this dilemma.

In a series of experiments one has to distinguish, as Liesegang remarked already, between the initial phase with accelerated diffusion ($d_w < 2$) and a final phase with impeded diffusion ($d_w > 2$). Within the framework of the approach outlined here, this behaviour, which combines the two approaches described above, can be explained as follows:

At the beginning of the diffusion process, when the external electrolyte is brought into contact with the internal electrolyte, very many small nuclei of the sparingly soluble salt are created spontaneously, i.e., on a very short time scale and in a very small region Δr direct at the contact zone, i.e. at the phase boundary solution / gel due to a situation of local over-saturation therein. Their radius is so small that they can still diffuse into the gel at a sufficient rate.

On the other hand, they are thermodynamically unstable because of their smallness, provided that their radius has not yet reached the critical radius for autonomous crystal growth.

In this way an accelerated diffusion ($d_w < 2$) is to be expected at the beginning of the Liesegang experiment. In the later course the diffusion is then taken over more and more by the ions.

Of course, one has to take into account the fact that some stable nuclei are formed over time, which may have a considerable influence on the diffusion during the course of the experiment so that the average Stokes radius gradually increases again after a sufficiently long period of time. For this assumption, however, it must be assumed that very small crystal nuclei can already be stabilized.

Starting from the results of a group of theoretical physicists in Rostock around Ulbricht et al. [19], it is justifiable to assume that even very small nuclei can be thermodynamically stabilized by blocking them in sufficiently small cavities. Due to its porous structure, a gel has a large number of differently sized pores, including those which are sufficiently small to stabilize the very small nuclei. In this way we can also describe the slow growth of the Stokes radii in the inhibited diffusion.

In the fractal description of these phenomena, these stabilized small nuclei serve to produce more and more dead-end roads, so that the diffusion of the ions and of the small crystal nuclei on the percolation cluster is more and more hindered.

3.4.1 Processes in the Precipitation Front

We find out a lot about the diffusing particles and the structure of the space in which the diffusion takes place by use of the Einstein–Smoluchowski relation of the mean shift of the precipitation front in diffusion processes and the Stokes–Einstein interpretation of the diffusion coefficient. However, the question remained open to what role the nucleation plays for the formation of the visible aggregation cluster in the area of the shifting front some time after the beginning of the experiment.

Diffusion of the small ions of the external electrolyte, e.g. the iodide ions, is faster than the diffusion of the nuclei and also much faster than the diffusion of the stable crystals or even the aggregation clusters, due to the lower Stokes radii, as we have seen above. For example, if such a crystal or cluster would have a radius of about $100 \cdot 10^{-8}$cm, its diffusion coefficient would be smaller by a factor of 10^{-2} than that of the ions. It is therefore reasonable to assume that large nuclei or crystals diffuse, if at all, only slowly, so that they are not important for the frontal shift. Moreover, the relatively few small nuclei that develop in the precipitation front, as long as they do not dissolve immediately on account of their small size, migrate only so long as they come together with other particles and are localized in a kind of agglomeration cluster or crystals in the agar–agar gel—for example in the dead-ends of the agar–agar percolation cluster, or even forming new dead-ends.

The very small nuclei that are created in the area of the front have only one a relative short lifetime in which they contribute to the shift of the front next to the ions. In space, far behind the fast-moving front area these small nuclei are caught by the already existing germinating aggregates.

Moreover, in the space behind the front, there are hardly any free ions of the internal electrolytes, so that, if nuclei do not make cause for the formation of ions again, there is scarcely any nuclei formation, so that this space can be filled with the ions of the external electrolyte almost unimpeded.

For the displacement of the front, in addition to the diffusion of the nuclei, are only those diffusion processes of the ions of the external electrolyte of importance, which take place within the front in the direction of the concentration gradient of the internal electrolyte. Moreover, they are only of importance as long as they have not yet fallen into germination or cluster formation, which occurs immediately in front of the front and thus marks the front boundary of the front area.

In other words, only those diffusing particles contribute to shift of the precipitation front, which traveled the distance Δr in the in the mean time Δt within the precipitation front. Thus, the path from the starting zone of the experiment (i.e. from the front to the time $t_0 = 0$) just to the actual front is not covered by a single diffusing particle, but by a large number of particles, which travel this path in common only for a short distance Δr or a short time Δt long respectively—until their destruction as a nuclei. If Δr and Δt are small enough, the set of these many partial paths is exactly comparable to the motion of a particle on the backbone of a percolation cluster, or the Brownian motion of a particle.

3.5 Pattern Formation

The fractal patterns mentioned at the outset, which, like root splints, evenly pass through the test tube, are very illustrative examples of a diffusion of the nuclei which have led to the formation of structures by an aggregation mechanism. In the light of this discussion, it is also quite clear that fractal spheres are formed in a three-dimensional reaction arrangement (Fig. 3.12). If one wishes to understand the macro scopic structure formation, in particular also the formation of helical surfaces and periodic disks, it seems unavoidable to start from these aggregation clusters and the mechanism of their formation.

The concentration of the inner electrolyte determines to some extent the concentration of the nuclei of the sparingly soluble salt which are formed in the front region of the diffusion. In a critical region of the number of these nuclei, formation of aggregation clusters occurs due to the diffusion of the nuclei. These clusters, which are formed by the mechanism of "diffusion-limited aggregation" (DLA mechanism), also

(a) (b)

Fig. 3.12 Fractal PbJ_2 Liesegang ball created in 5% agar–agar gel at 20 °C using a test tube of 3 cm in diameter; **a** left: at the beginning; **b** right: after almost one week, so that one can observe the formation of diffuse "rings" and granulated collections of crystals. (*Photo* … P.J.Plath)

stabilize the still very small nuclei when they cling to the clusters [20]. These aggregation clusters then form a more or less dense, random fractal network in the front region [21]. Depending on the prevailing concentration ratios in the front area and the mobility of the nuclei in the gel, the formation of largely "homogeneous" webs can occur (see Figs. 3.7 and 3.9) or one gets an almost dense, precipitation without any periodic structure, for example if one works with a high concentration of the internal electrolyte (Figs. 3.10 and 3.13). One can also observe a granular precipitation structure of the only rarely occurring, locally strongly limited DLA clusters, which can

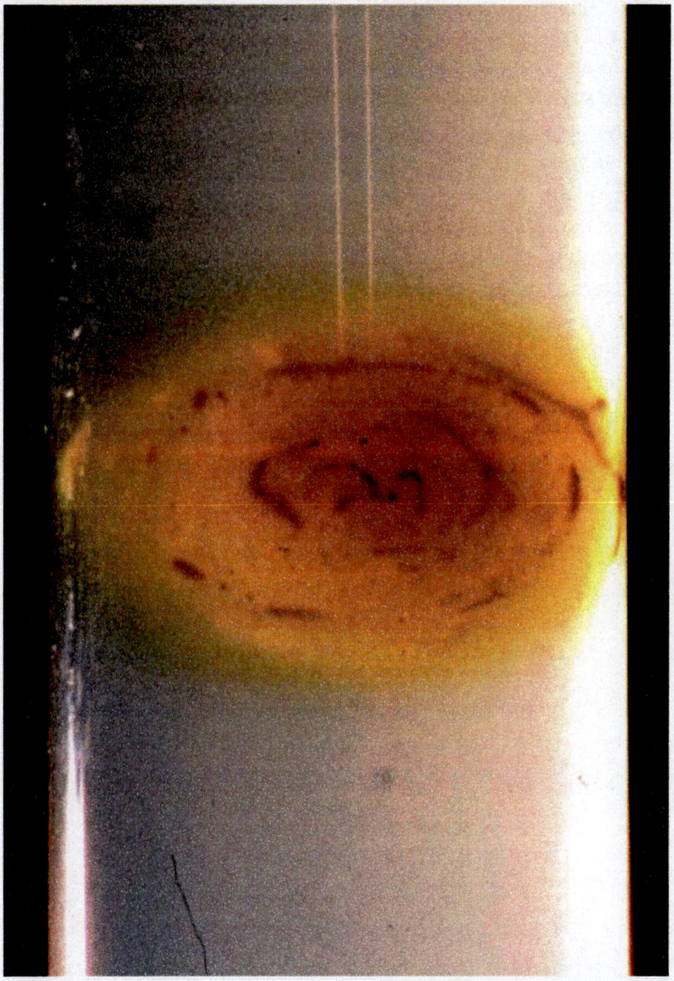

Fig. 3.13 Fractal onion-shaped Liesegang sphere of PbJ_2. The outer electrolyte diffuses by a capillary into the agar–agar gel. Care has been taken to ensure that there is no overpressure to flow of the outer electrolyte into the gel so that hydrodynamic effects can be precluded. (*Photo* … P.J. Plath)

be transformed into well-formed crystals under certain circumstances (Figs. 3.5 and 3.9 right).

Now, however, these fractal spheres usually consist of a multitude of delicate fractal spherical shells, which, like the shells of an onion or, more precisely, how the fractal leaves of the savoy cabbage surround the center of the diffusion, what needs to be explained as well (Fig. 3.13).

By the formation of the "diffusion-limited aggregation cluster" (DLA cluster) which stabilize the nuclei via adsorption, the area of the front depletes over time with free-diffusing nuclei. Only when a sufficient number of new nuclei have formed again at some distance a new formation of a layer of DLA-clusters of aggregating nuclei can occur (Fig. 3.14).

As a result, in a specific range of the concentration of the internal electrolyte, a spatially periodic structure formation occurs in the precipitation patterns in the form of concentrically successive ball shells. In the test tube experiment, segments of this series of ball shells are cut out, like cutting out the housing of an apple with the apple drill. Ball clusters are formed into disc-shaped segments which pass through the test tube in a periodic arrangement. However, the spherical shells themselves, or their corresponding segments, consist more or less two-dimensional of fractal precipitation patterns (see Fig. 3.15).

Interestingly, the three-dimensional test arrangement also shows that a series of helixes, helical surfaces or helicoids, which are becoming increasingly larger, extend from the diffusion center, on which one can pass from one fractal ball shell to the

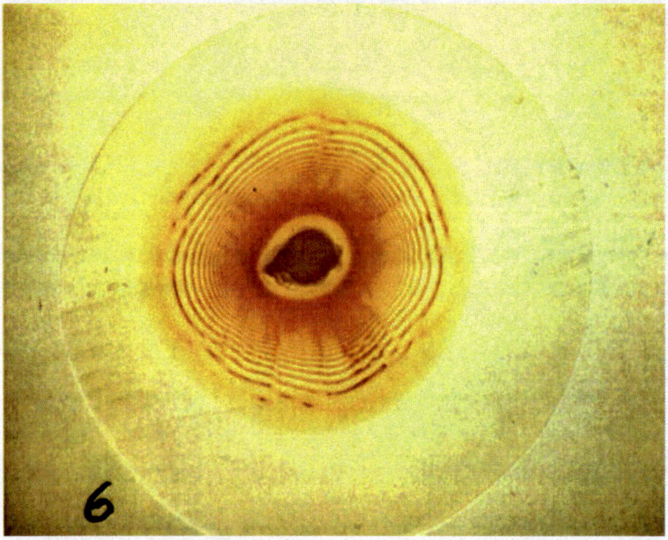

Fig. 3.14 A gel disc taken from the center of a Liesegang sphere (see Fig. 3.15) zig-zag patterns connecting neighbored ball shells. The experiment has been executed in a beaker glass of about 10 cm in diameter. The potassium iodide has been placed before in a small soluble capsule in the center of the beaker glass at the beginning. (*Photo:* S. Hollatz, T. Plikat, U. Sydow, P.J. Plath)

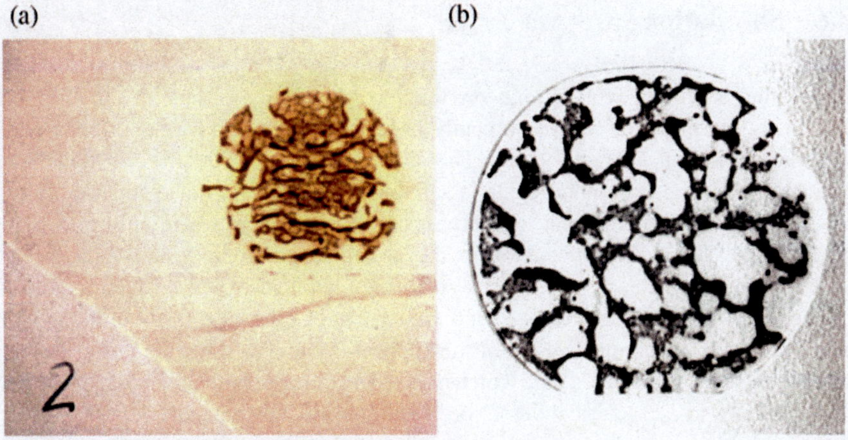

Fig. 3.15 **a** Left: Segment from a ball shell of a Liesegang ball, which reveals the fractal character of the concentric precipitation. This segment was obtained by a suitable cut from the periphery of the Liesegang sphere; **b** right: A disc-shaped segment produced by a test tube experiment. The fractal structure of this disc is clearly recognizable. (*Photos:* S. Hollatz, T. Plikat, U. Sydow, P.J. Plath)

next, as on spiral flights or spiral staircase in the tower respectively. On the more or less two-dimensional sections (Fig. 3.14) these spirals can be recognized on the sharp zig-zag patterns, which connect mutually displaced circular arcs. It appears that such helical surfaces are the result of random "disturbances" in the form of special accumulations of DLA clusters. The screw surfaces, as well as the disk surfaces, can be of a fractal struture (see Fig. 3.15), or even appear almost homogeneous to the observer at the macroscopic level. If the test tube is again used to cut sections from the three-dimensional structures, it may happen that a screw surface is more or less concentric with the axis of the test tube (Fig. 3.7). In the test tube such a structure can then form the dominant structure, because of the spatial restrictions which is very typically for structure formation in synergetic systems.

However, it is not always the case that only one or the other structure—helical surfaces, disk surface or granular precipitation—are formed, but these structures can be transformed into each other within only one test tube for example (Fig. 3.9). The distance from the starting point of the diffusion process acts as a bifurcation parameter with respect to the patterns. Of particular interest in this context are test tube experiments in which a transition from intertwined double spirals to single spirals is found.

3.6 Simulation

In all the previous explanations on the experimentally observed situation on Liesegang systems, the macroscopically observable structure formation was only explained in an unsatisfactory way. This is caused in the difficulties of modeling, which, if it is to be computationally verifiable, often have to make so simplifying assumptions that it is difficult to interpret chemically, or restricted only to certain partial aspects, e.g. the nucleation in the precipitation front. Whatever a model is designed today, the patterns observed in the chemical experiment must not be included in the modeling process, but these structures must also result in the model as a result of the self-organization of the system. In order to illustrate this, a simple simulation is to be presented here which is based on the experiments for the formation of a structure in a cylindrical vessel between two test tubes of different diameters (see Fig. 3.16).

For this purpose, let us imagine a cylindrically closed cellular automaton, where the state of a cell is described by a vector $\vec{z} = (J, Pb, PbJ)$ with three components. The temporal development of a cell is co-determined by the states of the cells which correspond in the square lattice of a von Neumann neighborhood. The components of the vector $\vec{z}$ represent the iodide and lead ion content as well as the amount of lead iodide nuclei in the spatial area of the cylinder mantle reactor which is represented

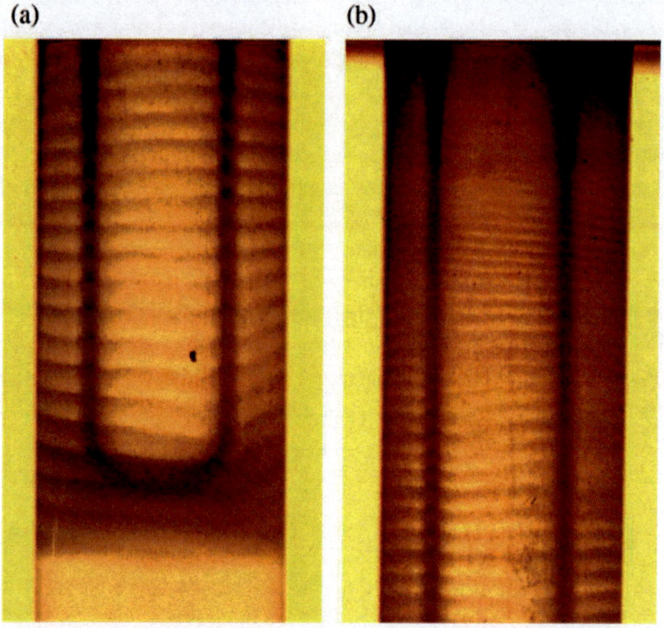

Fig. 3.16 Liesegang structure formation in the cylindrical vessel between two test tubes concentrically arranged about their longitudinal axis. **a** left: ring patterns; **b** right: spiral formation in the lower centre. (*Photos:* U. Sydow, R. Lipski, P.J. Plath)

by the cell: $J, Pb, PbJ \in \mathbb{N}^+$. This one vector automaton can also be understood as a superposition of three cylindrically closed two-dimensional scalar automata—an iodide, a lead and a lead iodide automaton—with one-component (scalar) states of the cells.

In the so-understood scalar "lead automata", the cells are randomly and homogeneously filled with lead ions, which then can diffuse almost unhindered on this two-dimensional automaton. Diffusion is understood here to mean the transition of the lead component of a cell from the value $Pb(t)$ at time t to the value $Pb(t) - 1$ at the time $t + 1$: $Pb(t + 1) = Pb(t) - 1$ when during this time step the state $Pb_{adj}(t)$ of a randomly selected, adjacent "von Neumann cell" changes by one with the probability $w_{Pb,adj} \leq 1$: $Pb_{adj}(t + 1) = Pb_{adj}(t) + 1$.

In the scalar "iodide automaton" the iodide ions "diffuse" only from one side (the upper) in the corresponding way "upside down" with the probability $w_J = 1$ into the "empty" automaton. When all cells of the lead and the iodide component have carried out the diffusion process at time t, a question is asked whether the two components J and Pb have values different from zero in each cell. According to the minimum of the two components in each cell, the third component PbJ of the state vector (J, Pb, PbJ) now receives an additional assignment by the value $min\{J, Pb\}$ at time t. If a cell has thus obtained a non-zero value of the PbJ component, it also undergoes diffusion in the next time step, provided the PbJ value is still small enough. The probability w_{PbJ} of the diffusion of the PbJ component decreases with increasing value of this component and becomes zero when a randomly selected threshold value $PbJ_c = crystal$ is reached which corresponds to the transition from the unstable nucleus to the stably growing crystal.

The acceleration of the diffusion at the start of the experiment is simulated by a larger "jump distance" during the first diffusion step of the iodide component away from the starting line.

When a "crystal" has arisen in the cell, i.e. when the PbJ component over-stepped the threshold of crystal formation: $PbJ \geq cryst$ this cell represents an insuperable obstacle for the diffusion of Pb and J component. However, if a PbJ component with $0 > PbJ < cryst$ joints a cell with $PbJ \geq cryst$ it will agglomerate with a certain probability.

It is of particular interest that in this simulation, which is based on a simple model, structure formation in the form of rings and spirals on the cylindrical automaton perpendicular to the main diffusion direction of the iodide component can be observed (Fig. 3.17). In addition, if the front-shift of the PbJ component is measured as a function of time, the simulation results surprisingly with scaling exponents dw d_w of a similar size to the chemical experiments (Fig. 3.11).

The role of the particular, fractal structure of the gel is neglected here, or is taken over by the formation of the PbJ cluster components. This raises a very interesting question from a chemical point of view, which has not yet received sufficient attention: to what extent does the fractal of the resulting DLA cluster of the PbJ_2 lead iodide agglomerates influence the reaction diffusion process in the chemical experiment?

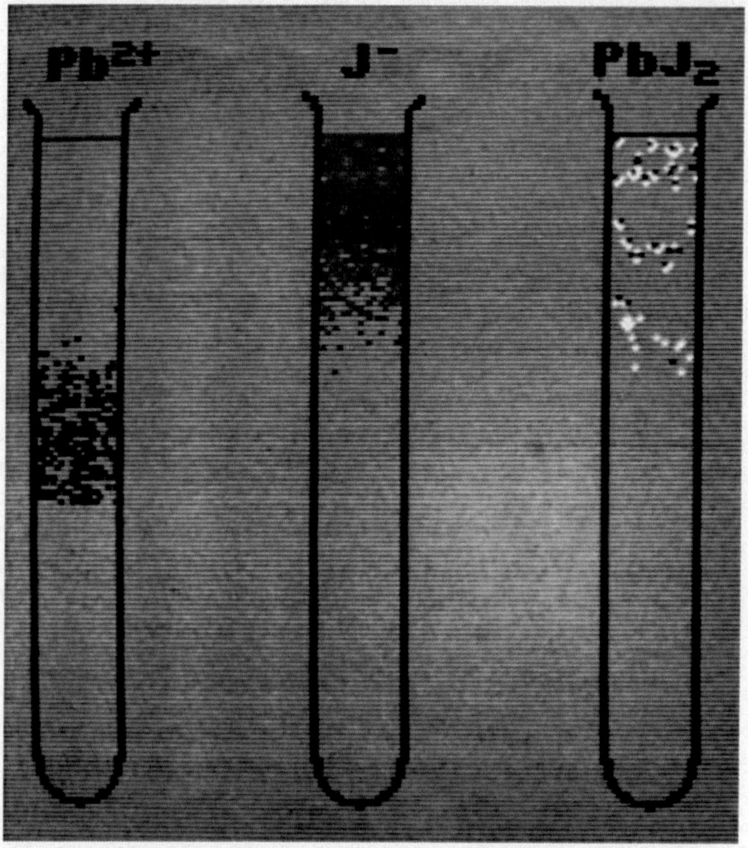

Fig. 3.17 Formation of spirals composed of DLA clusters in the simulation of experiments with cylindrical symmetry by means of a two-dimensional, cylindrically closed three-component vector automaton. The black points in PbJ_2 test tube represent bigger Clusters of PbJ2, which are already fixed in the gel, whereas the white points reflect small cluster which can diffuse yet, up to the moment where they touch each other or the bigger clusters

3.7 Twisted Scroll Waves—Much Too Early is Already Too Late

In 1988 I came together with my wife directly to Eisenach from the "8th Winter-Seminar at Zeinisjoch" (27.2.–6.3. 1988) to take part at the workshop "Dynamics of Networks" (6.3.–11.3. 1988). We were invited by Prof. Werner Ebeling from the Humboldt University of Berlin.

The Eisenach workshop was extraordinarily impressive for us in every respect.

On the rice to the nearby wintry Wartburg it was icy cold, slightly snowy and partly very slippy. Up on the Wartburg I met Harald Engel again and we agreed to continue our joint work on the Liesegang structure-building phenomena. With a joint lecture at the 8th Winter Seminar on the Zeinisjoch, this work had already begun to be promising. Harald Engel assumed that the model of Panfilov et al. [22] on the instability of *twisted scroll waves* might be a way to understand and describe mathematically the Liesegang spirals and screw surfaces in the test tubes (Fig. 3.18) I found this idea quite interesting and liked to go to his offer of further cooperation because I had known Harald Engel for several years, among others, of his stay in 1986 in my working group in Bremen.

For me, however, this meant that we first of all had to greatly expand and improve the experimental base. However, only chemistry students of the 7th and 8th semester were available to me in their research internships. The work dragged on. As the study progressed, students' interests changed and most of them did their diploma theses in other fields of research. Thus, the experimental know-how could not be accumulated fast enough to be able to trace the formation of the spiral formation sufficiently precisely and repeatably, as hoped.

When I finally gained the above-mentioned insights in 1992 and submitted them to the popular science journal *Wissenschaft und Fortschritt*, which was continued after 1990 by H. Haken and W. Ebeling as the new editors. However, the science-political agenda in Germany changed completely. The magazine ceased its publication. The GDR did not existed anymore since 1990, its scientific organizations were dissolved, my local cooperation partners now worked in precarious employment relationships.

But finally, I got a diploma student—Claudia Müller, who was able to devote herself to the topic of the Liesegang structural formations. But she found no spirals, but precipitation patterns, reminiscent of slightly rolled leaves. She called that tongue formation. She worked out everything experimentally very carefully, but now we

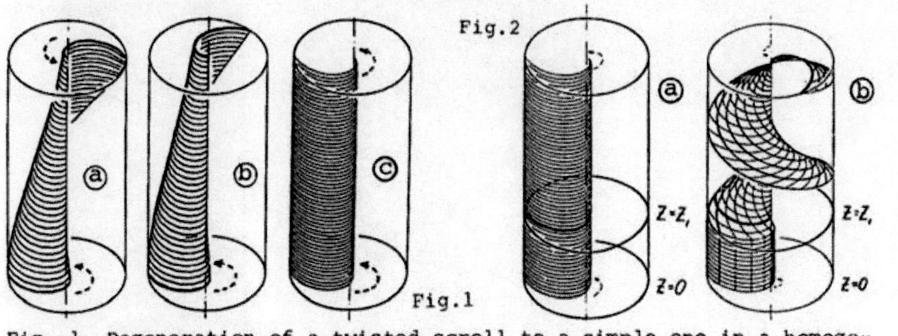

Fig. 1 Degeneration of a twisted scroll to a simple one in a homogeneous medium (g_f=0.9).
a - the initial shape of the scroll (t=0), the phase shift in the upper and lower cross-sections being equal to π. Dashed line is the wavefront, the arrow indicates the direction of rotation.
b - t=29.4, the phase shift reduced to π/2.
c - t=59.3, no phase shift is observable

Fig. 2 Generation of a twisted scroll.
a - initial conditions (simple scroll). At $z=z_1$, parameter τ is increased stepwise from τ = 1 to τ = 2.
b - the established shape of the twisted scroll

Fig. 3.18 Panfilov's Models for twisted Scroll waves

lacked the cooperation with the theoretical physicists, especially with Harald Engel. Panfilov's work had been forgotten, or simply it did not seem to be applicable. His twisted scroll waves did not move like our Liesegang leaves and above all they did not "grow" in the Liesegang way. In our opinion there was no explanation at that time for the observed phenomena and thus no prospects for further work in this area. The works of K-H. Jacob had not yet appeared. Claudia Müller went after her diploma thesis quite other ways.

However, we had found with the "tongues"-structures a whole new kind of pattern formation in precipitation reactions in gels (Figs. 3.19 and 3.20)!

The idea for her diploma thesis originated with the preliminary experiments in the delivery systems potassium chromate and potassium dichromate respectively as external electrolyte and lead nitrate as inner the electrolyte.

Fig. 3.19 Example for the new precipitation pattern in a test tube: tongues-formation: inner electrolyte 0,21 M $Pb(NO_3)_2$; external electrolyte: saturated potassium chromate $KCrO_4$ solution; maturing time of the gel < 10 h. **a** side view: splitting of the tongue, **b** frontal view: the same tongue viewed from a perpendicular position. (*Photo:* C. Müller, P.J. Plath, 1993)

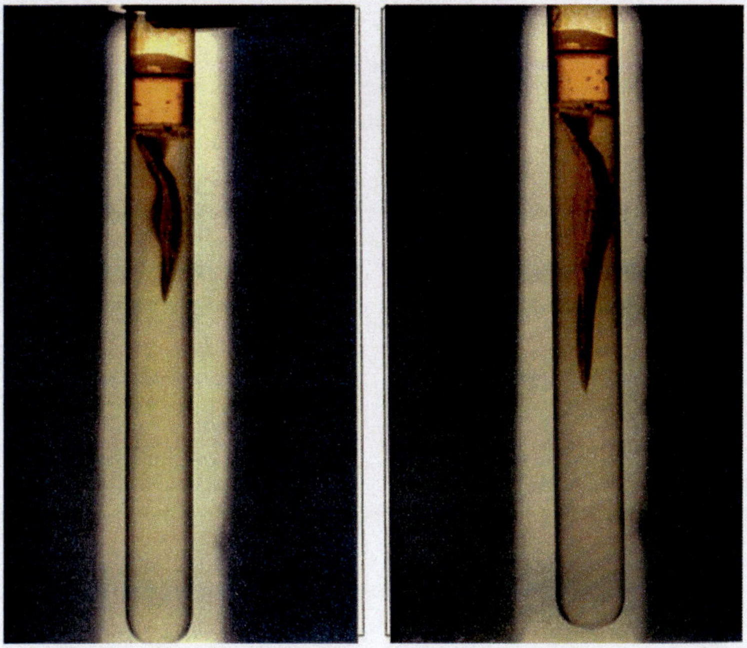

Fig. 3.20 Examples for twisted "tongue"-structures; Photos (5) und (6), experiment A.1.10a; inner electrolyte: 0.29 M $Pb(NO_3)_2$, external electrolyte: saturated potassium chromate solution; maturation time of the gel: $t < 10$ h; (*Photos* C. Müller, P.J. Plath, 1993)

It turned out that a very high concentration of lead nitrate in agar–agar led to completely novel patterning of lead chromate precipitates.

Compared with the known ligaments or spirals in gels, so-called "tongues" appeared here, which exerted not only influence on the agar–agar, which was dissolved in places, but showed completely different growth behaviour.

The experiments with "tongue" formation lasted longer than 60 days, so they last longer than those at lower levels of lead nitrate.

Thus, the question arose as to which lead nitrate concentration the transition from the known bands to the new "tongue" structures occurs and how the pattern behaviour is in the nearer concentration environment. Furthermore, the dependencies of growth rate and different "tongues" species on the concentration of lead nitrate should be investigated.

In the experiments, the main focus was on the differentiation of structure formation in the case of potassium chromate or potassium dichromate as external electrolyte.

The only experimental parameter that varies to a greater extent and which has been shown to be dependent on structure formation is the concentration of the internal electrolyte lead nitrate.

Other parameters such as the water temperature of 18° Celsius, the concentration of the outer electrolyte, the gel strength and the light conditions were kept constant, only the maturation time of the gel was between one and two days.

In the test series with **potassium dichromate** as external electrolyte, the lead nitrate concentration was between 0.1025 M and 0.13 M, in addition, experiments were also performed with 0.2 molar solutions.

When using potassium chromate, the range was more diversified, the concentrations were between $5{\cdot}10^{-3}$ M and 0.35 M lead nitrate. In doing so, some concise observations could be made:

- The plots of the progressive diffusion front or the "tongue" tip versus time gave characteristic curves for banding with potassium chromate or potassium dichromate and for the formation of the tongues

  All potassium dichromate experiments show macroscopic growth of the lead chromate crystals.
- Both in the range of $5{\cdot}10^{-3}$ M to 0,13 M lead nitrate in the experiments with **potassium chromate** and in all other experiments with **potassium dichromate**, only bands developed. In no case rings were visible.

From a concentration of 0.12 M lead nitrate, the hitherto unknown "tongues" structures were formed with **potassium chromate** as the outer electrolyte.

This is truly a complex system in the sense that its dynamics cannot be described only by several coupled equations of chemical kinetics, but that also crystal formation and thus phase transitions play a role, and moreover, that the agar–agar gel which serves as a "solvent" is structurally altered by the reaction. For example, the fact that these "tongue" structures only occur when using potassium chromate, but not potassium dichromate, is a clear indication that the acid concentration in the gel plays a very crucial role:

$$\underset{yellow}{2CrO_4^{2-}} - 2H^+ \rightleftarrows \underset{orange}{Cr_2O_4^{2-}} + H_2O$$

The acidity determines the form in which the chromate ion is present, as chromate ion or as dichromate ion. In a slightly alkaline solution at pH = 8, especially CrO_4^{2-} ions are present, in very dilute, whereas in practically neutral aqueous solutions at pH = 6.2 almost exclusively the hydrogenchromate ion $HCrO_4^-$ is to be found. In strongly acidic solutions at pH values below two: pH < 2 also H_2CrO_4 are present. However, hydrogen chromate ions dimerizes easily by dehydration to the dichromate ion even in aqueous solutions at room temperature:

$$2HCrO_4^- \leftrightarrow H_2O + Cr_2O_7^{2-}$$

Thus, every chromate solution also contains dichromate ions $Cr_2O_7^{2-}$ and each dichromate solution in turn also contains chromate ions CrO_4^{2-}. The equilibria are distributed as follows [23]:

$$pH > 7 \rightarrow CrO_4^{2-}; 2 < pH < 6 \rightarrow HCrO_4^- \; and \; Cr_2O_7^{2-}; pH < 1 \rightarrow Cr_nO_{3n+1}^{2-}$$

The saturated solutions used for the experiments resulted a pH of 9.6 for potassium chromate K_2CrO_4 and the value for pH = 3.8 for potassium dichromate $K_2Cr_2O_7$ at room temperature.

Thus, the agar–agar gel is dissolved in the experiments with potassium dichromate already during the overlaying with the outer electrolyte. The germs, from which subsequently macroscopic crystals are formed, remain partly in the resulting larger cavities and find an environment conducive to their crystal growth.

However, the precipitation reactions

$$2Pb^{2+} + Cr_2O_7^{2-} + H_2O \rightarrow 2PbCrO_4 \downarrow +2H^+$$

and

$$Pb^{2+} + CrO_4^{2-} \rightarrow PbCrO_4 \downarrow$$

take place in a gel loaded with lead nitrate $Pb(NO_3)_2$ as the inner electrolyte. Aqueous solutions of lead nitrate, however, are acidic! (see Fig. 3.21).

In these experiments, we are dealing with a chemical system that itself already has a very complicated kinetics. In particular, at a higher lead nitrate concentration, the precipitation event will certainly be largely determined by the experimantally observed pH value of the internal electrolyte.

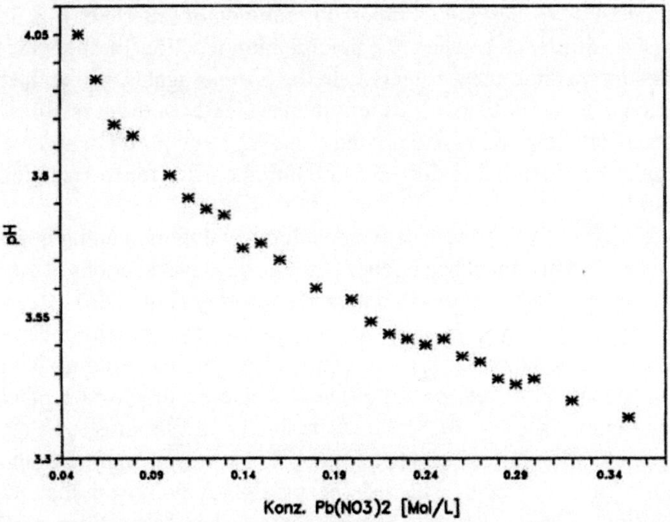

Fig. 3.21 pH values of aqueous $Pb(NO_3)_2$ solutions in the concentration range of the experiments from 0.05 M to 0.35 M

Here, in the case of potassium chromate as the outer electrolyte, the new "tongue" structures occur, at 0.21 M $Pb(NO_3)_2$ as straight growing tongues filled with potassium chromate solution and at 0.29 M $Pb(NO_3)_2$ correspondingly twisted, likewise filled tongues.

3.8 Agates and Other Mineralogical Liesegang Structures

As early as 1914, R. Liesegang [27] wrote his book on the agates, referring directly to his experiences with pattern formation in precipitation reactions in gelatin. The agates, consisting essentially of cavities enclosed in cavities, which may occur in the modifications chalcedony, quartzin, quartz or opal and silicate gelatin, must have been formed from aqueous silica sols:

> Die Kieselsäure kann nämlich ursprünglich eine gallertige Beschaffenheit gehabt haben und erst später in die dichte, wasserarme Form übergegangen sein. Diese Anschauung erleichtert dann zugleich die Erklärung der Bänderung. Denn inzwischen hat die physikalische Chemie festgestellt, daß solche gerade in Gallerten leicht entstehen können, wenn man chemische Prozesse darin sich abspielen läßt.

> The silicic acid may have originally been of a gelatinous nature and only later changed into the dense, water-poor form. This view facilitates then also the explanation of the banding. In the meantime, physical chemistry has established that such bands can easily be formed in jellies if chemical processes are allowed to take place in them.

As an example, he cites the precipitation reaction of silver chromate in gelatin and in ealso react with water to form hydroxilicxplaining the resulting pattern formation, he refers in detail to Wi. Ostwald's *supersaturation theory* of 1896 (Fig. 3.22).

The decisive difference between the precipitation reactions in the jellies and the pattern formation in the agates, especially in the fortress agates, is that the latter is a transformation of the various modifications of the silicates or the crystallization from silicate suspensions or galenics and not the classical, ionic precipitation reactions in which the gelantine is only the diffusion-inhibiting carrier for the reaction and the resulting germs.

Thus, the classical agates are usually not colored and their banding is only recognizable by the different optical behaviour of the various modifications of the silicates, which varies between white and different shades of gray (Fig. 3.23).

When comparing undyed agates with images of two-dimensional Liesegang precipitation patterns, however, it is also quite clear that the banding is due to the formation of "chemical waves" or "autowaves" which annihilate each other. In case the "fortress agates" (see Fig. 3.23) the essential processes seems to be crystallization or recrystallization instead of a precipitation. The waves migrate from the outer sphere of the agates into their inside. Liesegang already points out that, at the time of the formation of the agates, the available cavity had to be filled with not yet crystallized silica jelly [24].

(a) (b)

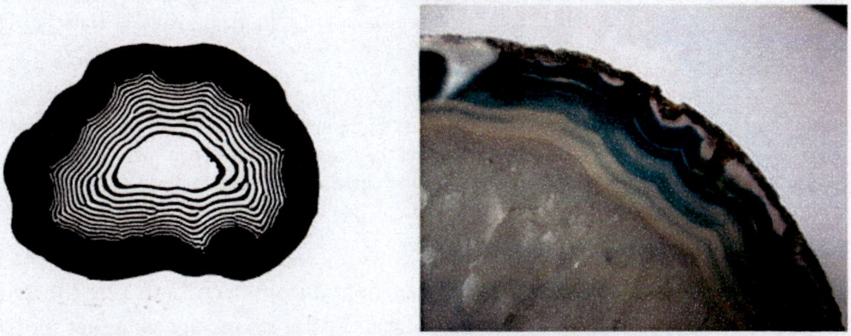

Fig. 3.22 **a** imitation of the agate structure with silver chromate on a glass plate; Copy from "Die Achate" by R.E. Liesegang (1915), inner electrolyte Ammonium dichromate $(NH4)_2Cr_2O_7$ in citric acid gelatin; outer electrolyte silver nitrate solution 25 percent. **b** blue colored agate (*Photo* P.J. Plath)

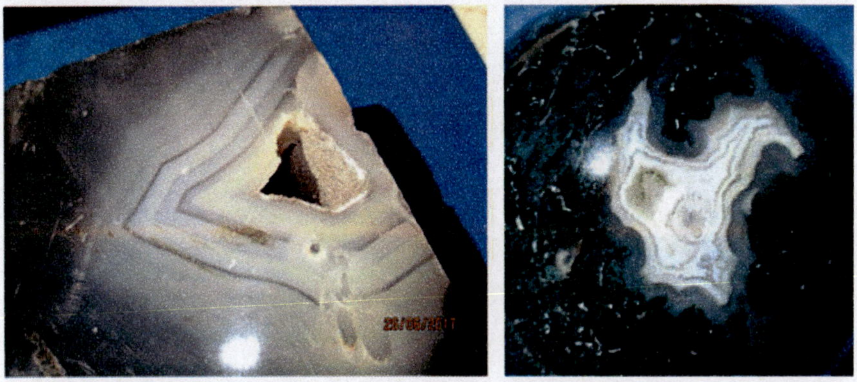

Fig. 3.23 Undyed agates of the type "fortress agate" ("Festungsachat"); left: with a cavity covered with quartz (Quarzdruse). (*Photo*: P. Plath)

Das Vorkommen von weicher Kieselsäuregallerte in den Hohlräumen von Gesteinen ist gar nicht so selten. Zuerst fand G. Spezia sie (1899) in einer Spalte des Gneises beim Bau des Simplontunnels. Dann wies besonders J.H. Levings (1912) auf ihr Vorkommen in einigen australischen Minen hin, als in einer Diskussion über die Gold-Quarz Arbeit von Hatschek und Simon daran gezweifelt wurde.

In den Blasenräumen der Basalte, welche den Melaphyren (Gestein vulkanischen Ursprungs und porphyrischer Struktur mit großen Einschlüssen, d. Autor) sehr nahe verwandt sind, findet man zuweilen Opale. Man nimmt gegenwärtig an, daß diese nur aus Kieselsäuregallerten hervorgegangen sein können. Von besonderer Wichtigkeit ist in dieser Beziehung die Tatsache, daß die Dreher-Sammlung des Mineralogischen Museums in Berlin zwei große Brasilianer Achate enthält, deren Inneres Opal ist.

The occurrence of soft silicic acid gallates in the cavities of rocks is not so rare. At first G. Spezia found them (1899) in a fissure of gneiss during the construction of the Simplon tunnel. Then especially J.H. Levings (1912) pointed out their occurrence in some Australian mines, when it was questioned in a discussion about the gold-quartz work of Hatschek and Simon.

In the vesicular cavities of basalts, which are very closely related to melaphyres (rocks of volcanic origin and porphyritic structure with large inclusions, author's note), sometimes opals are found. At present, it is assumed that they can only have been formed from silicic acid gallates. Of particular importance in this regard is the fact that the Dreher collection of the Mineralogical Museum in Berlin contains two large Brazilian agates, the interior of which is opal.

Silica gels now belong to the large group of hydrophilic colloids [25] the aging of which is due to dehydration. Intermolecular dehydration produces higher molecular weight silicic acids In this case, higher molecular weight silicic acids $H_{2n+2}Si_nO_{3n+1}$ are formed, for example ortho-di-silicic acid from ortho-silicic acid.

$$2H_4SiO_4 \rightarrow H_6Si_2O_7 + H_2O$$

Finally, the meta-silicic acids $(H_2SiO_3)_n$ with $n > 3$ which are formed as ring-shaped, linear, band or sheet-shaped polymers, and at the end, with complete elimination of the chemically bound water, quartz.

But how are the silicic acids inclusions in the igneous rocks formed? The double silicate or alumosilicate potassium feldspar or orthoclase $K(AlSi_3O_8)$ is the main constituent of the igneous rocks such as e.g. Basalt and porphyry. With molten alcalicarbonate (e.g., soda), it is easy to produce sodium silicate and carbon dioxide [26].

$$KAlSi_3O_8 + 3\,Na_2CO_3 \rightarrow 3\,Na_2SiO_3 + KAlO_2 + 3\,CO_2$$

Wässerige Lösungen von Alkalisilikaten reagieren stark basisch. Die Silikate sind in diesen Lösungen weitgehend hydrolytisch gespalten.

$$2\,Na_2SiO_3 + H_2O \rightarrow Na_2Si_2O_5 + 2\,NaOH$$

Säuert man die Silikatlösungen an, so wird zwar die Kieselsäure sofort in Freiheit gesetzt, aber sie fällt meist nicht gleich aus, sondern bleibt zunächst in Lösung. Erst nach längerer Zeit beginnt Ausflockung. Es beruht dies zum Teil darauf, daß die Kieselsäure in einer wasserlöslichen, monomolekularen Form auftreten kann, die sich je nach Versuchsbedingungen schnell oder langsam unter Wasseraustritt zu höhermolekularen, schließlich zu praktisch unlöslichen hochmolekularen Aggregaten kondensiert. Aber auch, nachdem die Kieselsäure vollständig in unlösliche Form übergegangen ist, braucht noch keine Ausfällung zu erfolgen, da sie kolloid in Lösung bleiben kann. Die Neigung der Kieselsäure, kolloide Lösungen (Kieselsole) zu bilden, ist außerordentlich groß. Sowohl in sauren, wie in schwach basischen Lösungen ist Kieselsäure in kolloidaler Form beständig. Durch Elektrolytzusatz erfolgt meist keine Ausflockung, sondern, wenn die Lösung nicht zu verdünnt ist, ein langsames Erstarren der gesamten Lösung zu einer Gallerte. Aus verdünnten Lösungen setzen sich schleimige Niederschläge ab.

Frisch dargestellte Kieselgallerte können bis zu 330 Mole H_2*O auf 1 Mol* SiO_2 erhalten. Bei längerem Stehen (an mit Wasserdampf gesättigter Luft) beginnen sie allmählich einzuschrumpfen, indem Wasser ausgepreßt wird, eine Erscheinung, die sich auch bei anderen Gallerten findet (z.B. bei der Synthese von Zeolithen aus Alumosilikatgallerten; Bemerkung des Autors) und die man als Synärese (see Remy p. 583).

In Form von kompakten Xerogelen (getrocknete Kieselgele; d. Autor) findet sich Siliziumdioxid in der Natur als Opal. Das in gewöhnlicher Weise gefällte Siliziumdioxid und dessen Gele liefern im allgemeinen keine auf kristalline Struktur hinweisende Röntgeninterferenzen. Entsprechendes gilt auch für durch Ausscheidung von SiO_2 aus Wasser bei gewöhnlicher Temperatur gebildete Opale. Aus heißen, magmatischen Wässern gebildete Opale dagegen geben, je nach ihrer Herkunft, für Christobalit oder für Quarz charakteristische Frequenzen. Chalzedon ist ein gealterter Opal. Er ist demgemäß wasserärmer als dieser (oft ganz wasserfrei) und weist schon mit gewöhnlichen Hilfsmitteln erkennbare kristalline Struktur auf. Bei mikroskopischer Betrachtung, besonders im polarisierten Licht, zeigt er faserigen Bau. Abarten des Chalzedons sind Achat, Onyx, Karneol, Heliotrop und Jaspis, ... sowie der Feuerstein, ... (Remy p. 580)

Aqueous solutions of alkali silicates react strongly basic. The silicates in these solutions are largely hydrolytically decomposed.

$$2\ Na_2SiO_3 + H_2O \rightarrow Na_2Si_2O_5 + 2\ NaOH$$

If the silicate solutions are acidified, the silicic acid is immediately freed, but it usually does not precipitate immediately, rather it remains in solution for some time. Only after a longer time flocculation starts. This is partly due to the fact that the silica acid can occur in a water-soluble, monomolecular form which, depending on the experimental conditions, condenses rapidly or slowly under water outflow to higher-molecular, finally to practically insoluble highly molecular aggregates. But even after the silicic acid has completely converted into an insoluble form, precipitation need not yet occur because it can remain colloidal in solution. The tendency of silicic acid to form colloidal solutions (silica sols) is extremely high. In both acidic and weakly alkaline solutions, silica is stable in colloidal form. The addition of electrolytes usually does not cause flocculation, but, if the solution is not too dilute, the entire solution slowly solidifies to form a jelly. Mucilaginous precipitates settle out of diluted solutions.

Freshly prepared silica gels can contain up to 330 moles of H_2O per 1 mole of SiO_2. During a longer period of standing (in air saturated with water vapor) they gradually begin to shrink by squeezing out water, a phenomenon which is also found in other gels (e.g., in the synthesis of zeolites from aluminosilicate gels; author's note) and which is called syneresis. (Remy -p. 583).

In the form of compact xerogels (dried silica gels; the author), silica is found in nature as opal. Silicon dioxide precipitated in the usual way and its gels generally do not provide X-ray interferences which indicate a crystalline structure. The same applies to opals formed by precipitation of SiO_2 from water at normal temperature. On the other hand, opals formed from hot magmatic waters provide frequencies characteristic of either cristobalite or quartz, depending on their origin. Chalcedony is an aged opal. Accordingly, it is less hydrous than the latter (often completely anhydrous) and exhibits a crystalline structure that can be recognized even with ordinary tools. Microscopically, especially in polarized light, it shows fibrous structure. Varieties of chalcedony are agate, onyx, carnelian, heliotrope and jasper, ... as well as the flint... (Remy p. 580).

So at the beginning we have a silicate jelly in the cavities (comp. Fig. 3.24). It is formed by hydrolysis of the salts of the adjacent igneous rocks and leads to an alkaline reaction.

The monomeric silicic acid anions also react with water to form hydroxilic cations and the formation of OH^- ions.

$$(SiO_4)^{4-} + HOH \rightarrow (SiO_3(OH))^{3-} + OH^-$$

The jelly or the forming gel ages, whereby the condensation process of the anions of the poly-acids with elimination of water is the crucial process. But especially the low-molecular meta-silicic anions can be hydrolytically split again by water. However, if it is already higher molecular weight poly-hydroxo-silicates, then an acidic reaction may occur:

$$Si_4O_{11}(OH)_2)^{8-} + 3\,HOH \rightarrow 4(SiO_3(OH))^{3-} + 4\,H^+$$

Fig. 3.24 An illustrative example of the filling of a cavity in the igneous rock with the silicate sol or gel respectively via the filling channel visible above. The agate formation process begins only after the filling of the cavity of all sides evenly

Due to this constant regression of the monomeric hydroxysilicate ions, this aging process is very slow!

The competition reaction is the condensation of the hydroxysilicate ions with the release of water, which leads to the polymerization:

$$2(SiO_3(OH))^{3-} \rightarrow (Si_2O_7)^{6-} + H_2O$$

Due to the fact that locally high levels of proton can be formed, the polymerization is enhanced, leading to a three-dimensional, fully amorphous and finally solid silicate gel, the opal.

This causes shrinkage of the gel and partly contributes to the formation of cavities in the interior of the resulting agate. The dehydration also happens in the presence of non-chemically bound water!

In order to convert the purely silicate gel into the amorphous, but partially crystalline chalcedony, it would require OH^- ions, formed almost exclusively in the "outside" of the igneous rock by dissolution of the alumo-silikates which are contained therein. Penetrating by diffusion into the interior of the opal, they may dissolve the amorphous structure of the opal "superficially" and in this way create low-molecular-weight silicates which diffuse themselves and can thus partially transform the amorphous structure into a crystalline structure, the chalcedony. This "recrystallization" then leads to the characteristic of the agates band structure.

Especially in the interior of the not yet fully polymerized jelly but caused by the very slowly progressing polymerization constantly new protons are created that neutralize the hydroxyl ions and propel the opalescence. This way enough water is formed so that OH^- ions can diffuse from the outside again.

If there are any other precipitable substances present in this whole event—or even afterwards—it may come even to colorings of the tapes, like R.E. Liesegang in his work "The Agates" describes in detail. The role of the local change in the pH value described here is likely to be of importance for the location of the respective precipitation reaction, which is very closely linked to the formation of chalcedony ribbons.

3.9 Rhythmically Banded Sandstone

In his book "The Agates" R.E. Liesegang also described the occurrence of rhythmic precipitations of iron compounds in sandstone.

> In der Nähe von Münzenberg (im Sandsteinbruch Rockenberg, in der Wetterau gelegen; d. Autor) findet man im Sandstein wundervolle braune, rote und gelbe Bänderungen, welche von mehr oder weniger wasserhaltigen Eisenoxyden bewirkt sind. Sie gleichen oft, wenn man von der viel gröberen Körnigkeit des Sandsteins absieht, ganz auffallend Achaten. Dabei sind sie auch chemisch sehr nahe mit diesen verwandt. Denn die Grundmasse ist bei beiden Kieselsäure, allerdings in verschiedener Verteilung. Und das färbende Prinzip sind Eisenverbindungen und zwar diese in gleicher Verteilung. Dieses Gestein sei deshalb

hier angeführt, weil es ganz besonders geeignet ist, uns zu zeigen, daß wenigstens in diesen "natürlichen Achat Nachahmungen" gar keine andere Erklärungsart möglich ist, als diejenige durch eine rhythmische Fällung des Eisensalzes. [27]

In the vicinity of Münzenberg (in the Rockenberg sandstone quarry, located in the Wetterau; the author) one finds in the sandstone wonderful brown, red and yellow bandings, which are caused by more or less hydrous iron oxides. They often look strikingly similar to agates, if one disregards the much coarser granularity of the sandstone. Thereby they are also chemically very closely related to them. This is because the basic mass in both is silicic acid, but in different distribution. And the coloring principle consists of iron compounds, and these in equal distribution. This rock is mentioned here because it is particularly suitable to show that at least in these "natural agate imitations" there is no other possible explanation than that of a rhythmic precipitation of the iron salt. [27]

These rhythmic precipitates of iron oxides in the sandstone are evidently also typical Liesegang structures, which are due to the fact that iron ions diffuse from the outside into the solid sandstone and precipitated as oxides in accordance with the respective prevailing pH values become.

The crucial difference to the agates is that it is a sedimentary rock with a proportion of at least 50% grains of sand, i.e. of grains which, according to the general definition of grain size, are between 0.063 and 2 mm in size. The grains of sand consist of various minerals, but mostly of quartz with additives, e.g. of feldspar and lime [28].

The Bentheim sandstone type Gildehaus is a fine to medium-grained quartz sandstone which consists of at least 90% quartz grains, and whose grain binding is largely caused by grain intergrowth. These sandstones can absorb a lot of water due to

Fig. 3.25 Liesegang's ring formation (60 to 80 cm in size of the ring shown here) in the Bentheim sandstone of the quarry near Gildehaus. (*Photo:* P. Plath) The "Gildehaus" type sandstone is known for its distinctive Liesegang structures

their porous structure (water absorption (atm.) 7.1 wt.-%;). Its yellowish Liesegang structures are due to precipitated Limonite (brown iron ore) [29, 30].

The granular but porous structure of the sandstone requires that the penetrating aqueous solutions must "flow" between the grains. These become in part "dissolved" on their surfaces or "re-crystallized". Thin superficial silicate gel layers similar to the agate, are it formed around the quartz grains. These layers are the actual environment for the Liesegang precipitation but also lead to grain intergrowth (Fig. 3.25).

References

1. Liesegang, R.E.: Spezielle Methoden der Diffusion in Gallerten. In: Handbuch der biologischen Arbeitsmethoden - Abderhalden, Abt. III, Teil 2 (1929); and G.F. Runge, R.E. Liesegang, B.P. Belousow, A.M. Zhabotinsky; Selbstorganisation chemischer Strukturen", Ostwalds Klassiker der exakten Wissenschaften, Band 272, (Eds. L. Kuhnert und U. Niedersen), Akadem. Verlagsges. Geest & Portig K.G. Leipzig (1987)
2. Jacobi, W.: Untersuchung kritischer Parameter Liesegangscher Ringe. Diplomarbeit, Universität Bremen (1984)
3. Sydow, U., Plath, P.J.: Structure formation during precipitation reaction in gels: New Liesegang patterns, Berichte der Bunsengesellschaft, Vol. 102, 11 (1998) pp. 1683–1688, and https://doi.org./10.1002/bbpc.19981021132 and Müller, C.A.: Neue Strukturen in Liesegang Systemen. Diplomarbeit, Universität Bremen (1993)
4. Krinsky, V.I.: Autowaves: results, problems, outlook. In: Self-Organization – Autowaves and Structures Far from Equilibrium, Poceedings of anInternational Symposium, Pushchino USSR July 18–23, 1983; (editor: V.I. Krinsky, series editor: Hermann Haken), Springer Series in Synergetics, vol. 28, pp. 9–19. Springer, Berlin, Heidelberg, New York, Tokyo (1984)
5. Raphael, E.D.: Liesegang. Verlag von Theodor Steinkopf, Dresden und Leipzig, Die Achate (1915)
6. Jacob, K.-H.: Ein neues Bild der Erde – Landschaftssteine fordern die Wissenschaft heraus, GEO Nr. 4 (1993) pp 160 – 178; and „Edle Steine im neuen Licht", GEO Nr. **5** (1994) pp 172 – 178, and „Liesegang-Ringe" in: Lagerstättenkundliches Wörterbuch, Clausthal, pp. 307–310 (1999)
7. Jacob, K.-H., Dietrich, S.: Electric field forces and selforganization – from common concepts to new insights. In: Proceedings of the 36th Interdisciplinary Workshop – The Expansion Evidence; St. Cwojdzinski, G. Scalera; 4./9. Oct. (2011) in Erice, Sicily, Italy, pp 407–419
8. Kuhnert, L., Niedersen, U., Runge, F.F., Liesegang, R.E., Belousov, B.P., Zhabotinsky, A.M.: Selbstorganisation chemischer Strukturen. Ostwalds Klassiker der exakten Wissenschaften Bd. 272, Akademische verlagsgesellschaft Geest Portig K.-G., Leipzig (1987)
9. Bradford, S.C.: Adsorptive Stratification in Gels. Biochem. J. **10**, 169–175 (1916)
10. Stern, K.H.: The Liesegang Phenomenon. Chem. Rev. **54** (1), 79–99 (1954)
11. Dhar, N.R., Chatterji, A.C.: Liesegangsches Phämomen und Niederschlagsbildung. Kolloid Z. **31**, 15–16 (1922)
12. Kai, S., Müller, S.C. and Ross, J.: Measurements of temporal and spatial sequences of events in periodic precipitation processes. J. Chem. Phys. **76**(3), 1392–1406 (1982)
13. Ostwald, W: Zur Theorie der Liesegangschen Ringe. Kolloid-Zeitschrift **36,** 380 (1925)
14. Feinn, D., Ortoleva, P., Scalf, W., Schmidt, S., Wolff, M.: Spontaneous pattern formation in precipitation systems. J. Chem. Phys. **69**(1), 27–39 (1978)
15. Ross, J., Müller, S.C., Kai, S.: Periodic precipitation patterns in the presence of concentration gradients. 1) Dependence of ion product and concentration difference. J. Phys. Chem. **86**, 4078–4087 (1982) and Kai, S., Müller, St., Ross, J.: Measurements of the temporal and spatial

sequences of events in periodic precipitation processes. J. Chem. Phys. **76**(3), 1392–1406 (1982)
16. Müller, S.C., Kai, S., Ross, J.: Curiosities in periodic precipitation patterns. Science **216**, 635–637 (1982)
17. Moelwyn-Hughes, E.A.: Physikalische Chemie, p. 468. Georg Thieme Verlag, Stuttgart (1970)
18. Pauling, L.: Die Natur der chemischen Bindung; Verlag Chemie GmbH, Weinheim/Bergstraße, vol. 962, p. 472
19. Ulbricht, H., Schmelzer, J., Mahnke, R., Schweitzer, F.: Thermodynamics of finite systems and the kinetics of first-order phase transitions. Teubner Texte zur Physik, Band 17, BSB B.G. Teubner Verlagsgesellschaft, Leipzig, p. 144 (1988)
20. Witten, T.A., Sander, L.M.: Diffusion limited aggregation: a kinetic critical phenomena. Phys. Rev. Lett. **47**, 14000–1430 (1981); and Phys. Rev. B **27**, 5686–5697 (1983)
21. Peitgen, H.-O., Jürgens, H., Saupe, D.: Bausteine des Chaos – Fraktale; Chapter 7.3: Zufalls-Fraktale in einem Laborexperiment. Springer, Berlin, Heidelberg, New York, and Klett-Cotta Stuttgart, pp. 437–444 (1992)
22. Panfilov, A.V., Rudenko, A.N., Pertsov, A.M.: Twisted scroll waves in three-dimensional active media. In: (Krinsky, V.I. (ed.) Self-Organization – Autowaves and Structures Far from Equilibrium. Springer Series in Synergetics, vol. 28. Springer, Berlin, Heidelberg, New York, Tokyo, pp. 103–105 (1984)
23. Holleman, A.F., Wiberg, E.: Lehrbuch der Anorganischen Chemie, pp. 508–511. Walter de Gruyter & Co., Berlin (1960)
24. ibid. p. 14
25. Hollemann, A.F., Wiberg, E.: Lehrbuch der anorganischen Chemie. Walter De Gruyter & CO., Berlin, p. 323 (1960)
26. Remy, H.: Lehrbuch der Anorganischen Chemie; Band I, 11. Aufl. Akademische Verlagsgesellschaft Geest & Portig K.-G. , Leipzig, p. 583 (1960)
27. Liesegang, R.E.: Die Achate, Verlag von Theodor Steinkopf, Dresden und Leipzig, pp. 12–13 (1915)
28. https://de.wikipedia.org/wiki/Sandstein
29. https://de.wikipedia.org/wiki/Bentheimer_Sandstein
30. http://www.geodienst.de/bentheimer.htm

Chapter 4
Runge Pictures

Peter J. Plath and Ernst-Christoph Haß

Fig. 4.1 Runge picture (Experiment executed by C. Baum, October 1995) [1]

"Runge pictures belong to the group of fascinating gadgets whose aesthetics are subject to the strictness of physical and chemical laws." With this slightly pathetic sentence, the former student Christof Baum at the University of Bremen begins his

P. J. Plath et al., *Imagery Synergetics*, Understanding Complex Systems,
https://doi.org/10.1007/978-3-030-95607-3_4

protocol to the experiments he carried out, with which he used his approach to the old experiments of F.F. Runge wanted to prove (see Fig. 4.1).

But of course, he is absolutely right, exactly this approach, to make these fascinating structures understandable with today's methods and the knowledge of synergetics that is our goal.

When Runge no longer carried out the precipitation reactions customary in inorganic chemistry for determining a wide variety of chemical compounds in test tubes, but instead started to carry them out on filter paper, he was also extremely fascinated by the patterns that occurred. In his book "Zur Farbenchemie [2]", published in 1850, he wrote [3]:

> Man macht diese Mischungen gewöhnlich in röhrenförmigen Glasgefäßen, die man Probegläser nennt, und hat besonders darauf zu achten, daß von dem einen oder dem anderen nicht zu viel oder zu wenig hinzugemischt werden, sonst kann es kommen, dass dem Beobachter etwas sehr Wichtiges entgeht und er einen Stoff nicht entdeckt oder auffindet, der bei abgeänderter Mischungsmenge sein Prüfungsmittel kenntlich gemacht haben würde.
>
> Da mir bei meinen Arbeiten diese Unsicherheit zuletzt zu unbequem wurde, so sann ich auf Abhilfe und fand sie im Wechsel der Gefäße oder vielmehr in Beseitigung jedes eigentlichen Gefäßes.
>
> Ich mischte nämlich das Aufeinanderwirkensollende nicht mehr in Glasröhren und gußweise, sondern tropfenweise auf Papier, und zwar auf Löschpapier.
>
> Hier zeigte sich nun mit einem Male eine neue Welt von Bildungen, Gestaltungen und Farbenmischungen, wie ich sie mir natürlich nicht gedacht hatte und die wohl auch nicht zu vermuten war, deren Wirklichkeit daher um so mehr überraschte.
>
> Bald lernte ich die Bedingungen kennen, unter welchen Bedingungen diese Bilder am schönsten und mannigfaltigstem nicht nur ausfallen, sondern auch wie es möglich ist, sie in willkürlicher Menge zu vervielfältigen. Dies zu ermitteln war mir besonders wichtig, denn dadurch gewann diese Entdeckung außer dem chemischen Wert auch noch einen für die bildende Kunst, und es wurde mir möglich, diese Bilder zu Tausenden als Musterbilder in die Welt zu schicken.

The most important sentence from this interesting train of thought for understanding the formation of the Runge pictures is:

> I no longer mixed what was supposed to interact in glass tubes and castings, but dropwise on paper, on blotting paper.

Ferdinand Runge has produced a multitude of wonderful pictures with his drop method. Although his methodical approach was born entirely from the idea of being able to carry out chemical detection reactions with less material and even faster, the "Musterbilder" and five years later also the pictures in his book "Der Bildungstrieb der Stoffe" were largely determined by aesthetic criteria (see Fig. 4.2).

In the book "Naturkunden No. 12" published by Judith Schalansky [4], all these and other Runge pictures are combined in excellent colors and well documented. There is also an extensive representation of the Runge pictures with much more chemicals in the book by Günter Harsch and Heinz Bussemas "Bilder, die sich selber malen—Pictures that paint themselves" [5] and a comprehensive project from the

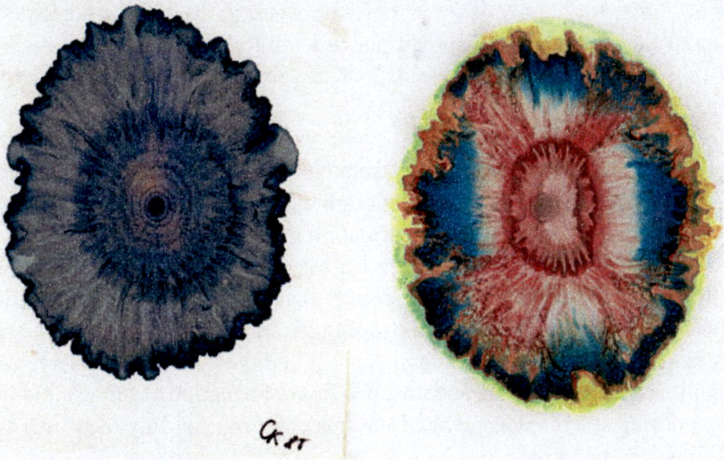

Fig. 4.2 Two examples for Runge pictures that have been produced mainly for esthetic purposes. **left** gift to the author by L. Kuhnert on the occasion of the Wartburg meeting in 1988 (*Photo* P. Plath); **right** produced by F.F. Runge himself. (Reproduced with friendly permission; "Der Bildungstrieb der Stoffe" edited by J. Schalansky, Naturkunden No. 12) p. 31) [4]

Institute Dr. Flad "Runge und Kapillarbilder" by Muriel Dupré and Mirjam Jaenisch under the guidance of Peter Menzel [6].

This aesthetic approach is undoubtedly very fascinating and suitable for approaching questions of pattern formation on a very emotional basis, but the underlying method of successively dripping different chemicals onto the filter paper makes understanding the underlying processes of pattern formation very difficult.

4.1 A More Analytical Approach

Drops on the filter paper can be of different sizes. They can lead to convection currents on the paper, especially since the moistened paper has a different tension than dried filter paper. This inevitably leads to valley and mountain formation, even if the filter paper has been carefully stretched on a frame beforehand. The resulting patterns are not influenced insignificantly by this.

In order to avoid all the associated problems, we have decided that the liquids are only drawn into the filter paper by the capillary forces of this. For this purpose, the liquid is continuously fed through a capillary that sits flat on the filter paper. On the other hand, the capillary is connected to a large reservoir of the liquid in such a way that the hydrostatic pressure at the point of contact on the filter paper can practically be neglected.

If we now let a solution of potassium dichromate soak up by the filter paper in this way, a more or less circular spot will appear. The paper will be dried at 100 °C afterwards. What will happen if we just give up water in the center of the spot in the same way? Will the water also spread out in a circle in the potassium dichromate stain on the filter paper? The answer is a clear "yes and no". We were very much surprised! For the right answer, see Fig. 4.3.

Obviously, one can observe a second cyclic spreading of the waterfront within the dried potassium dichromate stain! This front can be recognized by the change from the darker surroundings to the subsequent lighter, more yellow area. However, what was most surprising was the structure formation in the subsequent white area, which was completely washed out by the water.

This fingering pattern is very reminiscent of the images of frontal instabilities that arise when thinner liquids displace denser liquids in two-dimensional test arrangements, in so-called Hele-Shaw arrangements [7]. In Fig. 4.3 one can observe patterns which remind strongly to stable and unstable finger formation of immiscible Newtonian fluids during the displacement of the more viscose fluid by the much weaker one (see Figs. 4.3 and 4.4).

From this point of view, we can understand the filter paper as a kind of Hele-Shaw cell. From this point of view, we can understand the filter paper as a kind of Hele-Shaw cell. However, what are the two immiscible Newtonian liquids, of which the one with the lower viscosity replaces the one with the higher viscosity?

As already mentioned, when the water penetrates the dried potassium dichromate stain, a stable, almost cyclical front forms. This means that the resulting solution of potassium dichromate acts as a liquid of higher viscosity compared to the air in the still dry stain, which is already loaded with potassium dichromate crystals.

Fig. 4.3 Water spreads out in a circle in the potassium dichromate stain on the filter paper. However, in its rare pattern formation occurs which reminds strongly to fingering front instabilities in Hele-Shaw experiments, especially one can observe splitting of fingers

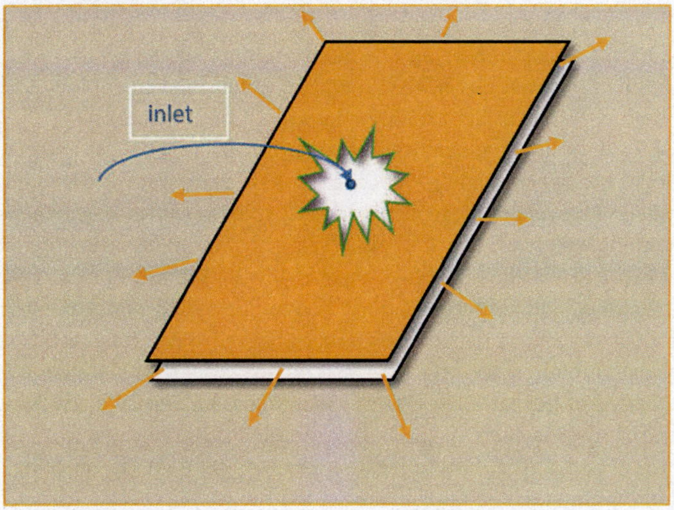

Fig. 4.4 Sketch of a radial Hele-Shaw arrangement [8]. The less viscose fluid from the inlet is displacing the more viscose fluid (yellow) between the two parallel flat plates separated by an infinitesimally small gap

The water that now constantly penetrates this salt solution is obviously of a much lower viscosity than the solution that was created by the re-dissolution of the potassium dichromate crystals in the filter paper. The consequence of this difference in viscosity and the immiscibility is the observed frontal instability, which we recognize in the fingering pattern.

The question that arises here, however, is whether this description fully describes what happened? Obviously, it does not understand that each of these fronts themselves are colored differently in their radial course. Thus, fronts must therefore be structured differently too.

Fig. 4.5 shows a photograph and Fig. 4.6 displays four corresponding transmission micro scans of the three important front processes discussed here. On the left-hand side of the graphs of the four micro scans, the sharp reduction in transmission shows the sharp front formation between the displacing water and the displaced solution of potassium dichromate. This is hard to understand, since water on the one side and an aqueous solution of potassium dichromate are by no means immiscible fluids, on the contrary, they are extremely easy to mix. Furthermore, the graph of the micro scan of the neighbored area of the first front, which should reflect the potassium dichromate solution, has a strong irregular structure. This does not fit the idea of a fluid in the filter paper.

Many questions arise; for example: What kind of front is it actually? What are the fluids that form a front here?

The small crystals in the dried stain are attached to the paper fibers. Along the fibers of the filter paper, the water is drawn into the paper by adhesion or by capillary forces between closely adjacent fibers. These swell when they are wetted with the water. The crystals dissolve and locally highly concentrated solutions are formed. However, ho w can we combine all this questions and statements?

It seems that the situations in the fronts are much more complex than we have discussed before. So, let us start with the first of all these questions.

Figure 4.7 shows how the penetrating water forms the sharp, structured front. In a sense, the "water" pushes an up-dividing wave in front of it.

The crystals in the rare of the front dissolve. A situation arises in which many small and the incoming water (see Fig. 4.8) pushes large crystal chunks together in a highly saturated solution. Here the pure water meets a locally limited, constantly renewing, highly concentrated and viscous solution. "A liquid in which a large number of small solid particles is suspended (suspension) can be treated as a homogeneous medium if the phenomena of interest are characterized by distances which are large compared to the dimensions of the particles. Such a medium has an effective viscosity η which is different from the viscosity η_0 of the base liquid." [9] (translated from German by the author) That is exactly the prerequisite for the formation of the observed structure formation of the front.

However, the question remains, what happens just behind the circular second front, which penetrates into the area of the previously dried potassium dichromate. It is reasonable to assume that the water, after passing the large front of the suspended crystals, now penetrates in the form of a highly concentrated solution along the fibers of the filter paper into the area that was previously dried but already loaded with potassium dichromate. In this solution, larger crystals can also grow at suitable points on the fibers, see Fig. 4.9.

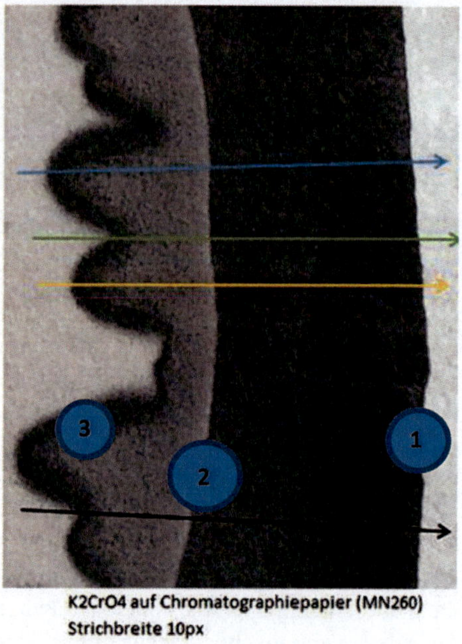

Fig. 4.5 Microscopic photograph of the front instabilities in case of the displacement of a potassium dichromate stain by water. The fronts are numbered in the order in which they were created. The water "flows" from the left to the right side. The arrows indicate the way in which the gray values of the photograph (see Fig. 4.6) are recorded. (*Photo* K. Guttmann and S.C. Müller, University of Magdeburg (2015))

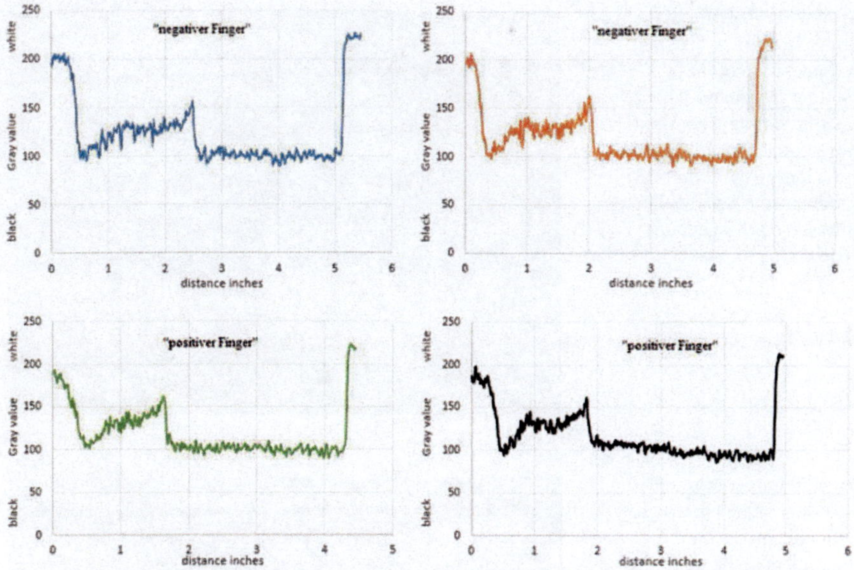

Fig. 4.6 Transmission micro scans at four different position of the fronts as indicated with the same colors by the as indicated in Fig. 4.5 (these scans have been executed by S.C. Müller and Katja Guttmann, University of Magdeburg, in 2015)

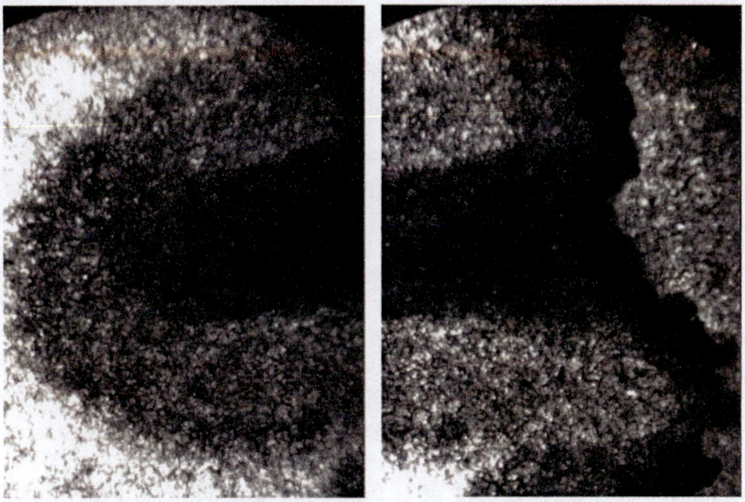

Fig. 4.7 Formation of the front (negative finger) between the penetrating water (coming from the left) and the area of the partially dissolved Potassium dichromate crystals. (*Photo* K. Guttmann and S.C: Müller, University of Magdeburg (2015))

Fig. 4.8 Dissolving of the former front of crystals and agglomeration of the resulting fragments. The water is coming from right above corner (magnification of the lens: 6.3). (*Photo* K. Guttmann and S.C. Müller, University of Magdeburg (2015))

Fig. 4.9 Larger crystals (dark dots in the photo) grow in suitable places on the paper fibers in the more concentrated solution after the water has passed the third front. (*Photo* K. Guttmann and S.C: Müller, University of Magdeburg (2015))

Above all, however, immediately behind the front penetrating into the dry area, the many small crystals that formed during the dry phase dissolve again, so that the filter paper becomes more translucent again. Similar to the case discussed above, the crystals that have just been loosened are only pushed onto one another directly in the front, as if melting pieces of ice were being pushed together to form larger and more compact chunks of ice. The result is a sharp drop in light transmission, which reveals itself as a front.

The experiment discussed here is, if we compare it with the beautiful Runge pictures, extremely simple: we let water penetrate into a dry spot of potassium dichromate on filter paper by capillary forces alone. In order to understand the observable phenomena, we have discussed the interplay of many different processes, such as the capillary flows along the fibers, the dissolution and growth of crystals and the formation and accumulation of large crystal agglomerates.

Fig. 4.10 **Left** 3D visualization of the macroscopic structure formation by gray values of micro scanning the **right** photograph of the Runge picture during washing out experiment of a dried potassium dichromate stain in the filter paper. The fronts are correlated with the very sharp changes gray values in transmission. (*Photo* K. Guttmann and S.C: Müller, University of Magdeburg (2015))

As in the Hele-Shaw experiment, the pattern formation observed during the migration of the unstable front, which ultimately also fascinates us so much in the Runge pictures, is largely only described by quantities such as the density of different fluids, their miscibility and the resulting surface tension. Figure 4.10 shows this macroscopic structure formation, which is based on complex microscopic processes, but which it makes sense to describe on a macroscopic level. This is exactly the subject of the science of synergetics and its great advantage.

4.2 Some Additional Measurements

When working with students, care must be taken to ensure that everyone wants to examine their own system. For C. Baum it was potassium ferricyanide (red prussiate of potash). (cf. Fig. 4.1) He investigated the speeds with which the displacing water penetrated into $K_3[Fe\ (CN)_6]$ loaded filter paper which had previously been impregnated with solutions of different concentrations (Fig. 4.11).

For sure, the potassium ferricyanide III $K_3[Fe^{III}\ (CN)_6]$ stain should look red but not green or blue. However, the time that has passed since these experiments were carried out (October 1995) until this chapter was written in 2020 has left its mark. The red potassium ferricyanide III, which is a light oxidizing agent, was partially reduced to the yellow potassium ferrocyanide II $K_4[Fe^{II}\ (CN)_6]$, which then reacted with the remaining iron II salt to give Berlin blue (Turnbulls blue) $K[Fe^{II}\ Fe^{III}\ (CN)_6]$. All this changes happened in the past.

It is noteworthy that the pattern formation in the third front differs slightly from that in the potassium dichromate experiment. One can detect a fracture-like shape

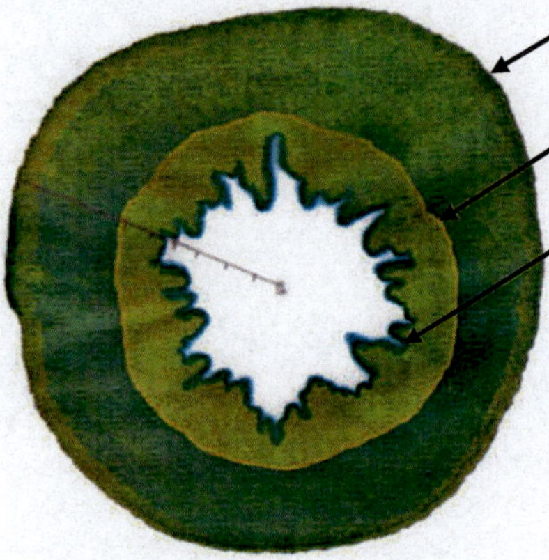

Fig. 4.11 Water spreads out in a circle in the potassium ferricyanide III ($K_3[Fe(CN)_6]$) stain on the filter paper. However, in its rare pattern formation occurs which reminds strongly to fingering front instabilities in Hele-Shaw experiments. The original concentration of the ferrocyanine solution with which the filter paper was impregnated was c = 1.32 mol/l

of the fingers.[7] This indicates that the front, which is displaced by the water, has visco-elastic properties. Van Damme observed something similar in the displacement of clay suspensions by water [10].

4.3 Impregnation of Filter Paper with Water and Potassium Ferricyanine III

4.3.1 The First Front

Up until now, the attention was clearly directed to the quite spectacular structure formations such as the fingering in the third front. But why is there a sharp formation at all on the first and second fronts? Shouldn't the solution diffusing into the filter paper gradually diffuse into the paper according to Fick's second law without a visible front formation?

To answer this question, let's take a closer look at the first step of our experiments—impregnating the filter paper. We can even go one step back by examining only the penetration of water into the filter paper, because this process is also already connected with the formation of a sharp front (Fig. 4.12). Besides, those who like to drink filter coffee and make it themselves can certainly confirm all of this observation through their own experience.

In order to be able to make quantifiable statements about the behavior of the fronts, we determined the temporal course of the respective front both for the water spot and for different concentrations of the potassium ferricyanide III $K_3[Fe^{III}(CN)_6]$

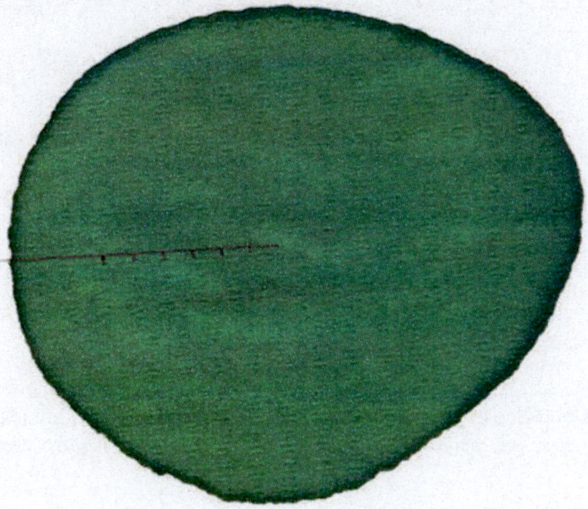

Fig. 4.12 First step in the experiment: Impregnation of the filter paper with potassium ferricyanide III $K_3[Fe^{III}(CN)_6]$ with concentration $c = 0.66$ mol/l. A sharp front (the first front) of the stain is formed, which over time spreads more and more without losing its sharpness and shape

spots. We cannot present here the water stain here because the water has evaporated meanwhile and nothing was left on the dried white filter paper to be photographed or scanned.

For our experiments, we used Selecta, Schleicher & Schüll brand filter paper for analytical paper chromatography of 185 cm in diameter.

Incidentally, it should be mentioned at this point that, as Harsch and Bussemann mention [5], Runge's "Musterbilder" [2] ultimately led to paper chromatography, an analytical method that is of great importance today in environmental and food chemistry as well as in clinical chemistry. After all, thin layer chromatography, which is widely used today, is closely related to it.

Using a graphite pencil, the paper scales from the centerat 0.5 cm intervals and fixed on a cork ring with pins afterwards. The filter paper impregnated with potassium ferricyanide III $K_3[Fe^{III}(CN)_6]$ was dried in a drying oven at 80° C for five minutes. The time is measured with a stopwatch when the liquid has passed the respective interval of 0.5 cm. The time is measured with a stopwatch when the liquid has passed the respective interval of 0.5 cm. The amount of liquid absorbed by the filter paper was determined by weighing the moistened paper whenever the radius of the stain had increased by a further 0.5 cm.

Assuming that the radius is proportion to t^α

$$R \propto t^\alpha \tag{4.1}$$

If one transform this relation to a linear one

$$lnR \propto \alpha lnt \tag{4.2}$$

one gets for α the value $\alpha = 0.6173$ see Fig. 4.13b.

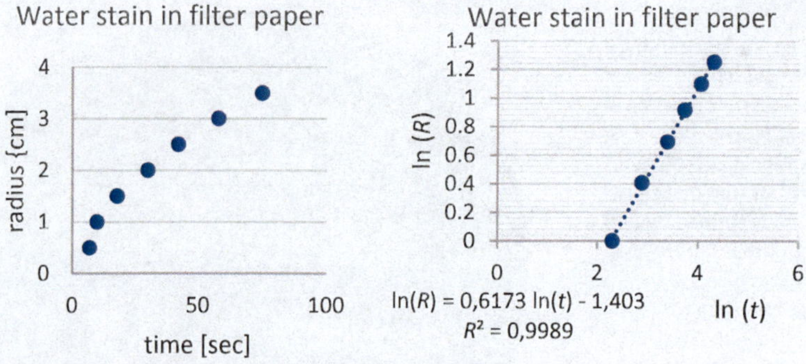

Fig. 4.13 **Left** The radius R of the circular water stain as function of time. **Right** double logarithmic plot of the experimental data: radius R versus time t with linear regression. The first point at $R =$ 0.5 cm was neglected because of the uncertainty of its measurement

Thus, we can write the relation:

$$R \propto t^{0.6173} \tag{4.3}$$

The velocity $v(R)$ of the radius R of the front can be estimated by differentiation.

$$v(R) = \frac{dR}{dt} \propto 0.6173t^{0.6173-1} = 0.6173t^{-0.3827} \tag{4.4}$$

The speed of the waterfront decreases with increasing radius or with advancing time respectively.

On the other hand, the amount m of water with density ρ absorbed by the filter paper with thickness h should be a quadratic function of the radius R.

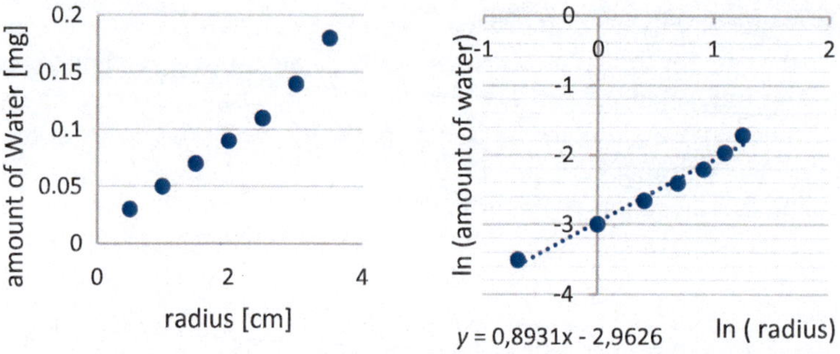

Fig. 4.14 **Left** Amount of water in the filter paper as a function of the radius of the waterfront; **right** double logarithmic plot of Fig. 4.14 left. The slope of the linear plot $\ln(m) = \beta \ln(R) + \ln(\rho\pi h)$ leads to $\beta = 0{,}893$

$$m = \rho \pi h R^2 \tag{4.5}$$

However, if we consider the amount of water absorbed by the filter paper as a function of the radius of the water spot, then there is no quadratic relationship according to Eq. 4.5. Instead of the integer exponent $\beta = 2$ we get a fractional exponent $\beta = 0.893$ (see Fig. 4.14).

$$m \propto R^{0.893} \tag{4.6}$$

In terms of water absorption, the filter paper does not act like a compact volume, but like a fractal. This is of great importance for understanding the dynamics of pattern formation in the Runge images. The water does not fill the spaces between the fibers of the paper, but as it penetrates the paper, it is bound to its fibers by adhesion or by their swelling.

Since the radius R of the waterfront itself is a function of time, we can also formulate the amount of water absorbed by the filter paper as a function of time.

$$m \propto \left(t^{0,6173}\right)^{0,8931} = t^{0.5513} \tag{4.7}$$

4.3.2 *Some Basics of Synergetics*

Some basics of synergetics may help to understand our simple experiments.

In the beginning of his book "Foundation of Synergetics I" [11]. Alexander Mikhailov describes the behaviour of *bistable media.*" Bistable media consist of elements that have two steady states which are stable under sufficiently small perturbations."

So let us describe the filter paper as a one-component bistable media (see Fig. 4.15). The filter paper itself acts as an active element in relation to the uptake of water. The dry state u_1 is the one stable state of the paper and the fully soaked paper (Fig. 4.12) is the other stable state u_3. If one disturb the dry state u_1 by putting a very small drop of water or any aqueous solution in the center of the paper the drop will vanish by diffusing into the paper and evaporation. However, if the perturbation is large enough, the system will get into its second stable state u_3 in this local area of the paper. In this case, the localized initial perturbation will evolve to a homogeneous steady state while forming a circular front that spreads out over the paper (Fig. 4.12).

The setup of our experiments is done in a way that the water or solution supply practically does not change during the time of the experiment "because its store is so large, that we can neglect the decline due to consumption." [11] This way we create a sufficient large perturbation of the state u_1 of the dry paper at the beginning of the experiment. The consumption of the water or the solution is associated with wetting of the filter paper with this aqueous solution. We know from experience

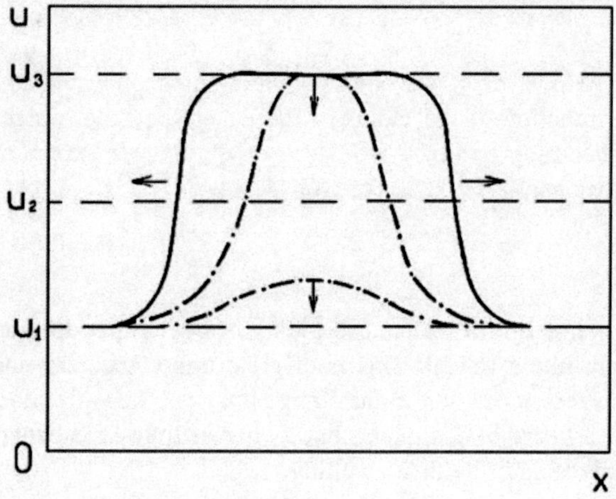

Fig. 4.15 Evolution of localized initial perturbations of a homogeneous steady state in a bistable system. The stable states are u_1 and u_3, whereas the state u_2 is instable. This figure is taken from A. Mikhailov's book "Foundations of Synergetics I" [11] and has been redrawn

that especially in hot summers with long dry seasons, the parched soil can absorb practically no water. The same applies to the filter paper.

4.4 Water Treatment of Dried Filter Paper Impregnated with Potassium Ferricyanide III

4.4.1 The Second Front

What we have just discussed for the first front also applies in principle to the second front. By drying the filter paper soaked with the aqueous potassium ferricyanide III $K_3[Fe^{III}(CN)_6]$ solution in the oven at 80° C, our system has become "excitable" again, because it returned in its dry state u_1 again (Fig. 4.15).

But now, something has changed compared to the previous condition of the untreated filter paper. The paper treated with the iron solution and dried now contains iron crystals of various sizes that adhere to the fibers of the filter paper. These crystals have to be dissolved, which affects the swelling of the fibers and certainly also the front speed of the now highly saturated and viscous solution. The partially dissolved crystals form a highly viscous suspension behind the advancing front, which only slowly dissolves in the water that follows. This can be observed particularly well when the concentration of the iron solution with which the filter paper was previously impregnated was particularly low (see Fig. 4.16).

If one compares the speeds of the first and the second front, it can be seen clearly that the second front of the water penetrating the impregnated filter paper is significantly slower than the first front of the solution or the pure water penetrating the untreated filter paper (see Figs. 4.11 and 4.12).

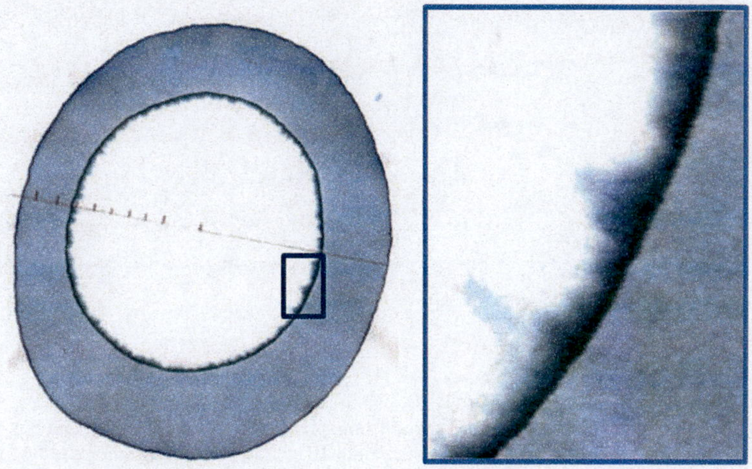

Fig. 4.16 Formation of the second front: **left** water penetrates the dried filter paper, which has previously been soaked with a very dilute solution of an aqueous potassium ferricyanide III, $K_3[Fe^{III}(CN)_6]$ with $c = 0.04$ Mol/l; **right** enlarged section from the adjacent figure. A sharp front can be seen with a diffuse, irregular decay in the area of the water pushing into the filter paper

In the case of the impregnation into dried filter paper which was previously impregnated with potassium ferricyanide III solutions of low concentrations (e.g., $c =$ 0.04 mol/l, see Fig. 4.17), the time development of the radius of the aqueous second front follows the approach $R \propto t^{\alpha}$ with $\alpha < 1$ as in the case of the first front or the simple water front, so that the front speed decreases as the radius of curvature increases.

A slight deviation of the approach $R \propto t\alpha$ with $\alpha < 1$ occurs when considering the variation of the velocity of the aqueous second front, where the filter paper

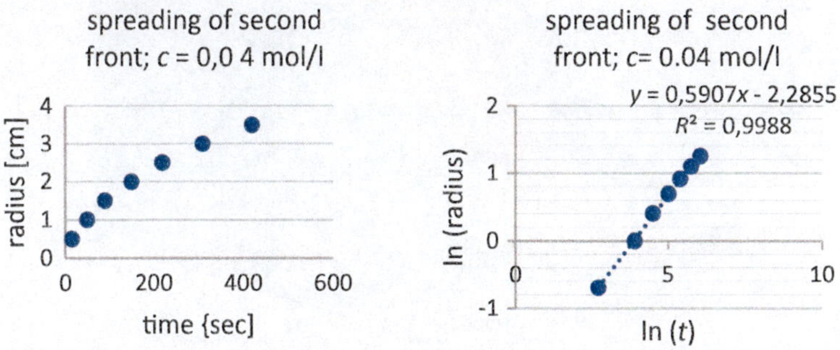

Fig. 4.17 Temporal development of the second front spreading in the filter paper which was previously impregnated with aqueous potassium ferricyanide III $K_3[Fe^{III}(CN)_6]$ solution of concentration c = 0.04 mol/l. **left** Radius of the front as a function of time. **right** Double logarithmic plot of the radius versus time. Concerning the ansatz $R \propto t^{\alpha}$, α gets $\alpha = 0.5907$

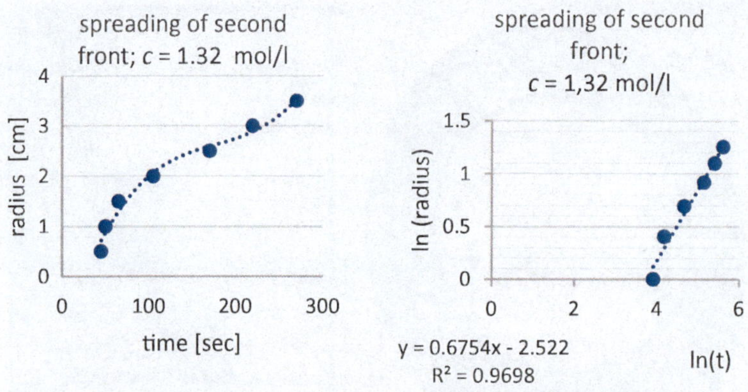

Fig. 4.18 Temporal development of the second front spreading in the filter paper which is previously impregnated with aqueous potassium ferricyanide III solution of concentration c = 1.32 mol/l; **left** the radius R of the front as a function of time with a polynomial trend line of third degree; **right** Double logarithmic plot of the radius R versus time. Concerning the ansatz $R \propto t^{\alpha}$, α gets $\alpha = 0.6754$ using linear regression, if the value pair for the smallest radius is not taken into account. Otherwise, the spreading velocity is much higher because of the development in the earlier stages

was previously impregnated with a high potassium ferricyanide III concentration of 1.32 mol/l (see Fig. 4.18).

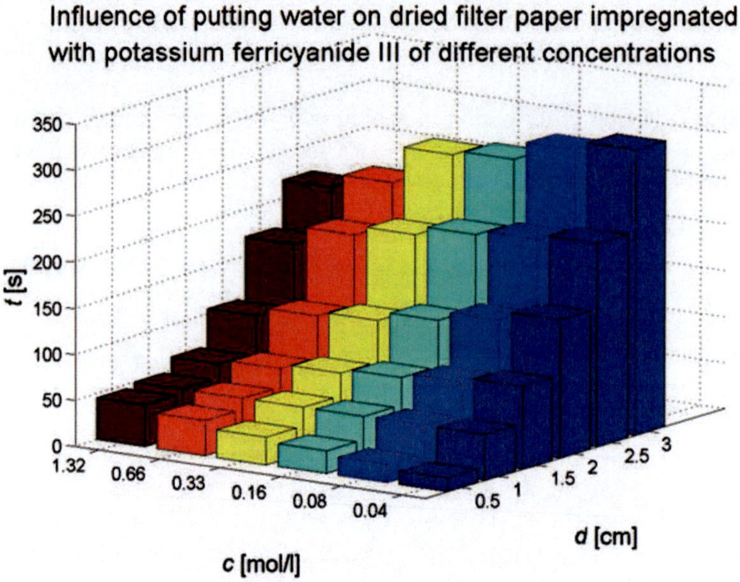

Fig. 4.19 3D bar diagram of the time dependence of the aqueous second front upon the impregnating potassium ferricyanide III concentrations c and the distances d from the centre of filter paper

To summarize the experimental results, Fig. 4.19 illustrates in a three-dimensional representation the time development of the second front of spreading water depending on the impregnation concentrations and the distances from the center of the filter paper. From this, it can clearly be seen that in the case of a very low concentration the second front moves fast in the beginning and slows down at increasing distances, while in the case of higher concentrations this behavior reverses, i.e., the second front is comparatively slow in the beginning and moves relatively faster at larger distances.

4.4.2 The Third Front

Behind the second front, an area that is getting wider and wider is created, which is separated from the purely watery area by the third, strongly structured, sharp front (Fig. 4.20). The generation of this third front, which we discussed in detail at the beginning in the potassium dichromate experiment, is therefore a direct consequence of the increase in concentration during the previous impregnation.

If we look again at Fig. 4.11 in this context, then with the concentration $c =$ 1.32 mol/l of potassium ferricyanide III used, not only a simple fingering of the third front can be seen, but the visco-elastic front brakes through in some places [10].

From the strong blue coloration of this third front, we can see that this is the poorly soluble Berlin (Prussian) blue or Turnbull's blue, of which the highly viscous suspension consists now.

However, if the concentration of the impregnating potassium ferricyanide III is increased, the patterns and the time dependence of the front shift change significantly. The second front is spreading faster and faster. Behind the second front, an area that is getting wider and wider is created, which is separated from the purely watery area by the third, strongly structured, sharp front (Fig. 4.20). The generation of this third front, which we discussed in detail at the beginning in the potassium dichromate

Fig. 4.20 Influence of the potassium ferricyanide III concentration with which the filter paper is impregnated on the structure formation of the third front. **Left** c = 0,16 mol/l; **centered** $c =$ 0,33 mol/l; **right** $c =$ 0,66 mol/l

experiment, is therefore a direct consequence of the increase in concentration during the previous impregnation.

4.4.3 *Chemical Interpretation of the Observed Colours*

The question that remains is what is happening in the area between the second and third fronts. The yellow-green color of this area today indicates that the formerly red iron salt $K_3[Fe^{III}(CN)_6]$ [12] had possibly changed into the yellow iron salt $K_4[Fe^{II}(CN)_6]$ [13] in the back of the second front (Fig. 4.21).

However, the aqueous solution of $K_3[Fe^{III}(CN)_6]$ is yellow in color itself and decomposes in the light to form iron (III) hydroxide $Fe(OH)_3$ [14] or $Fe_2O_3{\cdot}3H_2O$ and several oxy-hydroxides respectively. This can possibly also be responsible for the partly yellowish-brown color in the area between the first and the second front.

Today, after 25 years, it is difficult to judge what was taking place chemically when these experiments were carried out, since potassium ferricyanide III in particular changes over these periods and thus changes the colors. Lothar Kuhnert also draws attention to this fact [15].

> In einigen Fällen scheinen sich die Bilder (Runges Originale, Anm. der Autor) durch die lange Lagerzeit durch Luftoxydation verändert zu haben. Beispielsweise besteht Bild 1 der Musterbilder nach Runges Angaben nur aus einem Flecken gelben Blutlaugensalzes. Heute, nach 135 Jahren, hat sich dieser in Berliner Blau umgewandelt. … Bei der Berliner-Blau-Reaktion ist auch immer mit dem Auftreten eines grünen Niederschlages zu rechnen: Berliner Grün. Wie beim eigenen Nachgestalten einiger Bilder festgestellt wurde, hängt dies stark von den jeweiligen Arbeitsbedingungen ab.

> In some cases, the pictures (Runge's Origiale, author's note) seem to have changed due to air oxidation due to the long storage time. For example, according to Runge's information, picture 1 of the "Musterbilder" consists only of a patch of "yellow prussiate of potash" i.e., potassium ferricyanide II $K_4[Fe^{II}(CN)_6]$. Today, after 135 years, it has been transformed into Berlin Blue $K[Fe^{II}Fe^{III}(CN)_6]$. … In the case of the Berlin-Blue reaction, the occurrence of a green precipitate is always to be expected: Berlin Green. As we found out when we created some pictures ourselves, this depends heavily on the respective working conditions.

Fig. 4.21 **Left** potassium ferricyanide III $K_3[Fe^{III}(CN)_6]$; **right** potassium ferricyanide II $K_4[Fe^{II}(CN)_6]$ 17 mm × 15 mm und 9 mm; (*Photos* are made by Kathrin Götz [12] and Hubert [13])

Even if we start with potassium ferricyanide III $K_3[Fe^{III} (CN)_6]$, we have to consider a corresponding Prussian blue reaction (i.e. Berlin-Blue reaction). Baum [1] writes about his direct observations: "If water penetrates into an impregnated filter paper, an increasingly large ring forms around the penetration point with increasing concentration of potassium ferricyanide III. After a short time, three areas can be distinguished: (a) a central area empty of potassium ferricyanide III, (b) an approximately 0.5 mm light yellow zone and (c) a dark yellow zone. The dark yellow zone appears centripetally fingered with spikes of several orders. The light-yellow zone is sharply delimited against the dark yellow zone and blurred towards the center. It follows the lower order spikes."

If we look at today's coloring of the pictures with this description, then it is difficult to interpret them correctly, apart from the mention of fingering, which we identify with the third front. As in photography, we can regard the chemical reactions, which, after the actual experiment, lead to the colors and patterns we see today as the development of the original patterns.

It is particularly noticeable that the color change from blue to deep green in the area between the first and the second front depends on the concentration of the potassium ferricyanide III $K_3[Fe^{III} (CN)_6]$ originally used (see Figs. 4.11, 4.15, and 4.17). In other words: soluble Berlin Blue $K[Fe^{II} Fe^{III}(CN)_6]$ and Berlin Green [16] developed from pale yellow solution of potassium ferricyanide III $K_3[Fe^{III} (CN)_6]$. This presupposes that the trivalent iron in the potassium ferricyanide III has been partially reduced to the divalent iron in Berlin Blue $K[Fe^{II} Fe^{III}(CN)_6]$ or Berlin Green.

Potassium ferricyanide III $K_3[Fe^{III} (CN)_6]$ is known to be a weak oxidizing agent for organic synthesis [14]. Moreover, it has a strong oxidizing effect, especially in alkaline solutions, whereby it is reduced to the more stable potassium ferricyanide II $K_4[Fe^{II} (CN)_6]$ [17]. Since it is easy to see that Berlin blue and Berlin green appear in our experiments, we must assume that the $K_3[Fe^{III} (CN)_6]$ is reduced to $K_4[Fe^{II} (CN)_6]$, for example by a slight oxidation of parts of the filter paper, and then with the remaining $K_3[Fe^{III} (CN)_6]$ molecules react to the beautiful, brightly colored connections.

In our opinion, all of these reactions take place to some extent in the beginning while the experiments are being carried out. Therefore, it is not only the dissolution and crystallization processes and the sponging out of the loosened crystals that are important for the pattern formation, but also the associated chemical reactions.

4.5 Runge Pictures—Chemical Reactions Going on in Filter Paper

4.5.1 *Treatment of Impregnated Filter Paper First with Iron III and then with Copper Sulfate*

If, instead of pure water, one chooses a solution that reacts with potassium ferricyanide III $K_3[Fe^{III}(CN)_6]$, then this should result in a change in the second front speed and perhaps also in the shape of the front. In any case, this increases the complexity of the reaction in the filter paper and approaches the typical Runge images.

Since the filter paper was originally impregnated with Potassium ferricyanide III $K_3[Fe^{III}(CN)_6]$, it makes sense to use an iron II salt solution such as iron sulfate $FeSO_4$ as the second reagent. As is well known, the charge transfer complex of the soluble Berlin blue $K\left[Fe^{II}Fe^{III}(CN)_6\right]$ is formed during the reaction of the two iron salts.

$$K_3\left[Fe^{III}(CN)_6\right] + Fe^{II}SO_4 \rightleftarrows K\left[Fe^{II}Fe^{III}(CN)_6\right] \tag{4.8}$$

Iron sulphate $Fe^{II}SO_4$ penetrates the filter paper impregnated with potassium ferricyanide III $K_3[Fe^{III}(CN)_6]$ ($c = 0{,}33$ mol/l) and forms a dark green front, which is followed by a strong blue rear made of insoluble Berlin blue $Fe^{II}\left[Fe^{II}Fe^{III}(CN)_6\right]_2$ (see Fig. 4.22).

Subsequently, blue copper sulphate $Cu^{II}SO_4$ of high concentration (hot saturated) enters the filter paper from the center and forms a two-front system: a very narrow,

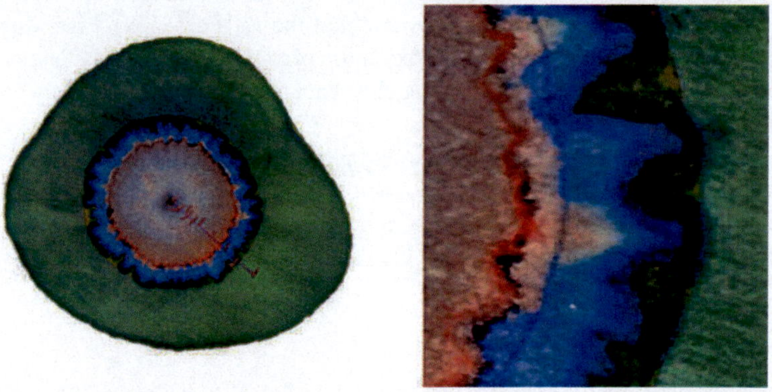

Fig. 4.22 **Left** Iron sulphate penetrates the filter paper impregnated with Potassium ferricyanide III $K_3[Fe^{III}(CN)_6]$ ($c = 0.33$ mol/l) and forms a front and a strong blue rear made of insoluble Berlin blue. When copper sulfate subsequently flows into the filter paper, a two-front system is formed: a very narrow, almost circular purple front and a red-brown-purple, slightly modulated rear, behind which an almost uniform red-brown tinted spot remains. **Right** Detail from the adjacent picture

almost circular front and a red-brown and purple, slightly modulated rear, behind which an almost uniform red-brown tinted spot remains.

Only then, does blue copper sulphate of high concentration (hot saturated) penetrate again from the center into the filter paper and forms a two-front system: a very narrow, almost circular front and a red-brown and purple, slightly modulated rear, behind which an almost uniform red-brown tinted spot remains.

The formation of two-front systems, which occurs when water or an aqueous solution penetrates a previously impregnated filter paper, appears to be a very general phenomenon. In the case of the penetration of water into a filter paper impregnated with potassium dichromate solution or a solution of red Prussiate of potash, this phenomenon was described in detail at the beginning. New in the example discussed here is that the penetrating $FeSO_4$ solution reacts chemically with the previously added impregnating compound $K_3[Fe^{III}(CN)_6]$ to form the soluble Prussian blue $K\left[Fe^{II}Fe^{III}(CN)_6\right]$ and even forms a poorly soluble complex, namely the insoluble Prussian blue $Fe^{II}\left[Fe^{II}Fe^{III}(CN)_6\right]_2$.

In general, it can be said that the higher the concentration of the impregnating solution, the more crystals and the larger crystals of the impregnating agent are present in the dry impregnated filter paper. These crystals are partially solved by the newly penetrating aqueous solution. This creates an increased concentration and in some cases a suspension in the front, so that the front has an increased viscosity and, in keeping with the Hele-Shaw experiment, penetrates the dried filter paper with a smooth, unstructured front.

However, the dissolution process takes some time, the greater the larger and more crystals of the impregnating compound have to be dissolved. The front moves on in the meantime. The area between the second and the third front, the rear front, increases with increasing concentration of the impregnating solution. However, the liquid that is pushed into the filter paper is no longer afflicted with the dissolved products or the resulting suspension, i.e., less viscous and thus penetrates into the more viscous suspension with structure formation such as fingering, also in the sense of Hele -Shaw experiment.

Now chemistry also comes into play. During the dissolution process of the crystals of the impregnating compound, this reaction with the penetrating solution takes place in a very complex way. Soluble Berlin blue is formed. This only works if the crystals have already dissolved, or are partially or superficially dissolved. The unpolluted subsequent flowing solution of the iron II sulfate provides additional iron II ions, so that the insoluble Berlin blue can be formed from the soluble one. The suspension that forms in the process consists primarily of these newly formed, insoluble Berlin blue crystals. Since they are formed very quickly in large quantities, they are very small and can be sponged up to the original front when fingering if the visco-elastic front cracks (see Fig. 4.22).

Of course, these truly complex processes in the area between the fronts do not occur in all places at the same time and to the same extent. Mainly because diffusion processes are associated with it, and the filter paper is also fractally structured. So, it is not surprising that this intermediate area is heavily spotted in color.

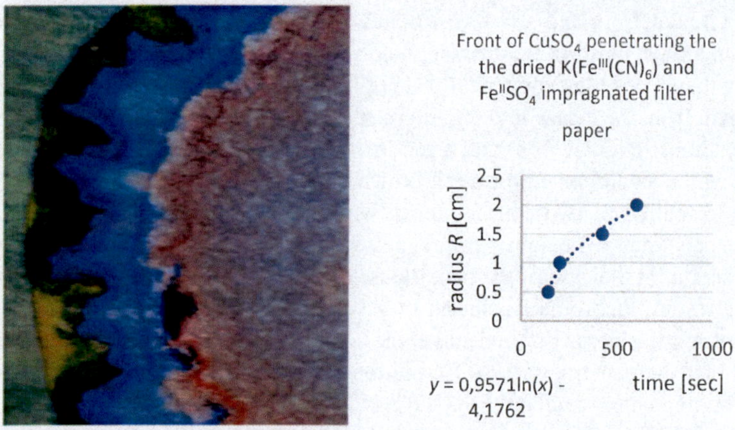

Fig. 4.23 **Left** Double front system of the copper sulphate solution, which penetrates the previously dried filter paper as the third solution. **Right** Radius of the most for-ward front of the copper sulfate front system, described by $R = 0.96\ lnt—4.18$ or $R \propto t^{0.96}$ according to the logarithmic approach

Let us now briefly look at the double front system of the copper sulphate solution, which now penetrates the previously dried filter paper as the third solution. This creates a very sharp blue-violet, largely smooth, circular front in contact with the iron (II) sulfate remaining after drying or the remnants of the soluble Berliner blue, (see Fig. 4.23 left). For the extremely slow expansion of the copper sulfate front system, the radius of its most forward front is described by $R = 0.96lnt - 4, 18$ or $R \propto t^{0.96}$ according to the logarithmic approach (see Fig. 4.23 right).

If we assume that the crystals remaining in the filter paper are only slowly dissolved and remain behind the moving front, we can expect a highly viscous suspension of copper-iron-cyanide complex compounds again. This is displaced by the penetrating copper (II) sulfate solution, which has a far lower viscosity. Fingering structure formation according to the Hele-Shaw experiment is the result.

Interestingly, a pattern also forms in the central area in front of the rear of the penetrating CuSO4 solution (Fig. 4.24). The patterns become more pronounced with increasing distance from the center. It appears that the pattern formation in the rear of the double front system of the $CuSO_4$ solution has already occurred here. It is however rather caused by the remains of the previous rear fronts, which repeatedly broke through, since they were not yet strong enough. Such breakthroughs through the rear front, even through the leading front of the double front regime of the $CuSO_4$ solution, can also be seen in the "last" front when the experiment was stopped. This observation confirms the assumption about the origin of the patterns in the central, inner area.

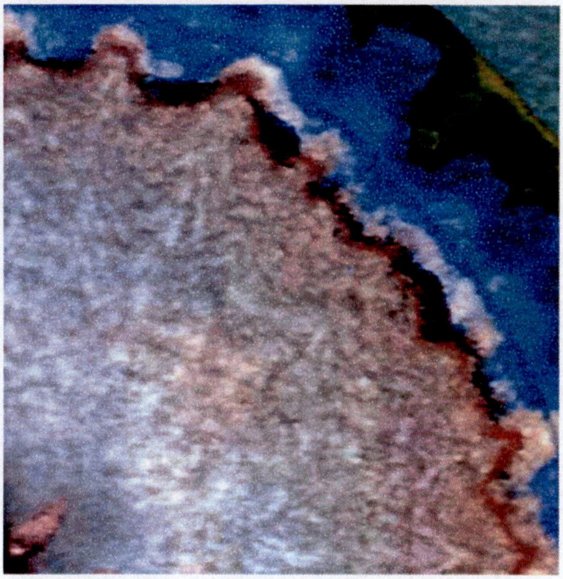

Fig. 4.24 Pattern formation in the central area. In the lower left corner, there is the starting point from which the solutions penetrate the filter paper

4.5.2 Modifying Runge's Drop Method to Understand Its Effect on Pattern Formation

In all of our experiments described here so far, we have carefully ensured that the individual liquids could only penetrate into the filter paper because of the absorbency of the filter paper. In order to understand Runge's drop method and its effect on the pattern formation, we will slightly increase the pressure with which the last solution flows out of the capillary and penetrates the filter paper. For this purpose, we lift the storage vessel, for example by 20 mm, in relation to the exit of the capillary. Figures 4.1 and 4.25 show examples in which a $CuSO_4$ solution is "pressed" in this way into the filter paper which was impregnated before at first with Potassium ferricyanide III ($K_3[Fe(CN)_6]$ (c = 0.33 mol/l) and afterwards with $FeSO_4$.

The effect of the pressure on the $CuSO_4$ solution, which is the last component to enter the filter paper, is mainly that the pattern on the rear side of the double-front system of the $CuSO_4$ solution is particularly strong. The front system also moves faster away from the center than if no additional pressure is applied. If the $CuSO_4$ front system approaches the front system of the previously penetrated $FeSO_4$ solution, it is slightly compressed and the pattern formation in its rear front is reinforced.

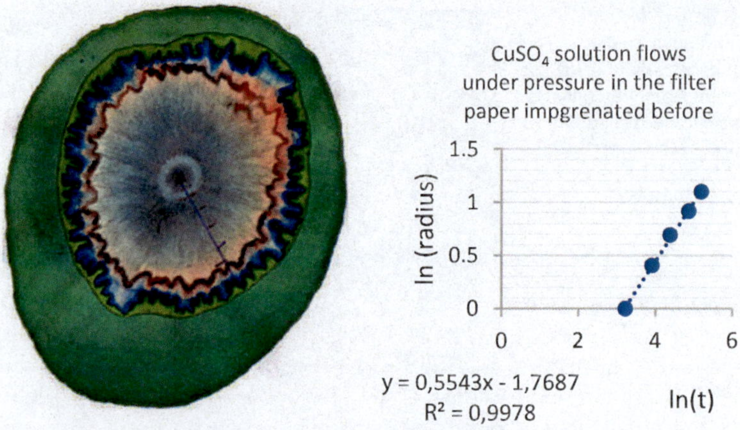

Fig. 4.25 **Left** A CuSO4 solution flows under a very small pressure in the filter paper which was first impregnated with potassium ferricyanide III ($K_3[Fe\ (CN)_6]$ (c = 0.33 mol/l) and then with $FeSO_4$. **Right** Double–logarithmic plot of the radius R of the leading front of the CuSO4 solution as a function of time neglecting the first point with $R = 0.5$ cm; $\ln(R) = 0.55\ln(t) - 1,77$; *or* $R \propto t^{0.55}$

4.6 Reversal of the Order of the Solutions Penetrating into the Impregnated Filter Paper

4.6.1 Treatment of with Potassium Ferricyanide III Impregnated Filter Paper with Only Copper Sulfate

Another interesting color option is to let a blue copper sulphate solution CuSO4 run into the filter paper that has been impregnated before with potassium ferricyanide III $K_3[Fe^{III}\ (CN)_6]$. Ch. Baum reports that a brown reaction product is formed. This reddish-brown spot is still clearly visible today, so it is quite stable, (see Fig. 4.26 left). In the literature [18], one finds a corresponding product with such a color only for divalent iron: $\left[Fe^{II}(CN)_6\right]^{4-} + 2Cu^{2+} \leftrightarrows Cu_2\left[Fe^{II}(CN)_6\right]$. Such a formulation would mean, however, that in an aqueous solution this compound would form a bluish hexaquo complex, i.e., it would by no means be brown in color. Brown and yellow colors of copper in connection with iron are known from the sulfidic copper ores, such as the copper pyrites $CuFeS_2$, the colored copper ore Cu_3FeS_3 and the copper luster Cu_2S. This reaction will involve the formation of a metal–metal Cu-Fe bond, whether with trivalent or divalent iron. A copper complex similar to the insoluble Berlin Blue would also come into question, for example: $Cu^{II}\left[Cu^{II}Fe^{III}(CN)_6\right]_2$.

$$K_4\left[Fe^{II}(CN)_6\right] + 2Cu^{II}SO_4 \rightleftarrows Cu^{II}\left[Cu^{II}Fe^{II}(CN)_6\right] + 2K_2SO_4 \quad (4.9)$$

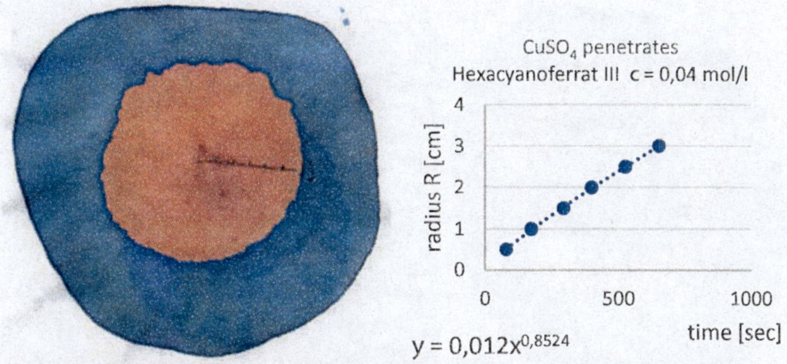

Fig. 4.26 **Left** Filter paper impregnated with potassium ferricyanide III $K_3[Fe^{III}(CN)_6]$ of concentration $c = 0.04$ mol/l and dried at 80 °C. A concentrated copper sulfate solution flows into the impregnated paper, and a brown complex is formed. A slightly irregular blue front of the penetrating copper sulfate solution can be observed; **right** Temporal development of the front of the copper sulfate solution

Anyway, it is very likely that this reaction presupposes that part of the potassium ferricyanide III $K_3[Fe^{III}(CN)_6]$ might need to be reduced. Therefore, we are dealing with a chemically extremely complex reaction process.

Similar to the attempt to displace the potassium ferricyanide III with water, with the very low concentration of $c = 0.04$ mol/l of the impregnating $K_3[Fe^{III}(CN)_6]$ solution used for impregnation, there is only a single front and no double front (see also Fig. 4.16).

The single blue front of the copper sulfate solution is formed slightly irregular. In addition, the front spreads very slowly with almost the same speed. The exact time dependence of the front spread is described with strong accuracy (correlation coefficient: $R^2 = 0.9996$) by the power function $R = 0.012t^{0.8524}$ where R is the radius of the circular front.

Wherever the copper sulphate has penetrated, it leaves an almost uniformly distributed brown color. This means that the impregnation with potassium ferricyanide III is not completely washed out by the copper sulphate solution, but that as soon as the contact with the potassium ferricyanide III material occurs, it is converted into the insoluble copper-iron complex and more or less precipitates on the spot.

4.6.2 Treatment of with Potassium Ferricyanide III Impregnated Filter Paper First with Copper and then with Iron II Sulfate

But if an iron sulfate solution is allowed to flow as a further component into the filter paper which was previously impregnated with red prussiate of potash and then

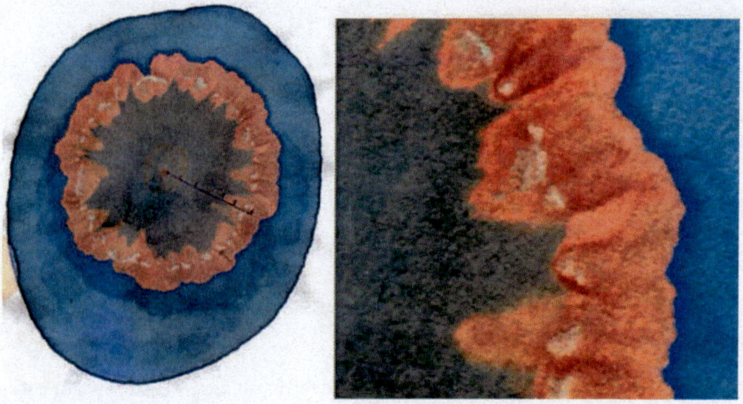

Fig. 4.27 An iron sulphate solution spreads in the filter paper, which was previously impregnated with red prussiate of potash and then with a copper sulphate solution; **left** Overview of the resulting Runge picture; **right** enlargement of the section which shows the double front system

treated with a copper sulfate solution and dried, we observe a very complex process (Fig. 4.27).

The result is a double front system in which the leading white front is heavily perforated and the rear is bizarre jagged. Trailing iron II sulfate $FeSO_4$ changes color to purple. Prussian blue, also known as Berlin Blue $K[Fe^{II}Fe^{III}(CN)_6]$ forms at the edges of contact with Potassium ferricyanide III $K_3[Fe^{III}(CN)_6]$, also called the red prussiate of potash.

A closer look at the double front system shows that the leading front surprisingly consists of white spots, around which dark brown plumes run (Fig. 4.27 right). This means that these white spots act as barriers to the spread of the $FeSO_4$ solution. The simplest explanation for this would be that these obstacles are white or colorless crystals, but of which compound? It cannot be the white potassium sulfate K_2SO_4, since its solubility in water is 111 g/l. These crystals form the front between the newly inflowing iron sulfate and the red-brown copper-iron cyanide $\left(Cu_2Fe^{II}(CN)_6\right)$ still present there. Furthermore, we should not forget that both the $CuSO_4$ ($CuSO_4$: 3,5 < pH < 4,5 at 20 °C and $c = 50$ g/l) [19] which has flowed in before and the $FeSO_4$ ($FeSO_4$: 2.5 – < pH < 3.8 at 20 °C and $c = 50$ g/l) react acidic [20]. Because of the large excess of Fe^{II} ions, the formation of the colorless $K_2[Fe^{II}Fe^{II}(CN)_6]$ complex could also be considered [21]. But also this statement should be taken into account: "By protolysis of hexa-cyanides II, the hexacyanide II acid $H_4[Fe^{II}(CN)_6]$ is obtained in the form of a white powder." [22].

In their very detailed report "Precipitation of Cyanide as $Cu_2Fe(CN)_6$ Compounds from Cyanidation and Detoxification Circuits" [23] Adams and Kyle discuss the many different copper-iron cyanide compounds. It is important in this context that there are other insoluble compounds in addition to the brown $Cu_2Fe(CN)_6$, such as e.g. copper I ferricyanide $Cu_3Fe(CN)_6$ and copper II ferricyanide $Cu_3\left[Fe(CN)_6\right]_2 \cdot 14H_2O$.

The back of this double front system shows a strong structure in the form of fingering. At the points where this visco-elastic front reaches and breaks through the leading, white front, the already mentioned flags of the poorly soluble, brown $Cu_2Fe(CN)_6$ appear. In the area between the source of the inflowing FeSO4 solution and the rear side, a dirty blue coloration remains in the filter paper. This indicates that only part of the original $Cu_2Fe(CN)_6$ reacts with the penetrating iron ions and is later oxidized to Prussian Blue by air.

4.7 Concluding Remarks

Today, after almost 25 years, it is difficult to reconstruct all the details of the complex chemical processes, which often works well, but sometimes remain vague.

Be that as it may, the synergetics view enable a fairly good classification of the pattern formations that occur, such as the fractality of the filter paper medium, the simple front formation, the fingering when water or corresponding solutions penetrates the impregnated filter paper and the non-uniform chemical reaction. The student Ch. Baum probably also sensed that all this was possible when he enthusiastically recorded his experiments on the Runge pictures.

References

1. Baum, C.: Versuchsprotokoll zur Lehrveranstaltung "Selbstorganisation in chemischen und biologischen Systemen" bei Peter Plath und Michael Vicker, SS 1995, Universität Bremen
2. Runge, F.F.: Zur Farbenchemie. Musterbilder für Freunde des Schönen und zum Gebrauch für Zeichner, Maler, Verzierer und Zeugdrucker, dargestellt durch chemische Wechselwirkung. Verlag E.S. Mittler & Sohn, Berlin (1850)
3. Kuhnert, L., Niedersen, U.: Ostwalds Klassiker der exakten Wissenschaften, Band 272. (F.F. Runge, R.E. Liesegang, B.P. Belousov, A.M. Zhabotinsky) – Selbstorganisation chemischer Strukturen, Akademische Verlagsgesellschaft Geest & Portig K.G. Leipzig (1987) p. 51
4. Schalansky, J.: Naturkunden No. 12. Matthes & Seitz Berlin (2014)
5. Harsch, G., Bussemas, H.H.: Bilder, die sich selber malen. Der Chemiker Runge und seine "Musterbilder für Freunde des Schönen." Anregungen zu einem Spiel mit Farben. Dumont Buchverlag Köln (1985)
6. Dupré, M., Jaenisch, M.: Runge und Kapillarbilder, Institute Dr. Flad, "Project work under the guidance of Peter Menzel", Lehrgang 62, Schuljahr 2012 /2013; https://www.chf.de/eduthek/projektarbeiten/Runge-Kapillarbilder.pdf (download 2020-09-11)
7. Van Damme, H.: Flow and Interfacial Instabilities in Newtonian and Colloidal Fluids (or The Birth, Life and Death of a Fractal). In: Avnir, D. (ed.) The Fractal Approach to Heterogeneous Chemistry. John Wiley & Sons, Chichester, New York, Brisbane, Toronto, Singapore (1989) 199–226; see also: Böckmann, M., Müller, S.C.: Coarsening in the buoyancy-driven instability of a reaction-diffusion front. Phys. Rev. E **70**, 046302 (2004)
8. Shaw, H.S.H.: Investigation of the nature of surface resistance of water and of stream-line motion under certain experimental conditions, Inst. N.A. OCLC 17929897 (1898) (https://www.worldcat.org/oclc/17929897).

9. Landau, L.D., Lifschitz, E.M.: Lehrbuch der Theoretischen Physik, vol. VI. Hydrodynamik, Akademie-Verlag Berlin, 3rd edn., p. 85 (1974)
10. Van Damme, H., Laroche, C., Gattineau, L., Levitz, P.: Viscoelastic effects in fingering between miscible fluids. J. Phys. France **48**(7), 1121–1133 (1987)
11. Mikhailov, A.S.: Foundation of synergetics – distributed active systems. In: Haken, H. (ed.) Springer Series in Synergetics, vol. 51. Springer-Verlag Berlin, Heidelberg, New York, London, Paris, Hong Kong, Barcelona (1990)
12. http://daten.didaktikchemie.uni-bayreuth.de/experimente/effekt/photo_kristalle.htm
13. https://www.mikroskopie-forum.de/index.php?topic=33128.15
14. https://de.wikipedia.org/wiki/Kaliumhexacyanidoferrat(III)
15. Kuhnert, L., Niedersen, U.: Ostwalds Klassiker der exakten Wissenschaften, Band 272. F.F. Runge, R.E. Liesegang, B.P. Belousov, a.M. Zhabotinsky – Selbstorganisation chemischer Strukturen, Akademische Verlagsgesellschaft Geest & Portig K.G. Leipzig, p. 101 (1987)
16. Gmelin, L.: Handbuch der Chemie: Handbuch der organischen Chemie. Organische Chemie im Allgemeinen, Organische Verbindungen mit 2 und 4 Atomen Kohlenstoff", Heidelberg, Jan., pp. 358–359 (1848)
17. Remy, H.: Lehrbuch der Anorganischen Chemie, Bd. II, 11th edn., Akademische Verlagsgesellschaft Geest & Portig K.G., Leipzig, pp. 339–340 (1961)
18. http://www.chemgapedia.de/vsengine/vlu/vsc/de/ch/6/ac/bibliothek/_vlu/kaliumhexacyanoferratii.vlu/Page/vsc/de/ch/6/ac/bibliothek/kaliumhexacyanoferratii/reaktivitaet.vscml.html
19. http://gestis.itrust.de/nxt/gateway.dll/gestis_de/000000.xml?f=templatesfn=default.htmvid=gestisdeu:sdbdeu3.0, Kupfersulfat
20. http://gestis.itrust.de/nxt/gateway.dll/gestis_de/000000.xml?f=templates$fn=default.htm$vid=gestisdeu:sdbdeu$3.0, Eisen(II)-sulfat
21. Röder, J.-K.: Zweikernkomplexe multifunktioneller Pyrazolatliganden als bimetallische Analoga von N-Chelatkomplexen - Synthese, Koordinationschemie, Eigenschaften", Dissertation, Ruprecht-Karls-Universität, Heidelberg, p. 15 (2001)
22. https://www.spektrum.de/lexikon/chemie/cyanoferrate/2114
23. Adams, M.D., Kyle, J.H.: Precipitation of Cyanide as $Cu_2Fe(CN)_6$ Compounds from Cyanidation and Detoxification Circuits. In: Conference: Minprex at Melbourne, Australia (September 2000). https://www.researchgate.net/publication/314286234_Precipitation_of_Cyanide_as_Cu2FeCN_6_Compounds_from_Cyanidation_and_Detoxification_Circuits

Part II
Fractal Structure in Chemistry and Biology

Chapter 5
Fractal Metal Zinc-Trees

Diffusion-Limited or Ballistic Aggregation?

Peter J. Plath

Fig. 5.1 Growth of the metal zinc-tree in a Petri dish at a voltage of 6 V. The 2 N zinc sulfate solution was covered with a layer of n-butyl acetate. For this picture the zinc tree has been removed from the Petri dish and placed and dried on a grey cardboard. (*Photo* taken by Thomas Rabbow)

5.1 Introduction

With the publication of his book "The fractal geometry of nature" in 1977 [1], which was soon translated into other languages, Benoit B. Mandelbrot succeeded in making his ideas of objects with a broken dimension (or fractional dimension)

P. J. Plath et al., *Imagery Synergetics*, Understanding Complex Systems,
https://doi.org/10.1007/978-3-030-95607-3_5

known to a wide audience. It had three English editions until 1983. A German translation appeared by Akademie-Verlag Berlin and Birkhäuser Verlag Basel in 1986 [2]. However, his book was mainly of a mathematical nature and difficult for non-mathematicians to understand. L. Pietronero and E. Tosatti published "Fractals in Physics" [3] in 1986 and thus considerably expanded the circle of readers. A big breakthrough came with Heinz-Otto Peitgen and Dietmar Saupe with their richly and colorfully illustrated book "The Sciences of Fractal Images" [4], which was even topped by their books "Fractals for the Classroom; Part I" (and "Bausteine des Chaos Fractals", German edition) 1992 [5]. In 1989 David Avnir published his book "The Fractal Approach to Heterogeneous Chemistry—Surfaces, Cololloids, Polymers" [6] and thus opened this wonderful idea of objects with broken dimensions even to very application-oriented chemists (Fig. 5.1).

In this book you can find the inspiring article of Mitsugu Matsushita on "Experimental Observations of Aggregations" from which we got the basic ideas of the experiments reported here.

If a carbon electrode is immersed in the interface of an aqueous zinc sulfate solution which has been covered with an immiscible liquid (n-butyl acetate) and a voltage of a few volts is applied so that the electrode becomes the cathode, zinc is deposited here off (Fig. 5.2). However, this happens in such a way that fractal metal deposition takes place in the boundary layer of the two liquids. A "zinc tree" is created. With the growth of the fractal structure, the current strength increases, so that one can infer the fractal dimension of the resulting structure from the dependence of the current strength or the "radius" on time.

The process described here can be described in broad areas according to the Witten and Sander model as "diffusion-limited aggregation" (DLA model). This model is based on a stochastic, spatio-temporal discrete process that takes place on a two-dimensional grid. A state moves randomly across the grid and is fixed as

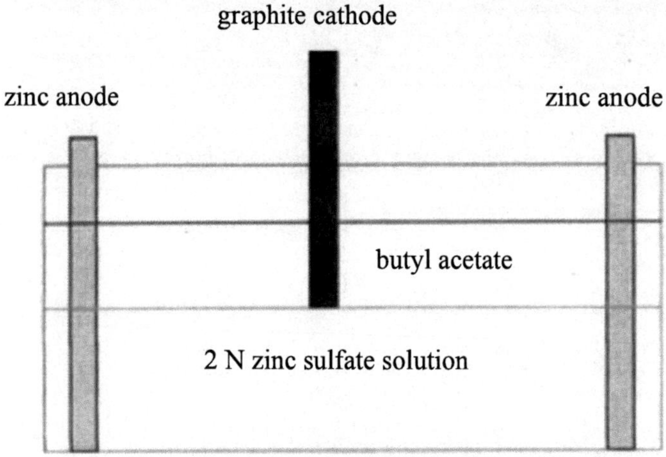

Fig. 5.2 Schematic representation of the experimental set up

soon as it touches the aggregation cluster. The dimension of the resulting structure corresponds in a very good approximation to the dimension of the zinc tree that can be experimentally determined in a chemical experiment. The aim of the experiment is to characterize the different structures of the zinc trees growing at different voltages by means of their fractal dimension and to simulate them using the DLA model.

5.2 Experimental Arrangement and Procedure

In this experiment metallic zinc in the form known as zinc metal-leaves was grown two-dimensionally. The experimental procedures used to grow zinc metal-leaves are as follows. A glass vat of diameter ca. 20 cm and depth ca. 10 cm is filled with 2 M $ZnSO_4$ aqueous solution (depth 4 cm), on to which butyl acetate $CH_3COO(CH_2)_3CH_3$ is floated to make an interface (Fig. 5.2). A tip of a carbon cathode (pencil core of diameter 0.5 mm) is polished carefully so as to make it flat perpendicular to the axis. The cathode is then set at the center of the vat so that the flat tip is placed just on the interface (Fig. 5.2). The electrodeposition is initiated by applying a d.c. voltage between the carbon cathode and a zinc ring plate anode of diameter ca. 17 cm, width ca. 2.5 cm and thickness ca. 3 mm placed in the vat.

A zinc metal-leaf grows two-dimensionally at the interfaces between the two liquids from the edge of the flat tip of the cathode towards the outside anode with an intricately branched random pattern (Fig. 5.1).

If the cathode tip is rounded or is immersed in the $ZnSO_4$ solution the deposit grows three-dimensionally into the solution.

Usually, the zinc metal-leaves grow to a size of 10 cm after about 10 min by applying a constant d.c. voltage of about 5 V. The temperature of the system was kept fixed, e.g. at 15 °C.[7]

If one looks at the fractal zinc structures that arise in these experiments, it is difficult for the author to compare them with the image of a tree leaf. During a pre-winter walk through the Uckermark in Germany, he saw defoliated oak trees and robidia, which reminded him very much of the fractal zinc deposits although they differ in their dimensions (Fig. 5.3). The author therefore suggests choosing the term tree and branches instead of leaves for the metal deposits structures observed.

5.3 Comments on the Theory

The procedure mentioned above looks very simple and can be executed easily. But indeed, one has to prepare the tip of the cathode carefully.

However, a lot of questions arise if one try to understand why one get nearly perfect fractal zinc metal-tree in the two-dimensional boundary layer between the two immiscible liquid phases. Why does the heavy zinc metal (specific weight 7.130 g cm^{-3}) does not sink down in the aqueous solution?

Fig. 5.3 Left: a defoliated oak tree in the foreground in pre-winter time which reminds the author on the fractal metal zinc deposition at higher voltages. right: hanging branches of a bare robidia which resembles the metal zinc-tree at low voltages. (*Photos* the author, January 2020)

First of all, if the cathode touches the water surface at first, we can assume a special *adhesion wetting* of its surface. For sure, the zinc ions deposit where the local field is largest. This is initially the case at the circular edge of the electrode because of its curvature, but not on its planar surface. The authoritative part of the electric field is orientated parallel to the interfacial layer. So, starting from the one-dimensional line of the cathode edge only those zinc ions, which are located in the intermediate zone of the interface, will be deposited at this circle line being discharged at the cathode.

This way the fractal metal zinc-tree is growing radially and spreading themselves along the two-dimensional interface layer.

The metal tree behaves like a steel needle on water—while growing gently it touches the water, but it does not set in the water.

But what is the role of butyl acetate in this game? The solubility of the butyl acetate in water is very low. Their volume fraction in water is only $7.300 \cdot 10^{-3}$, which is why both liquids can be regarded as immiscible to a good approximation. Furthermore, the interfacial tension between butyl acetate and water is $11.5 \cdot 10^{-3}$ N m^{-1}. In contrast, the surface tension of water, i.e. the interfacial tension of water and air, is $72.75 \cdot 10^{-3}$ N m^{-1}. The surface tension of von butyl acetate is $26 \cdot 10^{-3}$ N m^{-1}. When the aqueous zinc salt solution is covered with butyl acetate, the interfacial tensions of the water as well as butyl acetate are considerably reduced. In addition, the contact angle measurement on flat V4A steel for butyl acetate results in values of 99° and 83° for the advance angle θ_V and the withdrawal angle θ_R, respectively. All these data were measured by R. Magiera in his dissertation in 1995 [8]. But all these data reflect the equilibrium states at 20 °C. However, our growing system is indeed a complex electro-chemical reaction system in which the new phase of the metal zinc-leaf is generated and growing all the time. Hence, one has to look at it as a system far from equilibrium rather than trying to describe it in terms of equilibrium.

The metallic zinc, for which one can assume at the similar contact angle of the butyl acetate, now pushes into this interface during its growth. The zinc ions deposit where the local field is largest. This is initially the case at the circular edge of the electrode. In the course of the experiment the ions will be deposited only at the edge of the zinc-leaf growing in the two—dimensional space of the interface between the two liquids.

5.4 Experimental Estimation of Fractal Dimension of the Zinc Deposits

The structure thus created from the initial circle has a highly complex, fractal geometry, which can be characterized best following the history of its growth via regarding the development of the current over time, as is common in electrochemistry. If the voltage is kept constant, the current strength changes over time, since this is proportional to the amount of zinc ions deposited per unit of time.

According to the classic description of electrochemical processes, at constant voltage the current is proportional to the size of the area on which the discharge of the ions takes place.

Now, as we have just determined, the deposition does not take place so much on the surface as on the edge of the zinc-tree, so the current intensity should be proportional to the length L of the edge rather than to the size of the electrode surface F. However, that would be a statement that does not conform to the well-founded, classic, electrochemical view.

Due to the complex geometry of the zinc tree, we cannot specify the length of the edge of the growing cathodic zinc electrode or its surface at any time without prejudice, i.e. without specifying the scale used. On the other hand, we can measure the change in radius r of the smallest circle (radius of gyration, see Fig. 5.4) over time, which concentric to the origin, in which our graphite cathode is located, also includes the most distant branch of the zinc tree. This radius r must then grow over time.

$$r = f(t) \tag{5.1}$$

If the entire circular area described by r were completely filled with zinc metal, it would be well known that the circumference U is

$$U = 2\pi r \tag{5.2}$$

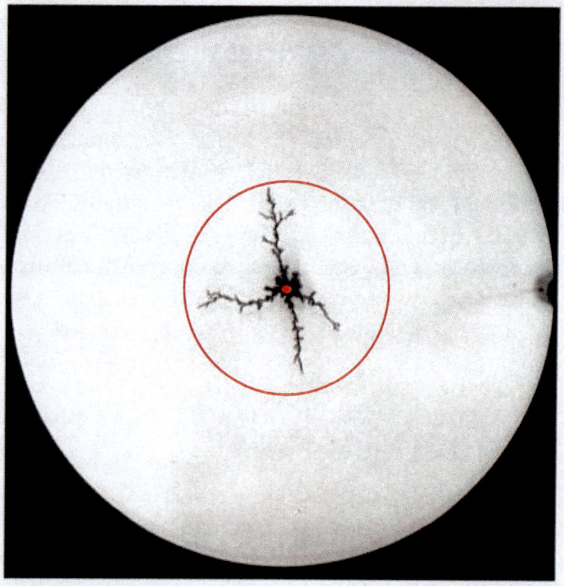

Fig. 5.4 Fractal metal zinc-tree together with its surrounding radius of gyration (red circle) and the centered graphite electrode (red dot in the center). The experiment has been executed at 3 V. (*Photo* Plath / Koblitz 30.3. 1994)

and the circular area F

$$F = \pi r^2 \tag{5.3}$$

This is certainly not the case in the case of the highly branched, "two-dimensional" zinc tree, since its edge is certainly longer than that of the circle circumscribing the zinc tree, and its area is certainly smaller than the area enclosed by this circle.

We now want to mathematically write the terms circumference and area somewhat more generally, by assuming from a flat, geometric object—be it an arbitrarily shaped, edged structure—that a size L assigned to it is proportional to the Dth power of the radius r of the surrounding circle:

$$L \propto r^D \tag{5.4}$$

where the still unknown exponent D has the character of a dimension. We want to admit that D need not be an integer, but can also be a fractional number. In this case, according to B. Mandelbrot (1977), D is called a fractional dimension and the structure characterized by this number is called a fractal. If $D = d_{top}$ is an integer, one speaks of a topological dimension. If for example $d_{top} = 1$, then the size L is the circumference as it is well known from the circle; accordingly, $d_{top} = 2$ means that we are dealing with a two dimensional surface that is connected everywhere, like a circular disk.

A simple transformation of relation (5.4) now leads to the double logarithmic form

$$\log L \propto D\, logr \tag{5.5}$$

The discussion ended earlier with the question of whether the current strength I is more proportional to the edge length of the 2D zinc tree or rather to the surface area of the zinc tree electrode. We already know from the experiment the function $I = g(t)$ which describes the temporal dependence of the current. We want to assume that it is of the form

$$I \propto t^{\alpha} \tag{5.6}$$

or, after appropriate transformation, the expression results:

$$\log I \propto \alpha \log t \tag{5.7}$$

We have also measured the function r = f (t) of the development over time of the radius of the circle comprising the zinc tree (radius of gyration) (Fig. 5.4). Here, too, we want to assume an exponential relationship, so that the following applies:

$$r \propto t^{\beta} \tag{5.8}$$

and

$$\log r \propto \beta logt \tag{5.9}$$

respectively.

With a simple transformation one gets:

$$\log I \propto \frac{\alpha}{\beta} \log r \tag{5.10}$$

or (Figs. 5.5 and 5.6)

$$I \propto r^{\alpha/\beta} \tag{5.11}$$

If we plot log$I$ as a function of log$r$ we get a straight line whose slope has the value $D = \alpha/\beta$. We can therefore decide the above mentioned question experimentally. We herewith connect directly the fractal Dimension D with the electrochemical deposition process, rather the width the geometrical shape of its resulting fractal metal zinc tree. The current intensity I is therefore largely proportional to the length of the edge, provided that D is much more similar to the topological dimension of the one-dimensional edge than the topological dimension $d_{top} = 2$ of a surface.

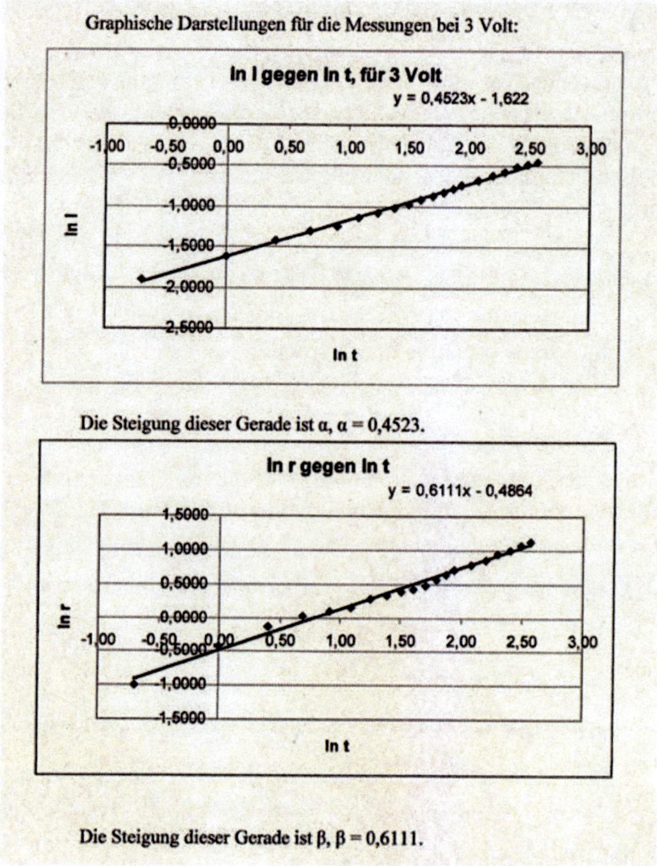

Fig. 5.5 An example: experimental estimations of the values α and β for the metal zinc deposition at the voltage of 3 V. The measurement has been executed by the students Olesya Kazakova and Marina Voropaeva during the practical training in physical chemistry—chemical synergetics under the guidance of Prof. Plath at the University of Bremen in 2005

The exponent D then expresses the accessibility of the locally one-dimensional electrode edge for the discharge of the zinc cations on the electrode (zinc cathode). The more or less accessible parts of the edge are, so to speak, perceived as gaps in the margin, which then takes on a fractional dimension, which is less than one: $D < d_{top} = 1$, the topological dimension of a "smooth", not fissured edge. The fractional dimension of the fractal object enclosed by this edge—the zinc electrode—is obtained by increasing D by one: $D_{electrode} = D + 1$. This additional dimension "one" corresponds to the local topological dimension of the circular edge of the graphite electrode from which the fractal growth started into the two-dimensional space of the interface (Fig. 5.7).

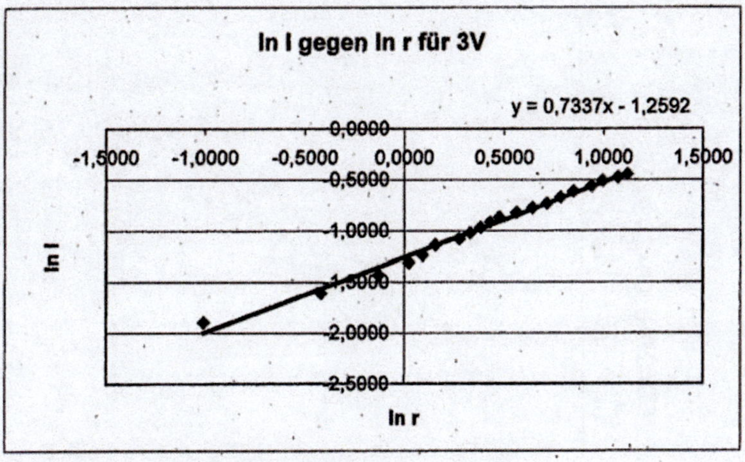

Also, d = 0,7337, dementsprechend, D = 1,7337.

Fig. 5.6 The example of Fig. 5.3: experimental estimations of the value of d and D respectively form the log–log plot of the current I versus the radius of gyration for the metal zinc deposition at the voltage of 3 V. The measurement has been executed by Olesya Kazakova and Marina Voropaeva during the practical training in physical chemistry—chemical synergetics under the guidance of Prof. Plath at the University of Bremen in 2005

Depending on the voltage applied between the zinc tree electrode and the zinc counter electrode there are different patterns for the zinc trees as is shown in Figs. 5.4, 5.8 and 5.9. The field strength at the electrode increases with the voltage and it is understandable at this level of argumentation that the patterns become increasingly dense. If one measures the dimension of the fractal zinc trees here, too, it shows that this remains constant $D_{electrode} = 1.66 \pm 0.03V$ up to a certain voltage $U_k \approx 8.2V$, but also increases approximately linearly up to $D_{electrode} = 2$, as Matsushita (1984) could show (Fig. 5.7). So there is a threshold U_k value of the voltage, which basically signals two different processes that take place below and above this threshold value. Above the threshold value U_k the discharge of the zinc ions also continues to take place at the edge of the electrode, so that D satisfies the condition: $0.66 < D < 1$.

5.5 Experiments with 4 and 8 V

Figures 5.9 and 5.10 shows clearly that the assumption on which Eq. 5.10 is based does not apply in the case of a constant voltage of 8 V. $\ln(I) = F(\ln(r))$ is by no means a linear function. But if we split the function into two parts, we can detect two linear functions intersecting each other (Eq. 5.9). This indicates two different processes taking place in the course of fractal zinc deposition (Figs. 5.4, 5.8, 5.9, 5.12, and 5.14).

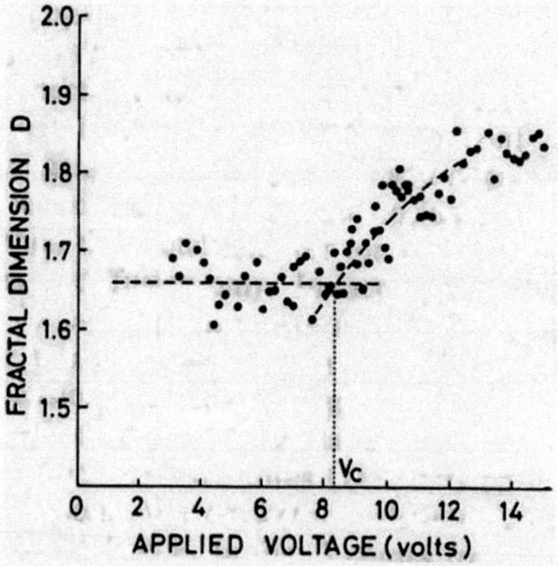

Fig. 5.7 Values of the fractal dimension of many zinc metal trees grown independently at various electrodeposition voltages. The solution depth and temperature are always fixed as 4 mm and 15 °C, respectively." (reproduced by friendly permission of Phys. Rev. Lett.) The data have been evaluated by measuring "the density–density correlation function $C(r) \sim r^{-A}$ where the exponent A is related to the fractal (Hausdorff) dimension $D = d - A$(d is the Euclidian dimension) [9]. $D = (d^2 + 1)/(d + 1)$. For this purpose Matsushita et al. analyzed scanned photographs of the metal zinc trees

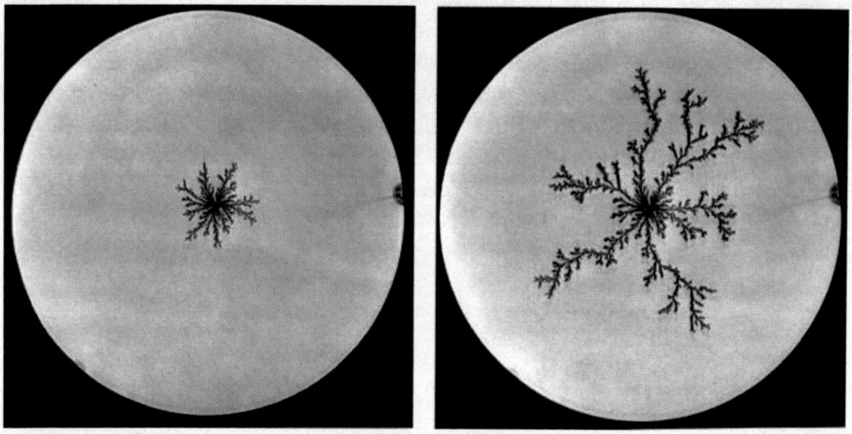

Fig. 5.8 Growing metal zinc-tree at 4 V; left: after 2:50 min. right: after 10:18 min. The fractal Dimension of the zinc electrode is $D_{electrode} = 1.698$ if one averages over the entire time of 875 s

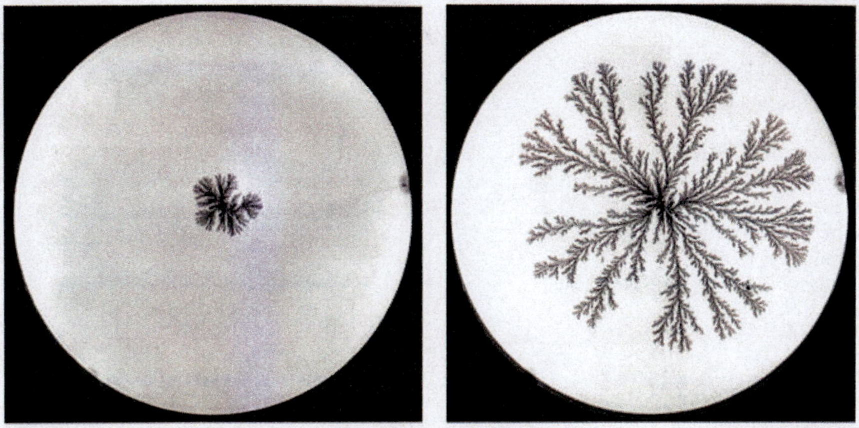

Fig. 5.9 Growing metal zinc-tree at 8 V; left: after 1:30 min. right: after 10:31 min. The fractal Dimension of the zinc electrode is is $D_{electrode} = 1.861$ if one averages over the entire time of 668 s

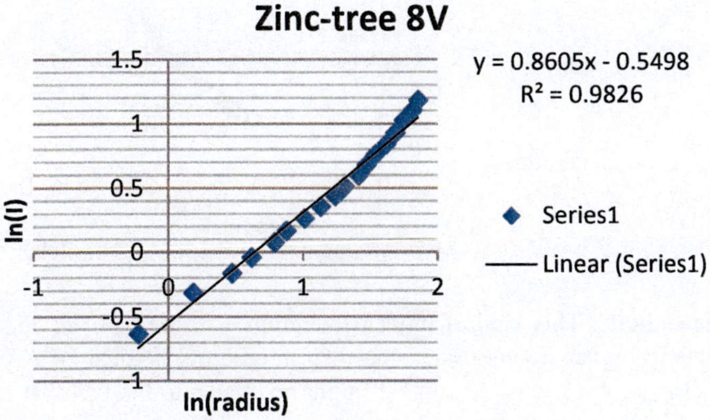

Fig. 5.10 Bi-logarithmic plot of current versus radius of gyration (see Eq. (5.10) at a constant voltage of 8 V

For lower radii the fractal dimension is $D = 0.707$ or $D_{electrode} = 1.707$ (Fig. 15.11a) and fits very well with the fractal dimension of the two dimensional Witten-Sander-Model [10] for diffusion limited aggregation (DLA) (D = 1.71) by randomly walking zinc ion particles as shown by Paul Meakin [11] theoretically.

However, if the zinc tree continues to grow, the fractal dimensions changes. For the 8 V-experiment the dimension becomes about $D = 1.1$ or $D_{electrode} = 2.1$ (Fig. 5.11b). Since the zinc ions don't aggregate during their walk to the fractal zinc cathode one can exclude a fusion of them before reaching the cathode being discharged. But a growing radius of gyration causes an increase of the electric field gradient, so that the ions move more and more in a random straight-line trajectories

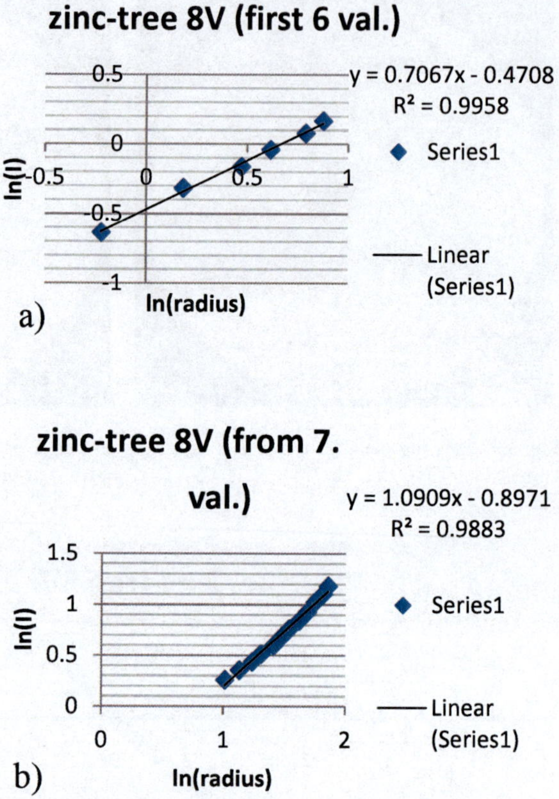

Fig. 5.11 **a** Low radii with is $D_{electrode} = 1.707$ if one averages over the time of the first 198 s and **b** larger radii is $D_{electrode} = 1.100$ if one averages over the time between 200 and 668 s; bi-logarithmic function of current versus radius of gyration

towards the cathode. This kind of movement reminds of the ballistic aggregation, which results in higher fractal dimension then in case of classical DLA processes [12]. That $D_{electode}$ oversteps the value 2 might be related to the fact that the three-dimensional Euclidian space has to be taken into account for aggregation if the tip of the tree approaches the wide anode.

5.6 Experiment with 12 V

Now, let's consider the results of the experiment at 12 V (Fig. 5.12). Taking into account the set of all data, the double logarithmic plot of the current versus the radii of gyrations results is a straight line whose slope is almost one, D = 1. This means that the envelope of the metal zinc fractal has practically the dimension one, so acts as a one-dimensional closed line with regard to the discharge of the zinc ions and their deposition as zinc atoms. Inspecting the photo of the zinc tree at 7:31 min one

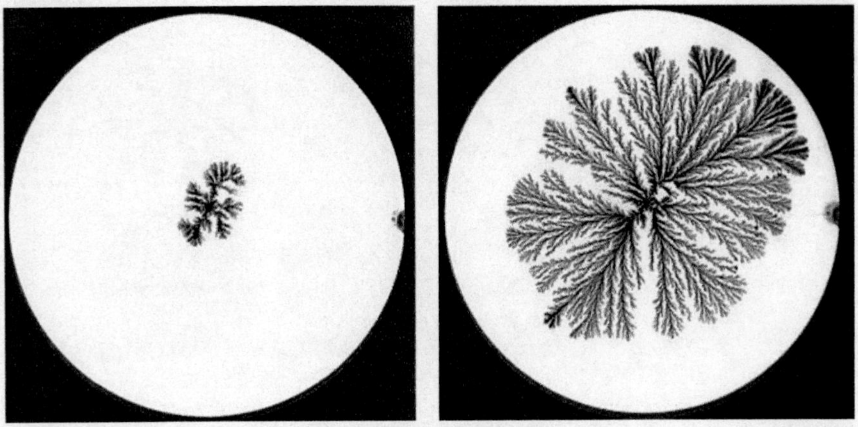

Fig. 5.12 Growing metal zinc-tree at 12 V; left: after 0:41 min. right: after 7:31 min. The fractal dimension of the zinc electrode is $D = 1.01$ or $D_{electrode} = 2.01$ taking the basis of all data up to 6:18 min. Only in the very beginning, the fractal dimension is $D_{electrode} \approx 1.7$

can perceive this conclusion. However, we know and see that this is only a rough approximation (Fig. 5.12).

R.C. Ball and T.A. Witten [13] emphasize that the way in which the particles move determines the fractal dimension of their aggregates. "Diffusion-limited aggregates, where the constituent particles move in random walks have a fractal dimension of about 2.4" in a space with the Euclidian dimension $d = 3$. "In ballistic aggregation, where the particles move in random straight-line trajectories" the fractal dimension becomes larger.

Due to the large field strength at the edge of the electrode, the zinc ions will hardly migrate erratically, but will follow the field lines and hit the edge in a kind of ballistic flight, if the applied constant voltage exceeds the critical value $V_c \approx 8, 2V$ as M. Matsushita et al. have shown. (see Fig. 5.7). However, the measured value of $D_{electrode} = 2.01$ for the 12 V experiment is unexpectedly high. Matsushita explained this situation referring to the theoretical work of Paul Meakin in 1983 [14].

> In fact, applying higher voltage ($V > V_C$) gives rises to a larger inner core with dense radial structure, which causes larger Hausdorff dimension, as seen in (Fig. 5.7, this work, the author). Recent computer simulations (see ref. 14, the author) demonstrate that the particle drift really entails uniform structures ($D \sim 2.0$) on the cluster formation. We conjecture, therefore, that the crossover at $V \cong V_C$ in (Fig 5.7 this work, the author) comes from the dominance of the drift effect.

In their paper "Aggregation in a mixture of Brownian and ballistic wandering Particles", S.G. Alves and S.C. Ferreira analyzed the scaling properties of a model that has as limiting cases the diffusion-limited aggregation (DLA) and the ballistic aggregation (BA) models [15].

> The particles added to the cluster (the fractal metal zinc cathode; the author) can follow either ballistic trajectories, with probability P_{ba}, or random ones, with probability $P_{rw} = 1$

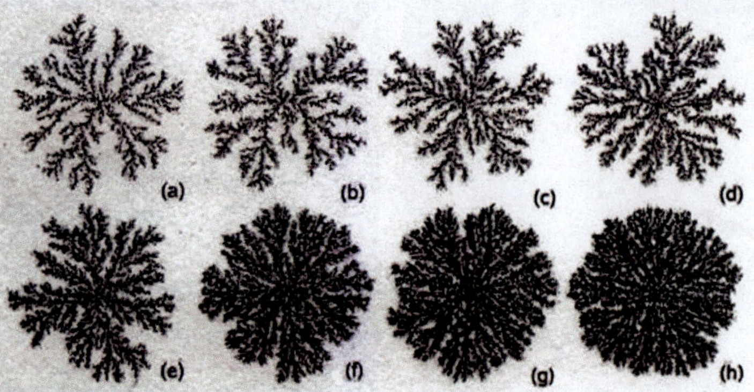

FIG. 2: Growth patterns for distinct values of the parameter P_{ba}: (a) 0.1, (b) 0.3, (c) 0.5, (d) 0.7, (e) 0.9, (f) 0.95, and (h) 0.99. All simulations were stopped when the cluster reaches the border of a region 750×750 centered at the origin. The number of particles varies from $N \simeq 3 \times 10^4$ ($P_{ba} = 0.1$) to $N \simeq 1.6 \times 10^5$ ($P_{ba} = 0.99$).

Fig. 5.13 Theoretical results of S.G. Alves and S.C. Ferreira [15]: Growth of fractal patterns for distinct values of the probabilities for random walk P_{rw} and ballistic walk P_{ba}. (with friendly permission of…Phys. Rev. E)

– P_{ba}. … The fractal dimension as a function of Pb_a continuously increases from $d_f \approx 1.72$ (DLA dimensionality) for $P_{ba} = 0$ to $d_f \approx 2$ (BA dimensionality) for $Pb_a = 1$.

However, the lacunarity and the active zone width exhibit a distinct behavior: they are convex functions of P_{ba} with a maximum at $P_{ba} \approx 1/2$. Through the analysis of the angular correlation function, we found that the difference between the radial and angular exponents decreases continuously with increasing P_{ba} and rapidly vanishes for $P_{ba} - 1ex$, in agreement with recent results concerning the asymptotic scaling of DLA clusters (Fig. 5.13).

5.7 Experiment with 14 V

The bi-logarithmic plot of current versus radius of gyration (see Eq. (5.10)) in the experiment at constant voltage of 14 V (see Fig. 5.15) shows a very surprising and dramatic event in the first seconds. Inspecting the photographs (see Fig. 5.14) one can detect large pieces of fractal metal zinc flowing just beneath the borderline between the organic phase and the aqueous zinc solution towards the anode.

In the vicinity of the fractal zinc tree, which was still small at the beginning, a very strong field gradient is formed when a high voltage is applied. As a result, particularly exposed parts of the fractally growing zinc tree are torn off like the branches of a tree in a storm. These small metal branches form the exposed parts of the metal zone

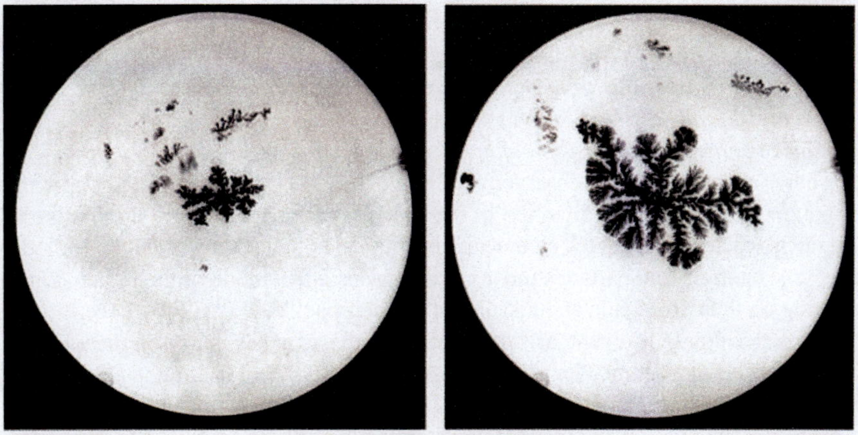

Fig. 5.14 Growing metal zinc-tree at 14 V; left: after 0:16 min. right: after 1:29 min. The fractal Dimension of the zinc electrode is $D = 0.709$ or $D_{electrode} = 1.709$

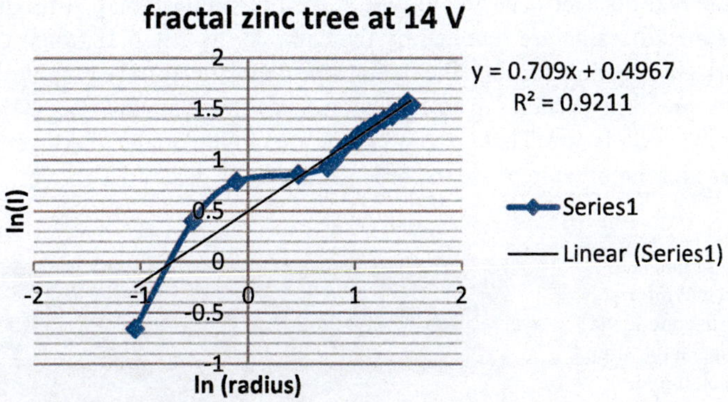

Fig. 5.15 Bi-logarithmic plot of current versus radius of gyration (see Eq. (5.10)) at a constant voltage of 14 V. The linear spline function for all data up to 128 s seems to result a dimension of $D_{electrode} = 1.709$. This dimension is the ideal value for a DLA process, but that is obviously misleading

tree will take with them an excess of electrons (negative charges) and thus they will move away from the negative charged cathode.

These free moving fractal zinc branches are polarized in the electric field. In the direction of the ring-shaped positive zinc anode, they become the cathode and continue to grow, while in the direction of the original negative graphite cathode, they themselves become the anode that supplies zinc ions.

In their article "Distinct dynamics on both sides of a metallic work piece electrode" M. Buhlert et al. describe such bivalent character of metal pieces in an electrochemical cell, which are unconnected to the external power supply [16]. Electrical connection of the unconnected metal sheet is established only via the electrolyte.

They wrote: "*Depending on the polarisation of an electrochemical cell, a metal workpiece will be either electroplated or metal will be electrochemically removed. Positioning the workpiece electrode between an anode and a cathode without connecting it to the external electrical circuit leads to removal of metal on one side of the workpiece and electrodeposition of metal on the other side.*"

That is exactly what we observed (see Fig. 5.16). We observed fractally growing metal zinc particles in the direction of the anode whereas we observe dissolution of the zinc particle on their back side which is directed to the original graphite cathode. This way the free zinc particles moves in direction of the anode of the system while growing on their front side and dissolving on their back side (see Fig. 5.16).

After the dramatic event of detachment of the exposed parts of the original fractal has been overcome, the typical DLA process of diffusion-limited aggregation continues. The typical fractal dimension $D_{electrode} = 1.723$ results (see Fig. 5.17).

This was the most surprising that despite the high voltage of 14 V, the dimension of the fractal zinc cathode returned to the value for the diffusion-limited aggregation!

To understand this, we look at the aggregation process very early on. First of all, what we have overseen looking at the bi-logarithmic plot of current versus radius of gyration, the real disaster occurred almost at the beginning. Small parts of the just aggregated metallic zinc are repelled by the cathode. In Fig. 5.18a one can detect them as dark grey shadows around the fractal zinc deposition on the graphite cathode. It looks like spreading particles in an explosion. These parts are growing at their front and dissolving in their rare. That is the way they are moving in the interlayer between the aqueous and the organic phase towards the anode as described above.

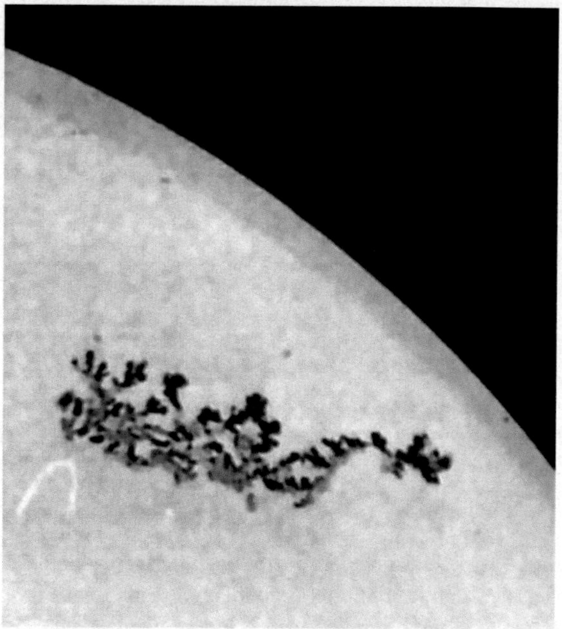

Fig. 5.16 Disconnected Zinc particle moving freely towards the ring anode via fractal growing at the front side and dissolving in its rare. Experiment with 14 V at 1:29 min after starting

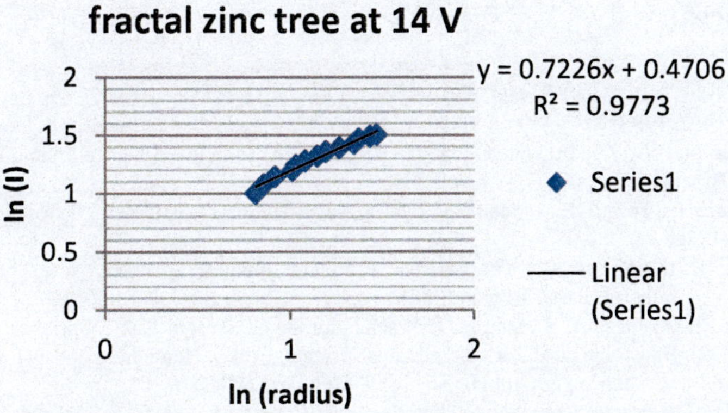

Fig. 5.17 Bi-logarithmic plot of current versus radius of gyration (see Eq. (5.10)) at a constant voltage of 14 V. The linear spline function for all data from 32 s up to 109 s the dimension $D = 0.723 and D_{electrode} = 1.723$ results again

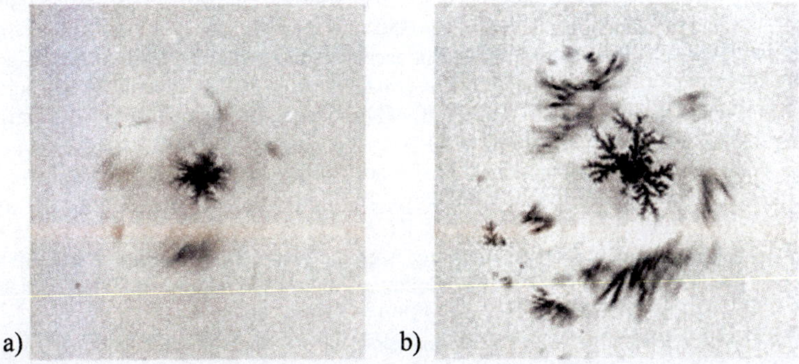

Fig. 5.18 Electrodeposition of metal zinc at 14 V just at the very beginning of the experiment: **a** after 3 s. **b** after 10 s

But we can also see that in the center of the radially moving zinc particles, the zinc stuck to the cathode develops in the form of a typical DLA cluster (see Fig. 5.18b). Obviously the corona of these zinc particles noticeably weakens the globally applied field opposite the cathode. This voltage drop at the zinc particle corona is the reason that the zinc deposition on the cathode also occurs practically completely in the further development over time by diffusion-limited aggregation (DLA), so that the typical fractal dimension of about D = 1.7 is measured.

References

1. Mandelbrot, B.B.: The Fractal Geometry of Nature. W.H. Freeman and Company, New York (1977, 1982, 1983)
2. Mandelbrot, B.B.: Die Fraktale Geometrie der Natur. Akademie-Verlag Berlin und Birkhäuser Verlag Basel (1987)
3. Pietronero, L., Tosatti, E.: Fractals in Physics. North-Holland Amsterdam, Oxford, New York, Tokyo (1986)
4. Peitgen, H.-O., Saupe, D.: The Science of Fractal Images. Springer, New York, Berlin, Heidelberg, London. Paris, Tokyo (1988)
5. Peitgen, H.-O., Jürgens, H., Saupe, D.: Fractals for the Classroom Part 1. Springer, New York Inc. (1992); Bausteine des Chaos – Fraktale. Springer, Berlin, Heidelberg und J.G. Cotta'sche Buchhandlung Nachfolger GmbH gegr. 1659, Stuttgart (1992)
6. Avnir, D.: The Fractal Approach to Heterogeneous Chemistry – Surfaces, Colloids, Polymers. Wiley, Chichester, New York, Brisbane, Toronto, Singapore (1989)
7. Matsushita, M.: Experimental observations of aggregations. In: Avnir, D. (ed.) The Fractal Approach to Heterogeneous Chemistry – Surfaces, Colloids, Polymers, pp. 161–179. Wiley, Chichester, New York, Brisbane, Toronto, Singapore (1989)
8. Magiera, R.: Trennung von flüssig-Flüssig-Dispersionen in durchströmten Faserpackungen. Dissertation at the Technische Universität München. Herbert Utz Verlag Wissenschaft, pp. 121 (1995). https://books.google.de/books?id=GnCZJukJUwQC&pg=PA126&lpg=PA126&dq=grenzfl%C3%A4chenspannung+wasser+butylacetat&source=bl&ots=d18trGnAlF&sig=ACfU3U0Y77XRAh7HVSOUnaP72LoOa5BrdA&hl=de&sa=X&ved=2ahUKEwjspcj5kI7nAhXioFwKHQylB8kQ6AEwAnoECAgQAQ#v=onepage&q=grenzfl%C3%A4chenspannung%20wasser%20butylacetat&f=false
9. Matsushita, M., Sano, M., Hayakawa, Y., Honjo, H., Sawada, Y.: Fractal structures of zinc metal leaves grown by electrodeposition. Phys. Rev. Lett. **53**(3), 286–289 (1984)
10. Witten, T.A., Sander, L.M.: Phys. Rev. Lett. **47**, 1400 (1981)
11. Meakin, P.: Simulations of aggregation processes. In: Avnir, D. (ed.) The Fractal Approach to Heterogeneous Chemistry – Surfaces, Colloids, Polymers, pp. 131–160. Wiley, Chichester, New York, Brisbane, Toronto, Singapore (1989)
12. Ball, R.C., Witten, T.A.: Particle aggregation versus cluster aggregation in high dimensions. J. Stat. Phys. **36**(5/6), 872 (1984)
13. Ball, R.C., Witten, T.A.: Particle aggregation versus cluster aggregation in high dimensions. J. Stat. Phys. **36**(5/6), 873–879 (1984)
14. Meakin, P.: Effects of particle drift on diffusion-limited aggregation. Phys. Rev. B **28**, 5221 (1983)
15. Alves, S.G., Ferreira, S.C.: Aggregation in a mixture of Brownian and ballistic wandering particles. Phys. Rev. E · (2006); PACS numbers: 61.43.Hv,05.40.Fb,47.53.+n,47.54.-r
16. Buhlert, M., Sydow, U., Plath, P.J.: Distinct dynamics on both sides of a metallic work piece electrode. In: Vernetzte Wissenschaften – Crosslinks in Natural and Social Sciences" (editors: Peter Jörg Plath , Ernst-Christoph Haß; Logos Verlag Berlin GmbH), pp. 65 – 86 (2008)

Chapter 6
The Fractal Character of Modified Zeolites

The Formation of Cobalt (II)-Phthalocyanine Fractals Inside Zeolites and the Dynamic Dimension of the Reverse Cobalt Ion Exchange

Peter J. Plath, Erwin Ignatzek, Ernst-Christoph Haß, and Uwe Hündorf

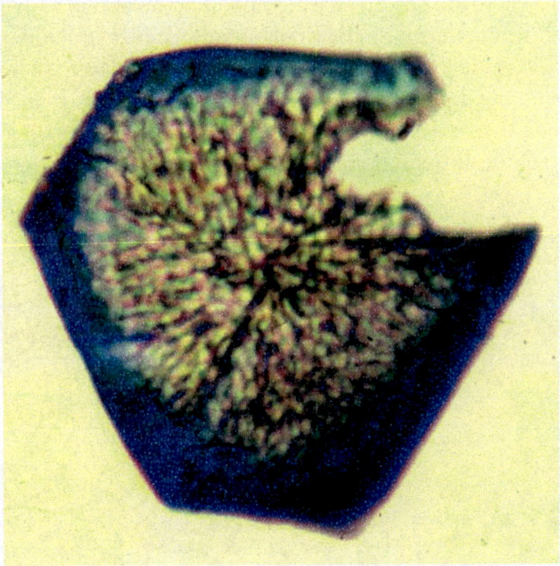

Fig. 6.1 Fractal river-like structure of CoPc in a zeolite-X disk (diameter about 70 μm) (Photo, H. Diegruber/P. J. Plath) A black and white version of this picture has been published by the authors in Z. phys. Chem, Leipzig in 1987

P. J. Plath et al., *Imagery Synergetics*, Understanding Complex Systems,
https://doi.org/10.1007/978-3-030-95607-3_6

6.1 Preliminaries, Which are Worth Knowing Beforehand

To introduce you to the topic (Fig. 6.1), we will briefly describe what a zeolite mineral is: zeolites form a group of alumino-silicate minerals with a microporous framework structure made of negatively chargedAlO_4^- and SiO_4 tetrahedra. The negative grid charge is compensated by freely moving cations. They have a regular arrangement of cavities and channels and in general an extraordinarily large inner surface, sometimes well over 1000 m^2/g. The special zeolite subgroup of faujasite (zeolite-X) $\left(Na, Ca_{0.5}, Mg_{0.5}, K\right)_x (Al_x Si_{12-x}) \cdot 12H_2O$ has an octahedral costume (shape) and an inner structure based on cages, super-cages and channels between them as shown in Fig. 6.2.

Such zeolite single crystals with a diameter up to 50 μ (Fig. 6.2b) have been synthesized in our laboratories under hydrothermal conditions from aqueous solutions of sodium silicate and sodium aluminate in the presence of triethanolamine keeping the gels under 80 °C [1]; the Si/Al-ratio was about 1.1 to 1.0 as determined by X-ray fluorescence analysis of the synthesized zeolite powder.

As we know from the textbooks, crystals are characterized by the translation group of their unit cells. These are described by the point group assigned to them. This also applies to the zeolite crystals with their rather complicated structure.

The various cavities in faujasite are of atomic/molecular order of magnitude a few Ångström. If we ignore the quantum character and see the system itself on this scale as a classic system, then we can say that there is a strictly deformed inner surface of the pores in the zeolites. But this deformation is by no means self-similar. However, if

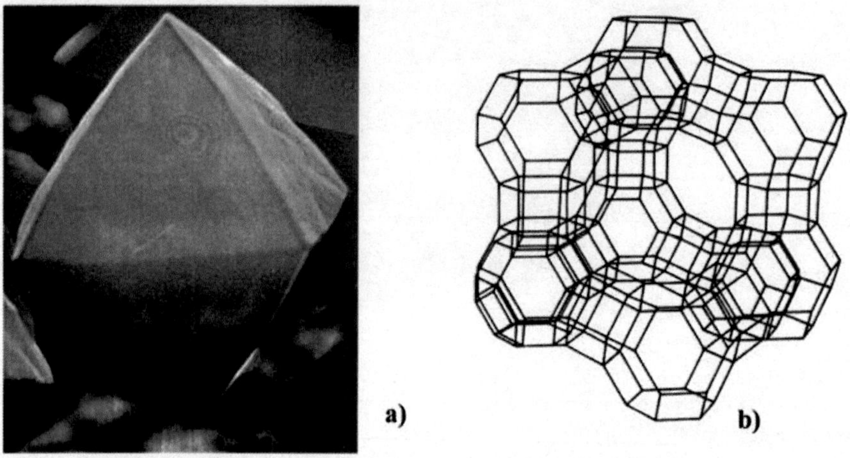

Fig. 6.2 **a** Faujasite crystal of about 40 μm; **b** Crystal structure of faujasite. The small cages are connected via hexagonal channels, forming a supercage. These supercages are tetragonally connected to each other. The edges of this graph represent the covalent bondings between the Si and/or Al atoms via oxygen atoms O in between. The vertices of the graph represent the Si or Al atoms

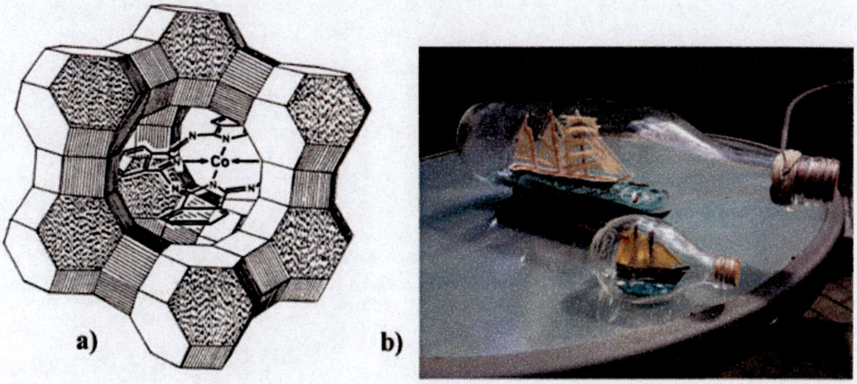

Fig. 6.3 **a** Schematic representation of a cobalt phthalocyanine (CoPc) complex in the supercage of a faujasite crystal; **b** two bottle-ships (*Photo* P. Plath)

we use molecules or atoms of different sizes, the measured surface size would differ indicating the roughness of the surface. Although real crystals, especially zeolites, always have dislocations, Pfeiffer and Avnir were able to show that ordinary zeolites do not have an internal surface that is fissured on all sizes. We were able to essentially confirm this statement [2].

But what happens if, after the primary ion exchange of the relatively free-moving sodium ions Na^+ of the faujasite with equally free-moving cobalt ions; are these Co^{2+} ions transformed to metal chelate complexes so that their mobility is then fixed in the zeolite lattice? If, for example, Dicyanobenzene (Dcb) is used as the reactant, the large, planar cobalt phthalocyanine CoPc is formed. It can be assumed that the large, bulky CoPc molecules each completely fill a large cavity—a supercage—of the zeolite (Fig. 6.3).

Such systems gained importance as so-called bottle-ship catalysts. For example, CoPc-loaded zeolite-X crystals catalyze the oxidation of ethyl benzene, while crystalline CoPc alone or supported on amorphous silicate carriers does not catalyze this reaction or at least not so effectively.

If one measures the inner surface of these CoPc-zeolite-X catalysts according to the BET-N_2 method (Brunauer-Emmet-Teller method) [3], as is usually done for the characterization of catalysts, it is found that this does not grow as expected, square or cubic with the edge length of the crystals however with an odd power. The dimension of the system can be determined from this dependency: it is $D = 2.71$ or, $D = 2.58$ depending on the synthetic method used. This has been described in detail in our previous work [2].

Figure 6.4 shows two images of CoX zeolites of 17 μm diameter which were dehydrated at different temperatures (60 °C and 122 °C, respectively). In the first case the crystal is whitish/grey translucent, but in the second case it shows a slightly bluish irregular coloration (compare the formation of cobalt blue glass). In both cases, however, there are some random fluctuations—due to the smallness of the crystals.

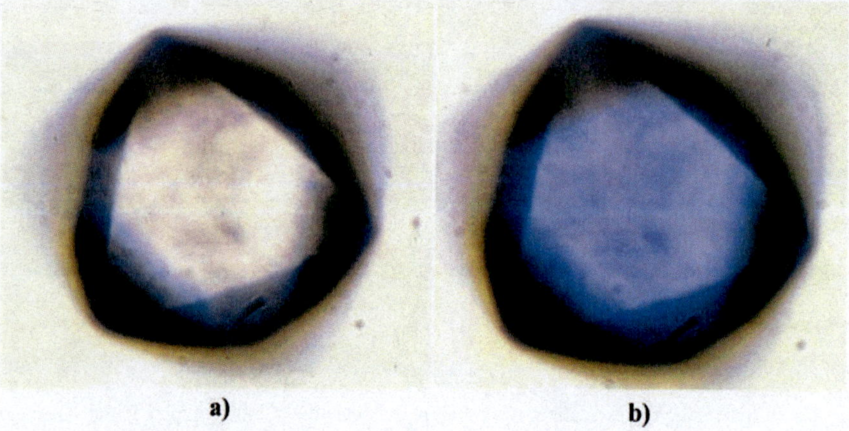

Fig. 6.4 Light-microscopic images of CoX zeolites of 17 μm diameter which were dehydrated at different temperatures: **a** at 60 °C and **b** at 122 °C. These photographs as well as those of Figs. 6.1 and 6.19 were taken by H. Diegruber and P. J. Plath in connection with the Doctoral thesis of H. Diegruber [4].

For CoX-zeolites, in which the Co^{2+} ions were not complexed with Dcb (Benzene-1,2-dicarbonitrile = Dicyanonobenzene), the measurement of the inner surface using the BET method results, as expected, in a dimension of $D = 2.97$, but not three ($D \neq 3$). In the picture of the fractal description of this phenomenon, one would say that the surface of the CoX is so fissured that it fills practically the entire three-dimensional volume of the crystal, whereby every partial volume is accessible for the N_2 molecules. This result is largely in line with the classic ideas. But a zeolite chemist would not talk about the inner surface of the zeolites in terms of fractal geometry.

However, if the CoPc-loaded zeolites have a very rugged inner surface of dimension, $D \approx 2.7$, how should one imagine the distribution of the CoPc molecules? Furthermore, is such a CoPc-loaded zeolite crystal still a crystal? In such a case, does there exist a crystallographic unit cell that is translation-invariant? Or, is it just the mimicry of an amorphous system that occurs in the costume of an octahedral crystal?

Most of the non-transparent CoPc-loaded zeolite crystals look uniformly blue or green from the outside and the intensity of the color distribution rarely shows a slight structure. From this, it was generally concluded that there is also a uniform distribution of the CoPc molecules loaded zeolites over the individual supercages, in accordance with the bottle-ship concept of the CoPc.

In the synthesis or growth of the zeolite-X crystals, however, it happens often enough that individual octahedra are not fully formed but only form crystal disks, since other crystals prevent them from growing regularly. Easily removable crystal twins are created in the form of disk-shaped crystals, which are only accessible from their freely accessible sides for Dcb molecules or Co^{2+} ions. As a result, they

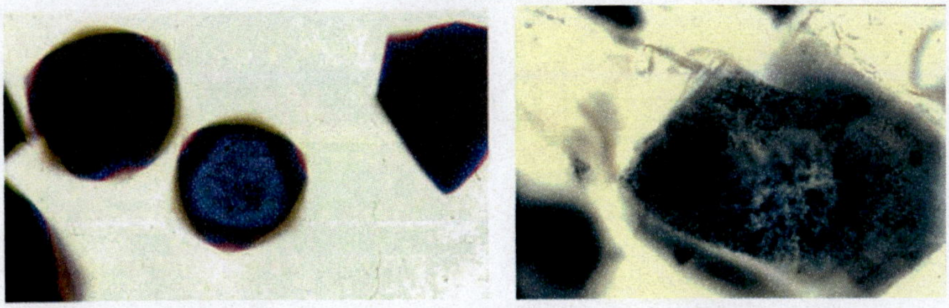

a) b)

Fig. 6.5 Light-microscopic view on CoPc-loaded zeolite-X crystals; **a** blue colored fully developed crystals; **b** zeolite-X disc (thickness of the microscopic cut: 5 μm) strongly patterned (stardust like) by CoPc

correspond to crystal disks that would have been cut out of the fully formed octahedral zeolite single crystal. So, one can see from them, like from a tomogram, what it looks like inside the CoPc-loaded zeolites. Such "cuts" by no means show an even distribution of the CoPc loading, but a wonderful pattern as can be seen in the pictures Figs. 6.1 and 6.5.

It looks like blue CoPc rivers which run through the Zeolite-X crystal. Their appearance, whether visible river structures or their cuts (stardust structure), depends entirely on the accessibility of the crystal interior during reactive loading. They make it understandable that a non-integer dimension is determined in the BET-N_2 measurements on CoPc-loaded zeolites.

If one looks at this flow structure of the CoPc then the question naturally arises, how can such a flow system be formed in the zeolite crystal? Where does it take the space from for its spreading? Obviously, the individual CoPc molecules are not isolated one from each other in its own supercage, but they form almost fractal structures of crystalline stacks of overlying CoPc molecules.

It is well known that faujasites are very sensitive to acids. If protons H^+ are formed during the formation of the cobalt-phthalocyanine, an increased acid concentration could occur locally, which leads to the partial dissolution of the crystal lattice of the faujasite. This would create the space necessary for the formation of the CoPc stacks. A similar process of locally dissolving the zeolite crystal lattice by acid formation in faujasites also takes place when palladium or platinum ions are reduced to larger metal single crystals within the zeolite [5]. They also urgently need more space than the individual super cages of the faujasite can provide.

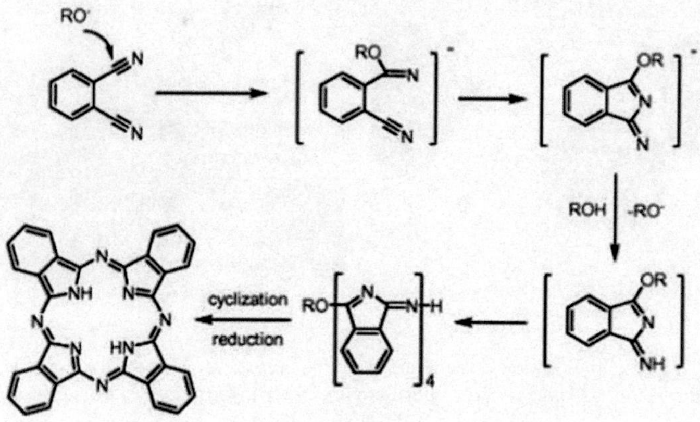

(6.1)

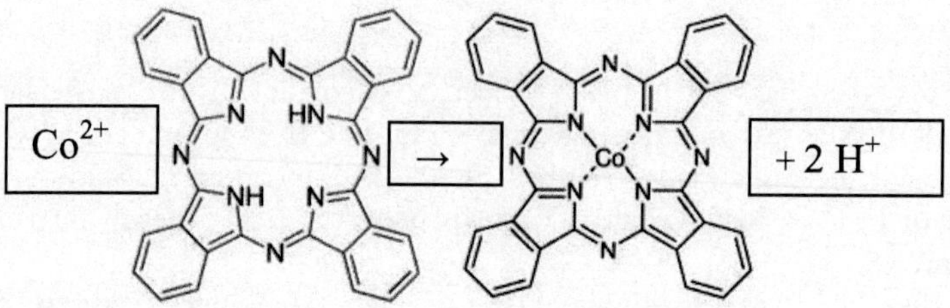

(6.2)

The reaction mechanism of Cheng-Hui Li, Xiao-Zeng You et al. [6] shown in reaction Eq. (6.1) for the formation of Phthalocyanine (Pc) from Dicyanobenzene (Dcb) was developed in a completely different context. However, if the substituent R in the formulas is replaced, for example by the hydrogen atom of the faujasite crystal water or by a docking point on the crystal lattice of the zeolite, then you get a very clear picture of the possible molecular reaction even in the case of this reaction in faujasite.

The reaction Eq. (6.2) shows formally, how the formation of the CoPc from Co^{2+} ions and Phthalocyanine Pc results in the hydrogen ion H^+ concentration necessary for the acid-related local destruction of the zeolite lattice. In this way, the resulting CoPc system creates the space it needs for its own creation.

If an ion exchange of the still free, not complexed Co^{2+} ions with Na^+ ions is carried out within these CoPc-loaded zeolites, this process also proves to be fractal: the dimension of the resulting CoPc-structure in the zeolite framework is $D = 2.71$ [2].

Now we are regarding the apparent total surface area A to be dependent on the average diameter of the crystals as $A \propto d^{D-3}$ or $\log A \propto (D-3)\log d + const$. For the CoX crystals, A does not depend upon d (see Fig. 6.6a). Therefore, the dimension becomes $D = 2.97$ (correlation coefficient $cc = -0.537$). In any case, all cavities of the zeolite can be totally filled with N_2 molecules.

In contrast to this, there is a strong dependence of the N_2 physisorption capacity on the crystal size in case of CoPc-loaded zeolites CoPcX (see Fig. 6.6b)). The observed dimension is $D = 2.71$ ($cc = -0.894$). This means, such systems actually form fractals. Therefore, one should expect areas of any size distributed stochastically in the zeolite framework, which are not

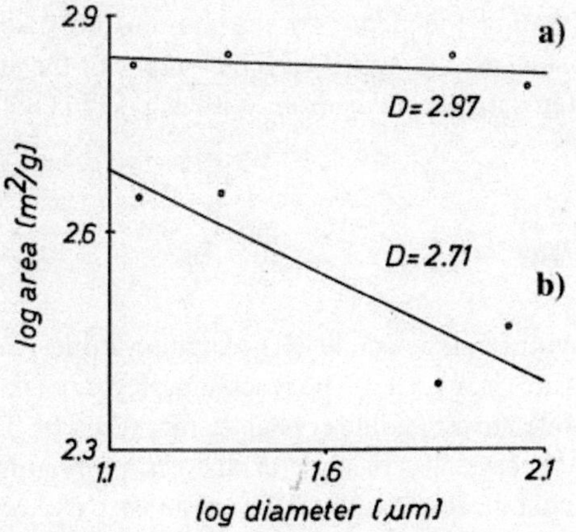

Fig. 6.6 Logarithmic plot of N_2-pysisorption data versus the diameter of the samples [2]; **a** in the case of CoX crystals; **b** in the case of CoPc-loaded zeolites CoPcX

> accessible for N_2-molecules. Large crystals should have even bigger inaccessible areas than smaller ones. Figure 6.1 shows the formation of systems of not self-avoiding CoPc-rivers in a faujasite crystal filled from the outer sphere into the middle of the crystal" [2].

Barnsley [7] now had a very interesting idea for creating fractal structures in the plane, which can be used to simulate a diffusion process.

He sets three striking attractive points on the plane and let a fourth point move between those attracted by these three points. This hiker between the three points uses the shortest route to the selected attractive point and stops at the location that divides the distance between his current location and his new destination in a certain predetermined ratio. The new point created in this way now becomes the starting point for iteration (see Fig. 6.7).

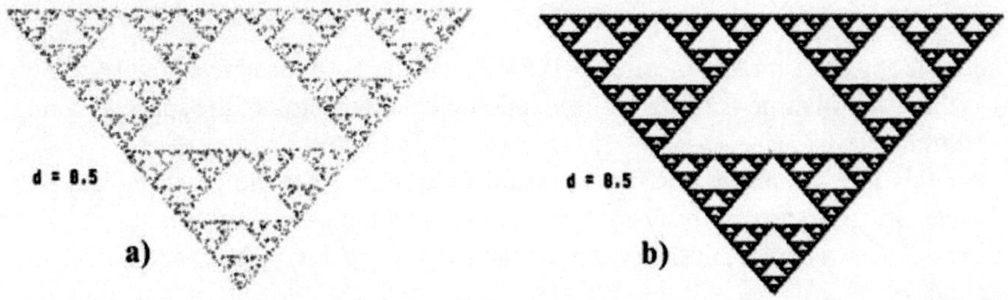

Fig. 6.7 Sierpinsky gaskets generated by an iterative random process. The division ratio between the randomly selected attractive corner point and the iterated, wandering point is $d = 0.5$; **a** with a low number of iterations; **b** with a high number of iterations, for which the set of iterated points has come very close to its attractor

For better understanding, the pictures of the non-uniform distribution of the CoPc inside the zeolite crystals are compared with the graphs of the attractors of different iterated function systems inside octahedrons with respect to the idea of Barnsley [7].

6.2 Introduction

A new method of synthesizing Cobalt (II)-phthalocyanine (CoPc) inside X-type zeolite crystals (faujasite) is reported. The reactants Dicyanobenzene (Dcb) and Co^{2+} ions are both incorporated in the zeolite crystals before reaction. The Co^{2+} ions, which remain unused inside the crystals, are ion re-exchanged afterwards. The reaction order of this reverse ion exchange depends upon the crystal size and reaches unexpectedly high values. This dependency can be expressed by the "dynamic fractal dimension" of the correlated ion exchange process.

In earlier works, especially by G. Zumofen, A. Blumen and J. Klafter, the idea of fractal geometry [8] has also been applied to chemical reactions [9]. To calculate the fractal dimension of faujasite type zeolites, D. Farin, D. Avnir and P. Pfeifer used data form adsorption measurements of small alkanes. They took the saturation volume of these n-alkanes as a function of their molar volume of bulk or liquid adsorbate, respectively. In this way, the dimension of the zeolite is about $D \cong 2$. They assumed a monolayer of linear alkane molecules forming parallel head-to-tail chains in zeolite cavities. The obtained value $D = 1.95 \pm 0.01$ is very close to $D \cong 2$, wherefore they concluded that the channel structure of zeolites may form an almost smooth two-dimensional sur-face [10].

However, some years later, they re-investigated experimental catalytic results and introduced the concept of a fractal reaction dimension. Their basic scaling law reads as follows [11, 12]:

$$a \propto R^{D_R - D_T} \tag{6.3}$$

where a is any kind of reaction rate, $2R$ is the size of the catalytic active metal particle, D_R is the reaction dimension and D_T the topological dimension of the space to which is referred.

For this interpretation, it is fundamental to assume, that the reaction rate is a characteristic measure for the catalytic reaction under the prevailing conditions. This means that the essential geometry of the reaction system does not change during the reaction.

But there are chemical systems in which the rate of the reaction changes while the reaction proceeds. Poisoning, for instance, will change the activity of the catalyst. In this case, one can usually spread the whole reaction into two widely separated time scales and for each of them one may estimate its own reaction dimension.

If of course both time scales are of the same order of magnitude, they are no longer separable. This is the case if the reaction creates and varies the fractal structure by which it can be characterized.

Studying the ion exchange of Na^+ ions versus the Co^{2+} ions which remain in the zeolite crystals after synthesizing Cobalt (II)-phthalocyanine inside the crystals, we obtained reaction orders n in the range between $n = 6.6$ and $n = 11.6$ depending upon the edge length of the zeolite crystals [2]. But, if the reaction order changes with respect to the size of this crystalline reactor, chemistry should also change with the crystal size. This can also be understood as a particle size effect of the reaction with respect to the zeolite crystal size. We have observed similar particle size effects using CoPc-loaded zeolites as a catalyst for the oxidation of ethanethiol [13].

To study the dependency of chemistry on the crystal size, we investigated the reaction of the Co^{2+} ions with Dicyanobenzene (Dcb) both inside the zeolite crystals. In large zeolite crystals Cobalt (II)-phthalocyanine is formed, whereas mainly the yellow by-product Tris (2-cyanophenyl)-1,3,5-triazine (in short: Triazine) is produced in the smaller crystals (see Fig. 6.8).

We have no desire to mention all possible fractal dimensions, but since the term *reaction dimension* is occupied by the definition of Farin and Avnir [11, 12], we propose the term *dynamic dimension* in order to express the dependence of the reaction order on the size of the crystalline reactor. The reaction order determines the mostly non-linear character of the dynamic system.

Two additional experimental observations are of great interest with respect to the fractal character of the distribution of the products. The first phenomenon is the river-like non–uniform distribution of the Cobalt (II) phthalocyanine in the zeolite crystal (see Fig. 6.1), which was first reported in our work in 1987 [2]. The second

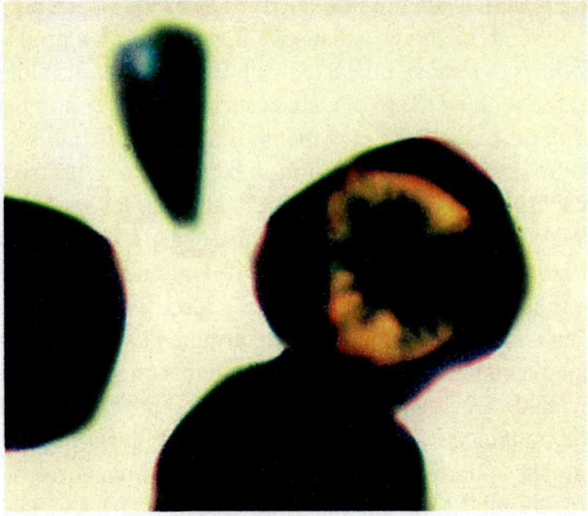

Fig. 6.8 Zeolite-X crystals in which the blue CoPc as well the yellow Triazine as a by-product is produced. In the big and bulky crystals, mainly CoPc can be observed (dark blue crystals) whereas in the small green–blue looking crystals both products are present. The eye-catching crystal disk shows the internal structure of the distribution of both products in one zeolite crystal

phenomenon is the local separation of the two products—the blue Cobalt (II) phthalocyanine (CoPc) and yellow Triazine insight the zeolite framework (see Fig. 6.8). For both phenomena, models are suggested to simulate the observations.

The models are founded on the idea that an iterated function system proposed by Barnsley in 1986 [7] might produce similar pictures as can be observed in photographs of the product-loaded zeolite crystals (see Figs. 6.5 and 6.21).

6.3 Experimental Part

6.3.1 Preparation of the Samples

In the present paper we are talking about Cobalt (II)phthalocyanine loaded faujasite crystals in a range of crystallite size from 15 to 105 microns using two different preparation techniques.

One of these preparation methods (named the **old method**) is already described in our previous paper in 1987 in detail [2]. For the synthesis of Cobalt (II)phthalocyanine loaded zeolites by the new method, the same Cobalt (II) ion-exchanged faujasites as for the old method are used. The Co^{2+} ion amount of the zeolite charges is 4.4 wt %, i.e., one Co^{2+} ion per supercage on average.

For the **new method**, mixtures of 8 mmol benzene-1,2-dicarbonitril (Dcb) and 1 mmol of the different charges of partially dehydrated Co/NaX zeolites (referred to as Co(II); 24 h, 573 K) are filled into glass ampoules which are sealed under special conditions [2]. After a so called “pre-heating” at 423 K for 3 h (10 °C/min) the glass ampoules are allowed to cool down to room temperature. At the end of this procedure the materials are put into the oven at 573 K for 18 h.

In both methods, exhaustive treatment of the products with acetone and pyridine follows to remove unreacted Dcb by-products and CoPc formed on the outer surface of the zeolite [13] (CoPc amounts and turnover numbers of the samples synthesized with the old and the new methods are listed in Table 6.1 A, B [14]).

To remove those Co^{2+}-ions which have not formed CoPc-complexes and in order to observe re-exchange kinetics, all samples are treated with 1 N sodium-acetate solutions. For this, 0.25–1 g of the Cobalt (II) phthalocyanine loaded faujasites (dried for 24 h at 475 K) are placed in 100 ml one-necked flasks and 50 ml portions of the 1 N sodium acetate solution are added. Then the mixtures are shaken thoroughly. The residual Co^{2+} ions were determined before the ion’s re-exchange procedure by complexometric measurements. For this purpose, the zeolite framework is destroyed in $HN0_3$. During the ion-exchange procedure, the Co^{2+} ion amounts which remain in the sample are determined by measuring the concentration of outcoming Co^{2+} ions into the sodium-acetate solutions with atomic absorption Spectroscopy (AAS).

The total time for the re-exchange is 10 h. The sodium-acetate solutions are renewed every two hours (see Table 6.2) [13]. To remove the acetate after completed re-exchange, the samples are washed 25 times with 50 ml portions of bi-distilled

Table 6.1 Physical characterization of the zeolite samples

	Old method sample	Diameter (μm)	Spec. N_2 surface (m^2/g) before after work off		Pc amount (w%)	Turnover (%)
A	1	14.67		445	1.5	5.1
	2	22.67		451	1.1	3.6
	3	71.74		245	0.5	1.5
	4	105.00		294	1.1	3.6
B	New method sample		Before after work off			
	5	14.67	10	449	0.3	1.0
	6	22.67	519	546	0.4	1.4
	7	71.74	361	483	2.9	9.5
	8	105.00	260	368	2.8	8.7

Table 6.2 Co^{2+} ion re-exchange with Na^+ ions after the formation of CoPc in the zeolites

Sample[a]	Residual amount of Co^{2+} ions in the zeolites (mg/g)					
	0 h	2 h	4 h	6 h	8 h	10 h
1	25.18	18.23	16.45	15.46	15.04	14.56
2	28.38	20.69	19.25	18.03	17.10	16.40
3	21.13	18.11	17.30	16.68	16.16	15.80
4	18.13	15.47	14.65	14.05	13.45	13.12
5	20.00	17.90	16.90	16.40	15.20	14.60
6	9.80	8.30	7.80	7.50	6.90	6.60
7	18.70	17.30	16.60	16.20	15.30	14.90
8	7.90	7.40	7.20	7.00	6.60	6.40

[a]The sample numbers correspond to the numbers in Table 6.1 A, B

water. Additionally, the re-exchange behavior of the Co/NaX charges are studied and reported [2].

6.3.2 Characterization of the Samples

The method for calculating the average diameter of the different zeolite charges used and their distribution functions are described in our previous work [2]. To characterize the samples, instrumental analytic investigations are carried out observing the morphology and surfaces of the crystals (scanning electron microscopy), the crystallinity of the products (X-ray diffraction spectroscopy), the distribution of the CoPc formed inside the faujasites (cross-cuts in Light-Microscopy), the amount of CoPc in the samples as well as the semi-quantitative amount of by-products

(UV/VIS-spectroscopy), the spectra of products inside the zeolite crystals after reaction (FTIR-spectroscopy) and the amount of Co(II) ions by AAS spectroscopy and complexometric measurements (UV/VIS range).

To study the arrangement of the cobalt chelates, EPR-measurements are carried out [13, 15]. Determinations of the nitrogen physisorption capacities of the samples are done by a dynamic method at 77 K, as described in the work of G. Schulz-Eckloff et al. in 1982 [16].

6.4 Results

During the 10 h of successive ion re-exchange, continuous non-linearly decreasing contents of Cobalt ions in the zeolite charges are observed. The rates of the re-exchange depend upon the diameters of the zeolites.

As in the case of N_2 physisorption for the specific surface area, we obtain decreasing values for totally exchanged ions with increasing diameter of the crystals of the catalyst charges (see Table 6.2 and Fig. 6.9). As in many other observations, the sample with the smallest diameter (about 15 microns) does not match this sequence. We explain this with the different morphology of this sample compared with the others, as we mentioned already [17].

Figure 6.9 shows an example for the "concentration" of the remaining Co^{2+} ions in the zeolites during ion re-exchange as a function of time.

The shape of this graph (Fig. 6.9) reminds us strongly to an exponential decay of the Co^{2+} ion concentration in the CoPc crystals. This encouraged us to describe the Co^{2+} ion re-exchange by a first order reaction kinetic (see Eqs. (6.4) and (6.6)).

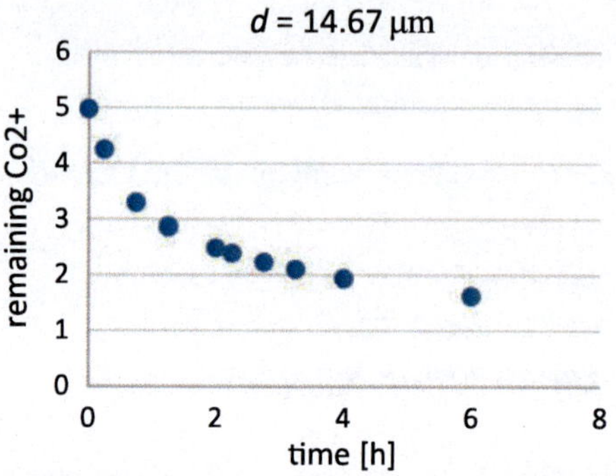

Fig. 6.9 "Concentration" of the remaining Co^{2+} ions in the small zeolites of the averaged diameter $d = 14.67$ μm (new synthesis method) during ion re-exchange as a function of time. The "concentration" of the Co^{2+} ions is calculated as the ratio of the weight of the Co^{2+} ions (in mg) and the related zeolite weight (in g). For better presentation, this ratio is diminished by the summand $s = -15$

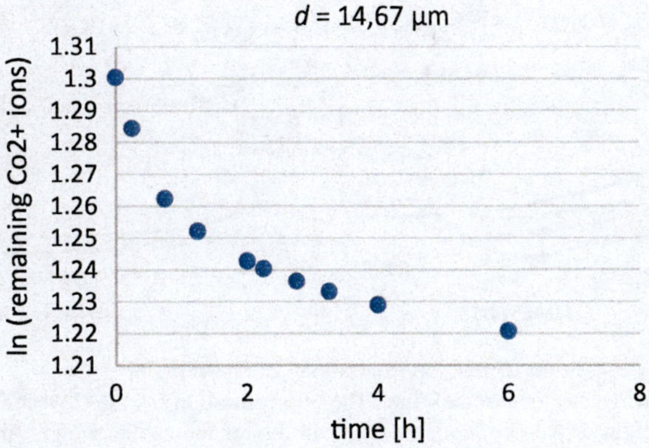

Fig. 6.10 The logarithm of the remaining Co^{2+} ion concentration as a function of time concerning Eq. (6.6). The mean crystal diameter of the zeolites of the sample (XXI) used here is $d = 14.67$ µm

$$\frac{dc}{dt} = -kc \tag{6.4}$$

The well-known solution of this linear differential equation is:

$$c = c_0 e^{-kt} \tag{6.5}$$

which can be transformed into the linear function $\ln(c) = f(t)$:

$$lnc = -kt + lnc_0 \tag{6.6}$$

Obviously, this is not a linear function (see Fig. 6.10) wherefore the approach of a linear differential Eq. (6.4) does not hold. This conclusion is not restricted to the sample with the smallest mean crystal diameter, but as Fig. 6.11 shows, is observed in all samples, regardless of whether the synthesis method used is the old or the new one.

For the other samples that were prepared with the new method, the total amounts of re-exchanged Cobalt ions after 10 h exchange time vary from 32.65% for the 23 microns sample through 20.32% for the 72 µm zeolites and to 18.79% for the 105 micron charges.

In contrast to these relatively low total exchange rates as mentioned above, nearly the whole Co^{2+} ion amounts could be exchanged back with the Na^{+}-ions in all the samples which were also used for the loading with Pc.

A logarithmic plot of the Co^{2+} ion amounts of the different samples as a function of time does not lead, as could be expected for "normal" ion exchange, to a linear function. This shows that the mechanism on which this particular ion exchange (CoPc- loaded zeolites) is based differs from the mechanism of the exchange with

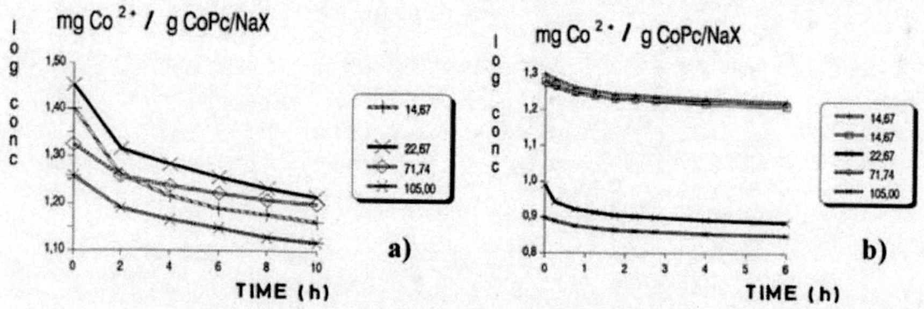

Fig. 6.11 Decadic logarithm of the "concentration" of the remaining Co^{2+} ions in the zeolites during ion re-exchange as a function of time. The "concentration" of the Co^{2+} ions is calculated as the ratio of the weight of the Co^{2+} ions (in mg) and the related zeolite weight (in g); **a** old and **b** new synthesis method

only Co^{2+} ion loaded samples as have been shown earlier [2]. Because of the non-linear logarithmic plot, we are not able to calculate the reaction order of the described ion re-exchange using the familiar equations, which can be taken out of basic text books on chemical kinetics.

XDS-measurements (Seifert ISO Debyeflex 1001) of the samples are carried out before and after the described synthesis of CoPcX. There is no detectable loss of crystallinity as E. Ignatzek has shown in his thesis [13]. Only those signals, which are typical for the zeolite framework can be found, but no CoPc signals occur. This is a hint that no significant amounts of crystalline CoPc are formed in the zeolite framework.

EPR-spectra (Bruker ER 200 D-SRC) of the CoPc-loaded faujasite taken in vacuum and in air show hyperfine structures. One can conclude that major parts of the CoPc are so-called mono-molecularly distributed. It is proposed that the arrangement of the macro cycles forms one-dimensional chains penetrating the crystals as it was already discussed for Ni-dmg (Ni-**dim**ethyl**g**lyoxime) in faujasites [18].

Additionally, it has been proven that the uptake of small molecules like O_2 and H_2O by the chelates is reversible.

6.5 New Synthesis Method

Taking up the idea of obtaining greater amounts of Cobalt-phthalocyanine in the zeolites by other synthesis parameters as mentioned in our earlier work, CoX crystallites mixed with Dcb (Dicyanobenzene) in sealed glass ampoules are pre-heated in such a way that the Dcb could get into the gas phase and diffuse in the faujasites, but could not cyclo-tetramerize with Co^{2+} ions to CoPc. The Dcb distributed in the zeolite supercages begins to react at the moment the filled glass ampoules are put in an oven at 573 K (new synthesis method).

As a result of quantitative CoPc determination, one should detect—in comparison with the values taken from the "old synthesis method"—higher CoPc amounts in the worked off reaction products. As shown in Table 6.1 A and B, the CoPc amounts are higher in the case of the bigger crystals, whereas in case of the small zeolites (15 and 23 microns) the values are smaller compared to those we obtain for the samples synthesized by the old method.

These data are applicable when looking at the pyridine UV/VIS spectra of the samples (the zeolite matrix was destroyed by stirring in 1 N $HN0_3$): The smaller the crystallites the greater the amount of substances absorbing in the shortwave region. Taking into account that, in general, extinction coefficient values of this region are in the order of 10^3 to 10^4 l mol^{-1} cm^{-1}, the amount of substances absorbing in the short-wave region should be higher by at least one order of magnitude in the case of the 15-micron crystallites than the CoPc-amount: $E(CoPc)_{654\,nm} = 1.19 \cdot 10^5$ l mol^{-1} cm^{-1} (thesis U. Hündorf) [14].

In the case of the small 15-micron samples, the substances which absorb in the short-wave range could be isolated and identified by FTIR-measurements and mass spectroscopy as tris(2-cyanophenyl)-1,3,5-triazine (Triazine: major by-product, yield more than 30 times as high as the yield of CoPc).

Usually, Triazine is synthesized from Dcb and catalytic amounts of water by cyclo-trimerization under strict exclusion of transition metal ions. The chemistry of the Triazine group was developed by Snow et al. [19] and Ross et al. [20]; in detail: if transition metal ions are present in the educt mixture even in very low concentrations, the metal ion traces lead to 100% turnover of these ions with Dcb yielding Pc by cyclo-tetramerization.

Contrary to this known reaction behavior in the case of the 15-micron faujasite crystals, we obtain Triazine as the major product although sufficient amounts of Co^{2+} ions are present in the applied zeolite compound. As was expected for all crystallite charges, the conversion of the Dcb to the reaction products is greater in the case of the new synthesis method as compared with the old one.

Likewise, for both the methods of synthesis discussed, the determined N_2 BET surfaces of the worked off CoPcNaX zeolite increases with decreasing crystal diameters. The 15-micron crystallites, synthesized by the new method, must be seen as an exception (as also mentioned for many things in the work of E. Ignatzek); before Co(II) ion re-exchange (see below), the N_2 physisorption value was 10 m^2/g and afterwards 44.9 m^2/g. For the N_2 molecules the previously non-accessible regions of the raw product become accessible after the Co(II)-ion re-exchange. Thus, the Co(II) ion re-exchange should be a very complex reaction. It might be better to understand this reaction as an extremely hindered diffusion reaction, induced by the fractal structure of the CoPc- loaded zeolites. This assumption will be discussed later on.

The relatively low N_2 BET surface of the 15-micron CoPcX zeolite sample seems to be causally related to their high Triazine loadings. For the other crystallite sizes, the difference between the N_2 BET surface values before and after the Co(II) ion re-exchange was not as remarkable as in the case of the 15-micron samples (see Table 6.1B). This correlates with the interpretation of the UV/VIS-spectra, which say that taking the 23-, 72- and 105-micron crystallites less or even indetectable amounts of by-products are formed, in comparison with the 15-micron CoPcNaX.

6.6 Catalytic Behavior of CoPcX

The preparation of the described samples is done following the concept of heterogenization of homogeneous catalysts [21]. Phthalocyanines are known to be good catalysts for oxidation reactions [22, 23] and the oxidation of thiols yielding disulfides (Eq. (6.7)) are important reactions in the oil processing industry [24], for which reason we have chosen the oxidation of ethanethiol in heptane as a model reaction for the following process:

$$2RSH + \frac{1}{2}O_2 \rightarrow RS - SR + H_2O \tag{6.7}$$

Into a 250 ml static reactor 100 ml heptane with 165 ppm ethanethiol is filled. The reaction temperature is 298 K. The total reaction time is set to 22 h. The amount of the added catalyst which is stirred into the solution is chosen to be $1.24 \cdot 10^{-5}$ mol of CoPc in the zeolite sample as described by Plath 1988 [17]. Blank experiments with NaX samples and with CoPc alone do not show noticeable amounts of turnover to the product after 22 h reaction time.

The values we received for the catalysts, which are prepared with the old synthesis method, are listed in Table 6.3 and the conversion of the thiol as a function of time is shown in Fig. 6.12. It can be seen that in catalysis, too, the reaction velocity is a function of the catalyst diameter. The crystals with the greater diameters show higher reaction rates. The turnover numbers listed in Table 6.3 are calculated using the converted amounts of ethanethiol at 0.5 h because this time could be taken as the initial reaction rate (see Fig. 6.12). Therefore, the fractal character of the catalyst, which is expressed in the BET values and the ion re-exchange is also viewed in its catalytic behavior.

Table 6.3 Catalytic behavior of CoPcX (old synthesis) on thiol oxidation

Sample	Turnover number at 0.5 [h^{-1}]
1	9.4
2	8.4
3	28.6
4	13.8

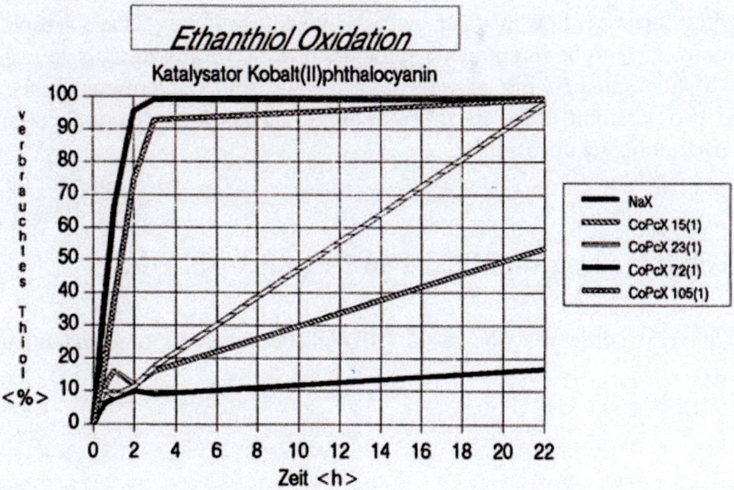

Fig. 6.12 Influence of the zeolite crystal size on the activity (thiol consumed in percent) of the zeolite catalysts (produced according to the old synthesis method) in the ethane-thiol oxidation in the batchwise stirred tank reactor. (There is practically no thiol conversion observable in pure NaX zeolites.)

6.7 Analysis of the Results

The amount of Co^{2+} ions remaining in the CoPc-loaded zeolite crystals after ion-exchange with Na^+ ions diminishes step by step with time (see Figs. 6.9 and 6.10, and the corresponding doble logarithmic plot in Fig. 6.13). Let us assume that time dependence for the decrease in the Co^{2+} content of the crystals is in accordance with a reaction law of first order as shown in Eq. (6.5).

However, the experimental well verified function $\ln c = \alpha \ln t$ is without any doubt a linear function in all cases, especially in the case of the new synthetic method.

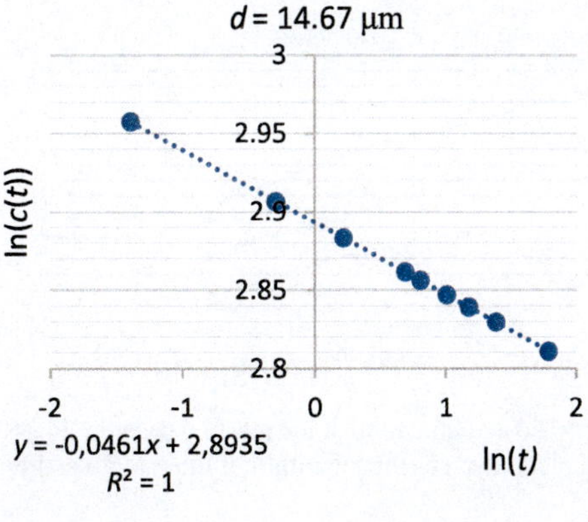

Fig. 6.13 Double logarithmic plot of the concentration of the remaining Co^{2+} ions in the CoPc-loaded zeolite crystals (new synthesis method) with the mean diameter $d =$ 14.67 µm as a function of time (compare Figs. 6.8 and 6.9). The result is the function $lnc_{Co^{2+}} = -0.0461\ln(t) +$ 2.8935, $R^2 = 1$

The question arises how the factor α in the linear function or in the corresponding relation $c \propto t^{\alpha}$ is to be interpreted. Since the linearity in Fig. 6.13 is so excellent ($R^2 = 1$), this must be the starting point of any further interpretation. For this purpose, two possibilities are discussed in the following: Interpretation via kinetics and interpretation via diffusion.

6.8 The Interpretation Based on Chemical Kinetics

Instead of the first order reaction law (Eq. (6.4)) let us take the more general reaction law:

$$-\frac{dc}{dt} = kc^n;\, n \neq 1 \tag{6.8}$$

with unknown reaction order n, to describe the decrease of the amount of Co^{2+} ions in the zeolite crystals with proceeding ion exchange. Solving this equation, one obtaines after a few steps:

$$c = [(n-1)k]^{-1/(n-1)} t^{-1/(n-1)} \tag{6.9}$$

The logarithmic transformation of c in Eq. (6.9) now gives the linear function $\log c = f(\log t)$:

$$\log c = -\frac{1}{n-1}\log t - \frac{1}{n-1}[(n-1)k] \tag{6.10}$$

Plotting the experimental data this way (Fig. 6.14), fairly good straight lines are obtained in the case of the old method and points which fit a straight line very well using the data of the new synthesis method. From the slopes m of these lines,

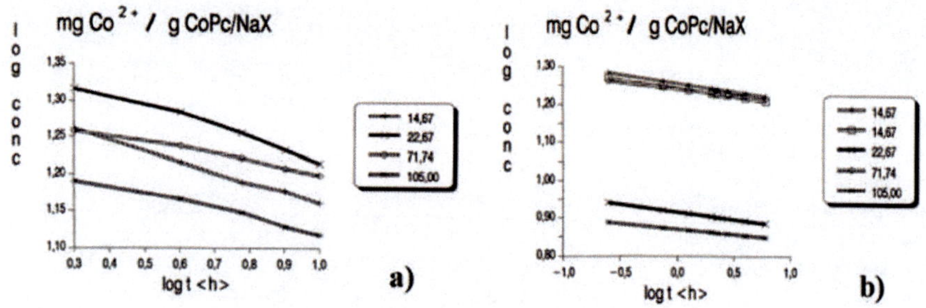

Fig. 6.14 Estimation of the reaction order in Eq. (6.10). Here, the logarithm of the "concentration" is plotted versus the logarithm of time; **a** old and **b** new synthesis method

Table 6.4 Estimation of the reaction order $n = -1/(m + 1)$ and the constant k of the formation of CoPc inside zeolite-X crystals according to Eq. (6.10). For comparison, the values for the CoNaX are also mentioned

Mean zeolite crystal size (μm)	Old synthesis method		New synthesis method		Unloaded CoNaX zeolite	
	Order n	Constant k	Order n	Constant k	Order n	Constant k
14.67	8.09	$8.24*10^{-11}$	22.94	$1.72*10^{-29}$	2.34	$2.34*10^{-2}$
22.67	7.92	$5.91*10^{-11}$	25.03	$3.81*10^{-24}$	1.67	$4.03*10^{-1}$
71.74	12.76	$6.31*10^{-17}$	29.16	$3.90*10^{-37}$	1.85	$1.81*10^{-1}$
105.00	10.70	$1.40*10^{-13}$	35.55	$2.21*10^{-32}$	1.98	$0.86*10^{-1}$

$$m = -1/n - 1, \text{ or } n = 1 - 1/m \tag{6.11}$$

one can estimate the reaction orders n concerning Eq. (6.8). The value of n obviously depends upon the size of the zeolite crystals (see Table 6.4). The smaller the crystal size, the larger is the reaction order n. The reaction constant k does not depend upon time any more, but it also depends upon the value of the zeolite crystal size, as expected.

The surprisingly high value of the reaction orders $n \approx 10$ and $n \approx 28$ for the old and the new synthesis method, respectively, reflects the complexity of the reaction inside the CoPc-loaded zeolites, which results in the observed ion exchange. In contrast, one gets a reaction order of about $n \approx 2$, as one could expect, for the ion re-exchange of the Co^{2+} ions in the case of the CoNaX-zeolites which are unloaded with CoPc. Although one can observe a dependency of the reaction rate on the crystal size even in this case, one has to recognize the dramatic change of the reaction order of the Co^{2+} ion re-exchange if CoPc is formed in the zeolite crystals.

Does it really make sense to think seriously about a reaction order of 10 to 30? Before answering this question, let us try to find any tendency which might be hidden in the sequence of these numbers. For this purpose, one may express the dependency of the reaction rate on the zeolite crystal size by its fractal dimension D.

Executing the log–log plot of the reaction order n versus the crystal size d, one can estimate a fractal dimension of $D = 2.763$ (cc $= -0.8606$) and $D = 2.786$ (cc $= -0.9838$) for the old and the new synthesis method, respectively (Fig. 6.15).

This dimension has been evaluated by regarding the change of the reaction order n for varying crystal sizes. For this reason, we call this dimension the dynamic dimension D of our chemical system. However, if the reaction order changes in such a dramatic way, chemistry should change as well. Therefore, chemistry also depends upon the size of the crystalline micro-reactors. This way one can understand that Triazine is the main product within the small crystals, whereas CoPc is the main product in the larger ones.

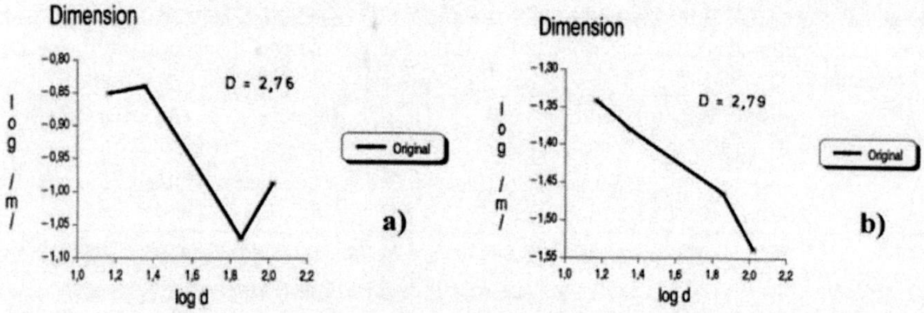

Fig. 6.15 Double logarithmic plot of the slopes of the straights shown in Fig. 6.14 and the mean crystal sizes in order to estimate the "dynamic dimension" D; **a** old synthesis method ($D = 2.763$, $cc = -0.8606$); **b** new synthesis method ($D = 2.786$, $cc = -0.9838$)

If we do not fix the reaction order, its resulting value reflects this temporal change of the molecular surroundings of the reacting species. But there is no unique traditional way of interpreting a reaction order of about 10 or 28. Does this really mean that ten or even thirty molecules have to come together? And what does it mean in case of the ion exchange procedure? Moreover, how can we understand this manner of dependence of the reaction order on the size of the zeolite crystals? We no longer believe that we can follow the trustworthy line of the traditional interpretation.

These results raise the question of a coherent interpretation of the underlying chemistry and physics. We are forced to ask the fundamental question: how do we have to define a molecule in a zeolite framework if we wish to understand the unexpected high values of the reaction order? To answer this question, let us go back to H. Haken's principle of slaving [25, 26]. In a dynamic system, one or only a few dynamic modes will slave all other possible dynamic variables, depending upon the constraints. In a chemical system, molecules are just the representatives of these leading modes [27]. Now, if we change the size of the zeolite crystals, we are modifying the constraints of the reaction and other modes may become the leading ones, whereas the former ones become slaved. Therefore, varying the crystal size from 14 to 105 microns, our system might pass many dynamic instabilities, undergoing several dynamic phase transitions. Different "molecules ", which are represented by different reaction orders—if we are referring to the concentration variables—characterize the various leading variables.

This consideration firstly leads to the conclusion that we cannot define a molecule inside the zeolite without specifying the correlated reaction and, secondly, even for the "primitive" ion exchange we have to consider non-trivial "molecules" of high complexity, represented by high values of the reaction order.

The question remains as to how we can imagine the manner in which the restriction of the volume of the crystalline reactor can influence the leading chemical reaction mode and the shape of the molecules. Of course, this can happen if the underlying space has a fractal structure. We have previously shown by BET measurements that CoPc-loaded zeolites are fractals [2]. Therefore, it should not be surprising that we can observe a change in the reaction order while varying the zeolite crystal size. This dependence will reflect the fractal character too.

6.9 An Interpretation Based on Fick's Second Law of Diffusion

We have seen that the approaches largely based on simple chemical kinetics to describe the temporal behavior produced completely unsatisfactory results in the case of Co^{2+} ion re-exchange. If we hold to the condition of a first-order reaction, then this inevitably leads to a very artificial time dependence of the reaction constant. If, on the other hand, we consider the reaction order $n > 1$ as the quantity to be determined, then completely unrealistically large reaction orders result.

However, if we abandon the assumption that the reaction rate depends on the concentration of the Co^{2+} ions present in the zeolite crystals, then a completely new possibility opens up.

Let us assume, for example, that the concentration of the Co^{2+} ions remaining in the zeolite only depends on the time according to relation (6.12):

$$c_{\mathrm{Co}^{2+}} \propto t^{\alpha} \tag{6.12}$$

Then, there is a linear relationship,

$$\ln\left(c_{\mathrm{Co}^{2+}}\right) \propto \alpha \ln(t) \tag{6.13}$$

which describes our observations excellently with unexpectedly good correlation coefficients, $R^2 = 1$, in all cases, as can be seen in Fig. 6.16b for example.

For the example shown in Fig. 6.16b, we get the linear equation.

$$\ln(c_{Co^{2+}}) = -0.0342 \ln(t) + 2.8582 \tag{6.14}$$

for the zeolites with the mean diameter $d = 71.74$ μm, where $\alpha = 0.00342$, according to Eq. (6.14).

Since this linearity is no exception, the question arises, what is the physical reason for these observations: Is the process of the ion re-exchange simply a diffusion of the Co^{2+} ions out of the zeolite crystals? Then, according to the 2nd Fick's law, the following should apply,

$$\frac{\partial c}{\partial t} = D\left(\frac{\partial^2 c}{\partial t^2}\right) \tag{6.15}$$

where D is the diffusion coefficient.

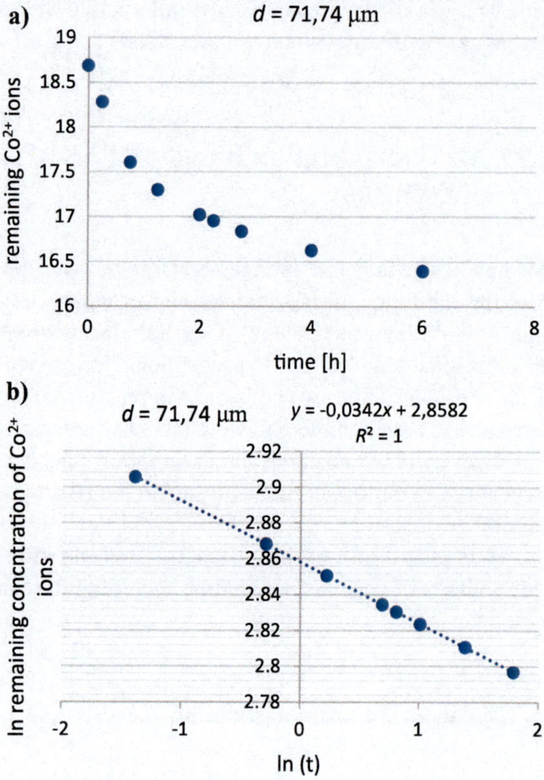

Fig. 6.16 **a** Concentration of the remaining Co^{2+} ions in the zeolite crystals with the mean diameter $d = 71.74\ \mu m$ as a function of time; **b** double logarithmic plot of Fig. 6.16a

For the sake of simplicity, let us consider an octahedral zeolite X crystal as a sphere. Let us also assume that the total Co^{2+} ion concentration in a zeolite crystal is concentrated at the origin of the sphere at time $t = t_0$. Outside this sphere of radius r_s, the concentration is zero at this time. Furthermore, the outer sphere should be in an infinite volume, so to speak. These assumptions allow a more or less simple solution of the partial differential equation of Fick's 2nd law:

$$c(r, t) = \frac{N_0}{8(\pi Dt)^{\frac{3}{2}}} e \qquad (6.16)$$

However, what we measure is not the concentration $c(r, t)$ of the Co^{2+} ions inside or outside the zeolite crystals, but the integral I of this concentration over the space within the crystal sphere of radius r_s after a certain time (see Eq. (6.17)). That is, what we are calling the remaining amount of the Co^{2+} ions in the zeolite crystals after re-exchange with Na^+ ions after a given time interval.

$$I = \int_{0}^{r_s} c(r,t)dr = \frac{N_0}{(4\pi Dt)^{\frac{3}{2}}} \int_{0}^{r_s} e^{-r^2/4Dt} dr \tag{6.17}$$

In order to solve the integral over the exponential function e, we develop this function into the well-known Taylor polynomial of order n at the expansion point $a = 0$, which is defined by,

$$T_n(f(x,a)) = \sum_{k=0}^{n} \frac{f^{(k)}(a)}{k!}(x-a)^k \tag{6.18}$$

where in our case $f(x) = f(r) = e^{-br^2}$ and $b = 4dt$.

If we restrict ourselves to the 2nd order in the Taylor series expansion, then we can write the integral expression (Eq. (6.17)) as follows:

$$I = \frac{N_0}{(\pi 4Dt)^{\frac{3}{2}}} \int_0^{r_s} (1 - br^2)dr \tag{6.19}$$

Emphasizing the dependence on time, the solution of this integral can be transformed into Eq. (6.20):

$$I = \frac{N_0 r_S}{(\pi 4D)^{\frac{3}{2}}} t^{-1.5} - \frac{N_0 r_s^3}{96\pi^{\frac{3}{2}} D^{\frac{5}{2}}} t^{-2.5} \tag{6.20}$$

In view of the very small crystal radii on the one hand, and in view of the strict linearity of the experimentally determined function $\ln\left(c_{Co^{2+}}\right) \propto \alpha \ln(t)$ (Eq. (6.13)) as well as the comparatively long times, it is justified that we limit ourselves here to the first term of the integral. Thus, we obtain Eq. (6.21) in double logarithmic representation:

$$\ln I = -1.5 \ln t + \ln r_s + \ln \frac{N_0}{(\pi 4D)^{\frac{3}{2}}} \tag{6.21}$$

The constant factor $-1.5 = -3/2$ in the logarithmic term $\ln t$ is the exponent of time $t^{\frac{3}{2}}$ (see Eqs. (6.20) and (6.21)) and reflects the assumption that this is a space with the topological dimension $d_t = 3$ in which the diffusion takes place. A factor of -0.0342 would mean that this diffusion would be based on a space with the fractal dimension $d_f = 0.0684$. This could be reminiscent of the diffusion of the ions in the gel in the Liesegang experiments (see chapter 'Liesegang structures' in this book) which takes place on the "backbone" [28] of the fractal gel body. There, the measured fractionated dimension was $d_f = 1.5$. But, a dimension as small as $d_f = 0.0684$ cannot be a hike on a fractally jagged path, because then it must be greater than one.

This is also the case if Euclidean obstacles or tremas, for example in the form of spheres, would partially block the path of the ions on the fractal [29].

A dimension between zero and one can only mean that it is a matter of "dust", i.e., an incoherent, possibly slightly clumped set of points or arbitrarily small distances. However, how can we imagine a migration of the Co^{2+} ions on such an incoherent dust? Again B. B. Mandelbrot gives mathematical examples of such processes. He refers to Lévy Flights. "A Lévy flight is roughly a sequence of jumps separated by stopovers (with Brownian trails; the authors). Only the latter are of direct interest in this chapter, but jumps are a necessary part of the construction. ... Decreasing of *D* (dimension; the authors) by subordination makes the Lévy clusters become increasingly separate" [30]. However, how do we generate a chemical process from these ideas that fits our experiment?

Let us assume that the formation of the CoPc molecules leads to a fractal that extends more or less over the cavities of the zeolite. Figure 6.1 illustrates this assumption., Whether or not the CoPc molecules are always isolated only as single ones in a supercage of the zeolite, i.e., whether or not the supercages are partially destroyed by the incorporation of the planar CoPc molecules, is not discussed further here. It is important that the typical color of the crystalline CoPc is due to the interaction of the co-molecules lying on top of each other. The CoPc molecules in the zeolite are also colored blue, but with a slight color shift, so that corresponding stacks or towers of CoPc molecules must have formed [31, 32].

If a free, isolated, not complexed Co^{2+} ion begins its "Brownian migration" during the ion re-exchange, it will soon encounter a branch of the CoPc fractal. Using the example of Newton's cradle, with which the conservation of momentum can be demonstrated excellently, it can be made clear what can happen to the Co^{2+} ions that hit a stack of CoPc molecules. Like the steel ball in Newton's cradle that hits the row of balls and at the other end repels the previously quietly hanging ball, the Co^{2+} ion can transfer its charge to the metal-like, "one-dimensional" ...–Co–Co–... chain of the CoPc molecule stack. Somewhere at any "end" of this fractal CoPc structure, a Co^{2+} ion is then formed and repelled. In this way one can interpret the "jumps" in the Lévy flight. Therefore, it's not one and the same ion that migrates, because its identity has changed; however, that does not matter here.

But this is where the analogy to the mechanical model of Newton's cradle ends, because in the case under consideration not only the charge is transferred from atom to atom, but also a mass transport or a shift of the central Co atoms from one CoPc to the neighboring CoPc molecule takes place.

Anyway, there is an ion that walks, and the fractal CoPc stack passes on the baton of charge to the next ion. In this way, the CoPc in the zeolite framework serves as fractal tremas in the sense of Mandelbrot's explanations [33]. They are "cutting out" the jumps in the underlying Lévy flight, so to speak. In this manner, an unbound Co^{2+} ion moves in the zeolite crystal, which is permeated by CoPc fractals according to the new synthesis method. The ion tumbles erratically in a stopover from one point of the Levy dust to another and then, changing its identity, jumps as another ion to another stopover.

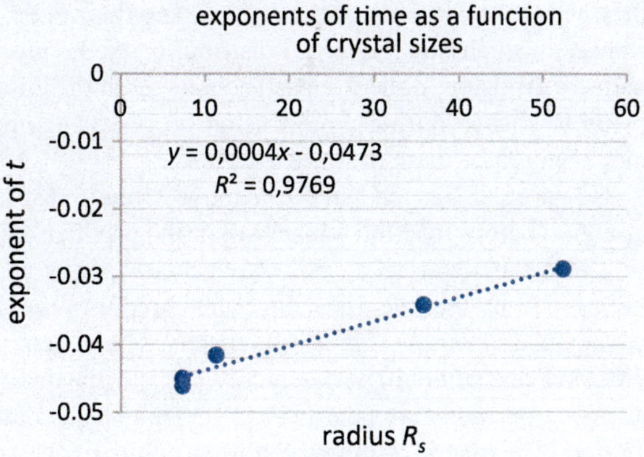

Fig. 6.17 The exponents of time in Eq. (6.12) are plotted versus the radius R_s of the spheres which represent the zeolite crystals

The result of this movement of the Co^{2+} ion re-exchange is a "migration movement" from point to point, the dimension of which is considerably smaller than one.

After all these explanations about the ion exchange of the Co^{2+} ions, it is not surprising that the dimensions of the space for this ion migration depend on the mean crystal size of the zeolites. Figure 6.17 shows that this dependence of the dimensions on the radius of the mean crystal size is linear in form $d_f = 0.0004$ and $R_s - 0.0473$.

6.10 An Iterated Function Model for the Formation of Fractals by Diffusion in Octahedral Zeolites

Besides the dynamics of the ion re-exchange, we have to understand the formation of the fractal shape of the CoPc distribution within the zeolite crystals. At the beginning of the synthesis, we should have randomly distributed Co^{2+} ions as well as Dcb molecules:

$$Co^{2+} + 4\,Dcb + 2\,H_2 \rightarrow CoPc + 2\,H_3O^+ \quad (6.22)$$

Therefore, the CoPc product should be randomly distributed as well. But this conclusion, which seems to be obvious, includes some strong assumptions about the diffusion process of the reactants inside the zeolite crystals.

If one assumes a Brownian-like diffusion, as it occurs in gases and liquids, where all directions in space are of equal probability, then the products shall also be equally distributed in a stochastic manner. However, there arises the question as to whether this assumption is justifiable in the case of zeolites.

On the contrary, in the zeolite crystals one should assume a diffusion process in which some directions in space are favoured, because of concentration gradients between the inner parts of the crystals and their outer surface. Under those circumstances, a stochastic and equal distribution of the products requires only very short paths of diffusion, if the stochastic character of the initial distribution of the reactants is not to change during the diffusion process. This means the reactants should not move but react where they are localised.

However, if the reactants can move "freely" for a while within the preferable direction before reaction, clustering of the products should occur.

These statements can easily be evaluated using an iterated function system [7]. Let us assume that the diffusion is favoured into the direction of the six corners of the octahedral zeolite (faujasite) crystals. The huge number of starting points of the diffusing reactants might be equally and stochastically distributed. Now an erratic diffusion process may begin [17, 18], This means that, in contrast to the truly free motion of a spheric symmetry, not all points of the sphere are target points of the diffusion, but only the six corners of the octahedron. These target points are randomly chosen in a coin toss algorithm.

The reactants only move directly one part of the total distance between their position $x(t)$ at time t and the chosen target point. This fraction of the distance depends on the attracting strength of the randomly chosen target point. This way we have established an erratic diffusion process based on an exponential law:

$$y_{i,m}(t-1) = |(i - y_i(t))|a^{-1}; a^{-1} \in [0, 1] \tag{6.23}$$

$y_{i,m}(t)$ vector of the iterated point at time t.
t discrete time,
i vector of the chosen target point and.
m starting point.

This can easily be evaluated, if i remains constant for a successive amount of time steps. Then Eq. (6.23) can be transformed into:

$$y_{i,m}(t+1) = \left|i - y_{i,m}(t)\right|a^{-1} \tag{6.24}$$

After a given number T of time steps t, the diffusion of the reactants may be finished by a successful reaction. The products should not diffuse any more. Let us mark these last points $y(T)$. If T goes to infinity: $T \to \infty$, the set of all points $y_{i,m}(T)$ forms the attractor of this erratic diffusion process. But if we have many starting points, $y_m(0)$; $m \to \infty$, we need only very few time steps to reach the attractor, i.e., to come sufficiently close to nearly all points of the attractor. Figure 6.18a–d shows some computed attractors for various fractions $1/a$. For small values of $1/a$; $(a \approx 2)$ one obtains only slightly structured distributions of the points $y(T)$, which reminds us of the product distribution of the CoPc inside the zeolite crystals.

Nevertheless, these numerical distributions are of fractal geometry and differ essentially from the stochastic and equal distribution of the starting points$y(0)$. If $1/a = 2$, one gets the 'three-dimensional' SIERPINSKY-like octahedron [8]; and if $1/a > \frac{1}{2}$, isolated clusters occur (see Fig. 6.18).

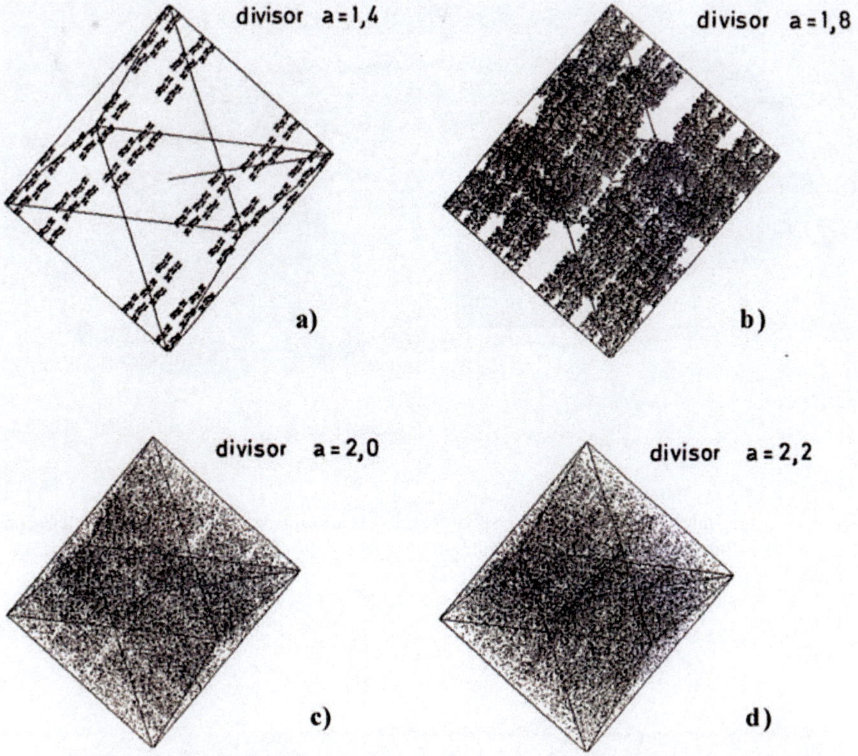

Fig. 6.18 Attractors for different iterated function systems (IFS)7 within an octahedron. The IFS's are characterized by various divisors a: **a** $a = 1.4, 1/a = 0.713$, **b** $a = 1.8, 1/a = 0.556$, **c** $= 2, 1/a = 0.5$, and **d** $a = 2.2, 1/a = 0.455$. The fractal structures within the octahedron can only be guessed at as very diffuse stripe-like structures in Fig. 6.18c, d

To get a visual impression of the CoPc distribution in the zeolite, fractal CoPc rivers in thin CoPcX zeolites are shown in Fig. 6.19a, b which are obtained by light-microscopic transmission photography.

To simulate real crystals and images of crystal disks, we have truncated the octahedron at the apexes of the two opposite pyramids (see Fig. 6.20a, b) or ground two opposite sites forming a thin slice in each case.

But these numerical pictures differ from the experimental ones; they do not show the river-like structure which we observed (Fig. 6.19) and discussed earlier [2]. To create pictures which correspond to the experimental photographs (c.f. Figure 6.21) to a greater degree, we have to change our iterated function system a little. We have to use affine mappings [7].

This means that we have to add a vector, which shifts $y(t+1)$ in a desired direction. This requires an additional chemical interpretation. One may imagine that the reactants are turned away from their straight path of diffusion by the tetrahedrally structured system of channels inside the zeolite crystals. Figure 6.22 shows the images

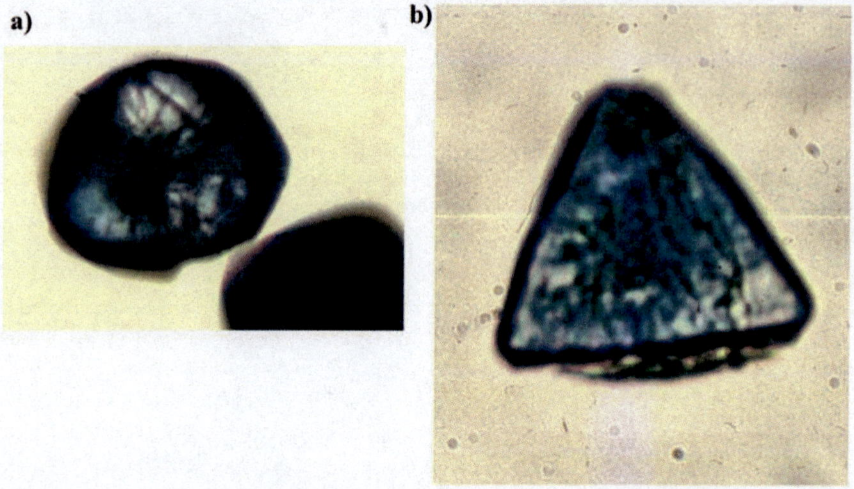

Fig. 6.19 Light-microscopic transmission photographs of CoPc-loaded zeolite-X-crystals; **a** fractal CoPc rivers within thin CoPcX crystals; **b** thin grown crystal with a fractal CoPc river-like structure. (*Photos* H. Diegruber/P. J. Plath)

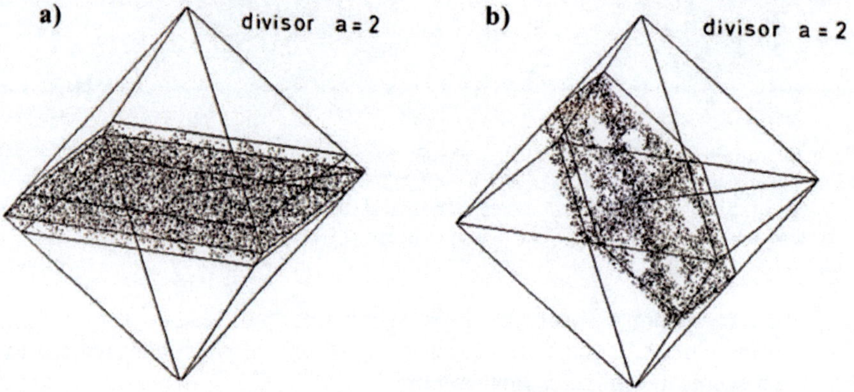

Fig. 6.20 Attractors of an iterated function system (IFS) [7] with the divisor $a = 2.0$ in different slices of an octahedron (**a**) and (**b**). The IFS works in the whole octahedron. So, these pictures may represent CoPc-loaded zeolite-X crystals which are ground after finishing the CoPc synthesis within the whole octahedral crystal

of such an iterated function system. These pictures resemble better the photographs of the CoPc-loaded zeolite crystals [2], showing the river-like structure of the product distribution of CoPc which is represented by the set of $y(T)$ points.

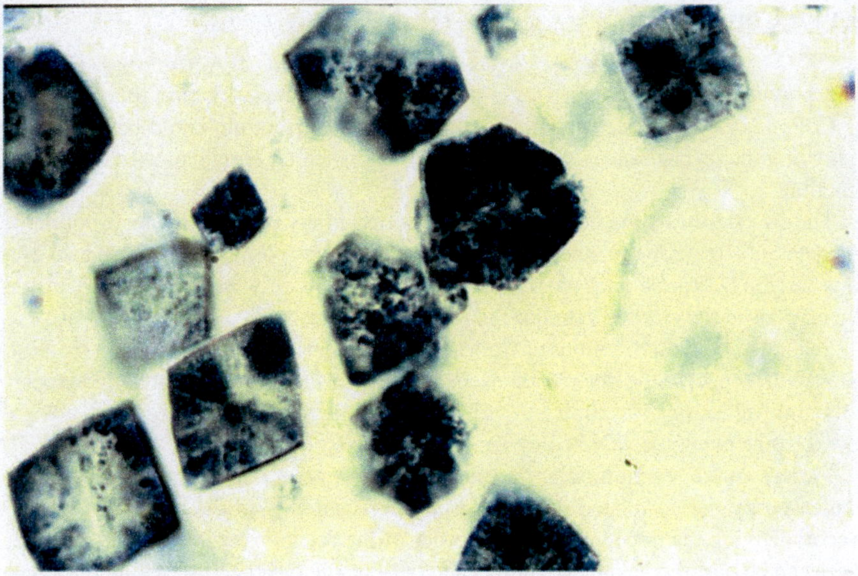

Fig. 6.21 Light microscopy transmission photograph of slices (thickness: 5 μm) of CoPc-loaded zeolite-X crystals with an average diameter of 22.7 μm (see Table 6.1) at 100 times objective magnification. (*Photos* H. Diegruber/P. J. Plath). Using this method, CoPc rivers are usually mapped as dots or small blue clouds as they are sliced by the cut formation

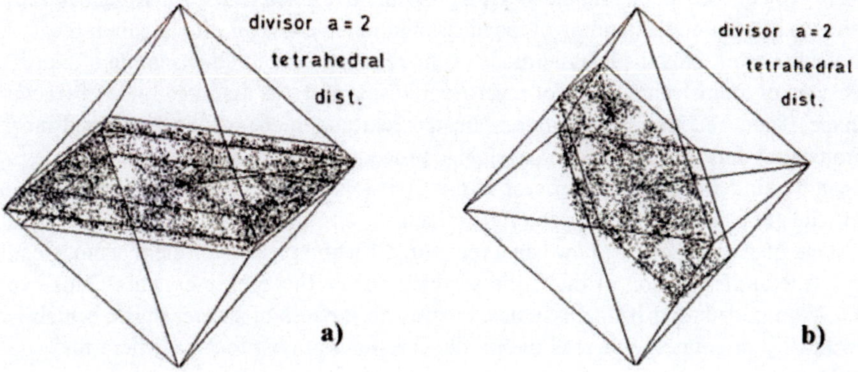

Fig. 6.22 Attractor of an iterated function system (divisor $a = 2$) with a tetrahedral shift. Again, the iteration takes place in the whole octahedron, where the slices (**a**) and (**b**) give an insight into the internal structure of the three-dimensional pattern of the attractor of the IFS

6.11 Conclusion

The previous results achieved by N_2-BET measurements [2], that the formation of CoPc inside zeolite crystals results in a fractal distribution, could be confirmed following the reverse ion exchange of the Co^{2+} ions which remain unused during the reaction.

The two synthesis methods which are reported firstly differ in the distribution of the reactants between the inner and the outer part of the zeolites: either both the Co^{2+} ions and the Dcb molecules are inside (new method), or only the Co^{2+} ions are inside the crystals at the starting point of the reaction (old method).

Secondly, both distributions of CoPc differ extremely in the order of their reaction rates with respect to the reverse ion exchange. However, both orders are of unexpectedly high values ($n_{old} \approx 10$; $n_{new} \approx 28$), whereas the reaction order for the Co^{2+} ion re-exchange using CoNaX-zeolites unloaded with CoPc is of about 2.

Moreover, the reaction order depends upon the size of the zeolites which can be expressed by the correlated dynamic dimension which is in the range of 2.7–2.8. Accompanying the dependence of the order of the reaction rate on the crystal size, there can be observed a change of the chemistry inside the crystals. Within the small crystals mainly Triazine is produced, whereas CoPc is the main product in the large crystals.

These results require a new interpretation of what a molecule inside a zeolite crystal means and of what a reaction order of the value of 10 to 30 means in this case. We are convinced that the slaving principle of Haken [25] and Wunderlin [26] leads to a better understanding of the ideas mentioned above. Following their ideas, a molecule represents the leading mode of the reacting structured ensemble of atoms. This set of atoms can even be of macroscopic size and it is assumed to be of fractal shape. Then, reaction means a complicated rearrangement of this structured unit. Translating this idea into the ion exchange procedure, this rearrangement becomes a strongly hindered diffusion process of the Co^{2+} ions and the Na^+ ions, respectively, and the great order of the reaction rate reflects the complexity of this rearrangement.

One question remains: how can a reaction diffusion process of the reactants lead to a fractal distribution of the CoPc products inside the zeolite crystals? This can really be understood if one assumes a diffusion process of the reactants, which is erratically orientated, whereas the products remain almost located where they are produced.

To evaluate this idea, we have simulated this process by an iterated function system in a three-dimensional space spanned by the six corners of an octahedron, which correspond to the octahedral symmetry of the zeolite-X crystals. There is a qualitative correspondence between the microscopic photographs of the CoPc-loaded zeolite crystals and the two-dimensional projections of the attractors of the iterated function systems in the octahedron.

References

1. Haß, E.C.: Chemische und physikalische Strukturen von Ni-dotierten Zeolithkatalysatorsystemen. Mikrospektralphotometrische Untersuchungen an benzol- und toluolbeladenen Ni2+-dotierten Faujasit-Einkristallen. Doctoral thesis, University of Bremen (1979)
2. Ignatzek, E., Plath, P.J., Hündorf, U.: The fractal character of zeolites; Part I: The fractal dimension of cobalt-(II)-phthalocyanine loaded faujasite. Z. Phys. Chemie, Leipzig **268**(5), 859–873 (1987)
3. Brunauer, S., Emmett, P.H., Teller, E.: Adsorption of gases in multimolecular layers. J. Am. Chem. Soc. **60**(2), 309–319 (1938)
4. Diegruber, H.: Reaktionen unter räumlicher Beschränkung—Charakterisierung von in Faujasit-Einkristallen immobilisierten Ni(II)- und Co(II)-Chelaten. Doctoral thesis, University of Bremen (1984)
5. Jaeger, N.I., Ryder, P., Schulz-Ekloff, G.: Concepts of reduction and disper-sion of metals in zeolites. In: Jacobs, P.A., Jaeger, N.I., Jirů, P., Kazansky, V.B., Schulz-Ekloff, G (eds.) Structure and Reactivity of Modified Zeolites, Studies in surface science and catalysis, vol 18. Elsevier Amsterdam, Oxford, New York, Tokyo, pp. 299–311 (1984)
6. Zheng, W., Wan, C.-Z., Zhang, J.-X., Li, C.-H., You, X.-Z.: Facile synthesis of phthalocyanine at low temperature with diisopropylamide anion as nucleophile. Tetrahedron Lett. **56**(30), 4459–4462 (2015)
7. Barnsley, M.F.: In: Barnsley, M.F., Demko, St.G. (eds.) Chaotic Dynamics and Fractals. Academic Press, Orlando, San Diego, New York, Austin, London, Montreal, Sydney, Tokyo, Toronto (1986)
8. Mandelbrot, B.B.: The Fractal Geometry of Nature. W.H. Freeman and Company, New York (1982)
9. Zumofen, G., Blumen, A., Klafter, J.: Reaction kinetics in disordered systems: hierarchical models in optimal structures in heterogeneous reaction systems. In: Plath, P. (ed.) Springer Series in Synergetics, vol. 44. Springer, Berlin, pp. 82–100 (1989); Blumen, A., Klafter, J., Zumofen, G.: Reactions in disordered media modelled by Fractals. In: Pietronero, L., E. Tosatti, E. (eds.) Fractals in Physics. North Holland Physics Publishing, Amsterdam, Oxford, New York, Tokyo, pp. 399–408 (1986)
10. Pfeifer, P., Avnir, D.: Chemistry in non-integer dimensions between two and three. I. Fractal theory of heterogeneous surfacesd. J. Chem. Phys. **79**, 3558 (1983); Avnir, D., Farin, D., Pfeifer, P.: Chemistry in noninteger dimensions between two and three. II. Fractal surfaces of adsorbents. J. Chem. Phys. **79**, 3566 (1983)
11. Farin, D., Avnir, D.: The reaction dimension in catalysis on dispersed metals. J. Am. Chem. Soc. **110**, 2039–2045 (1988)
12. Farin, D.: and Avnir, D."Structure Sensitivity Scaling Laws in Heterogeneous Ca-talysis on Dispersed Metals. 2. Fractal and Non-Fractal Interpretations of the Reac-tion Dimension". In: Phillips, M.J., Ternan, M. (eds.) Proceedings 9th International Congress on Catalysis. Characterization and Metal Catalysts, vol. 3, pp. 998–1005. Chemi-cal Institute of Canada, Ottawa, Ontario (1988)
13. Ignatzek, E.: Strukturuntersuchungen und Oxidationsreaktionen an trägergebundenen metallhaltigen Phthalocyaninen, Doctoral thesis, University of Bremen (1987)
14. Hündorf, U.: Synthese und Charakterisierung von trägerfreien und gebundenen Metallphthalocyaninen sowie deren Einsatz als Katalysatoren in der Ethanthiol-Oxydation, Doctoral thesis, University of Bremen (1987)
15. Iliev, V., Prahov, L., Andreev, A., Ignatzek, E., Schulz-Ekloff, G.: Proc. Vlth Symp. Heterog. Catalysis, Sofia Part 2, p. 79 (1987)
16. Briese-Gülban, S., Kompa, H., Schrübbers, H., Schulz-Ekloff, G.: Compara-tive characterization of monodisperse platinum, palladium and nickel in a faujasite X matrix by dynamic oxygen chemisorption and electron microscopy. React. Kinet. Catal. Lett. **20**, 7–12 (1982)

17. Plath, P.J.: Spatial and temporal fractals in heterogeneous catalysis. In: Velarde, M.G. (ed.) Proceedings of the Conference on synergetics, order and Cha-os. World Scientific, Singapore, New Jersey, London, Hong Kong, 331–348 (1988)
18. Plath, P.J., Jourdan, A., Diegruber, H.: Spectroscopic and magnetic properties of nickel-dimethylglyoxime within zeolites—possible one-dimensional metal-metal bonding. Z. Phys. Chemie Leipzig **271**(4), 665–681 (1990)
19. Snow, A.W., Griffith, J.R., Marullo, N.P.: Syntheses and characterization of heteroatom-bridged metal free phthalocyanine network polymers and model compounds. Macromolecules **17**, 1614–1624 (1984)
20. Ross, S.D., Fineman, M.: The Trimer of o-Phthalonitrile. J. Am. Chem. Soc. **72**, 3302–3304 (1950)
21. Herron, N.: A cobalt oxygen carrier in zeolite Y. A molecular 'ship in a bottle'. Inorg. Chem. **25, 26**, 4714–4717 (1986)
22. Born, J.: Katalytische Flüssigphasenoxidation von Ethylbenzol mit verschie-denen Kupfer-phthalocyaninen. Diploma thesis, University of Bremen (1983)
23. Hronec, M., Ilavsky, J.: Phthalocyanines in catalysis. Petrochemia **23**(2–3), 89–97 (1983)
24. van Streun, K.H., Piet, P., German, A.L.: Effects of 2,4-ionenes of different molar masses on the oxidative coupling of Thiol catalyzed by Cobaltphthalocya-nine. Eur. Polym. J. **23**(12), 941–946 (1987)
25. Haken, H.: Erfolgsgeheimnisse der Natur—Synergetik, die Lehre vom Zusammenwirken. Deutsche Verlagsanstalt, Stuttgart (1985)
26. Wunderlin, A., Haken, H.: Generalized Ginzburg-Landau equations, slaving principle and center manifold theorem. Z. Phys. B Condensed Matter **B44**, 135–141 (1981)
27. Plath, P.J.: Keyword: Chemie (chemistry). In: Sandkühler, H.J., Regenbogen, A. (eds.) Europäische Enzyklopädie zu Philosophie und Wissenschaften, Felix Meiner Verlag GmbH, Hamburg, Bd1 A – E, pp. 459–466 (1990)
28. Chhabra, A., Herrmann, H.J., Landau, D.P.: Fractal dimensionalities of backbones and clusters in a kinetic gelation model. In: Pietronero, L., Tosatti, E. (eds.) Fractals in Physics, pp. 129–131. North Holland Physics Publishing, Amsterdam, Ox-ford, New York, Tokyo (1986)
29. Mandelbrot, B.B.: The Fractal Geometry of Nature; Chapter: 33, Disc and Sphere Tremas, Moon Craters and Galaxies, pp. 301–309. W.H. Freeman and Company, New York (1983)
30. Mandelbrot, B.B.: The Fractal Geometry of Nature; Chapter: 32, Subordination; Spatial Lévy Dusts; Ordered Galaxies, pp. 288–300. W.H. Freeman and Company, New York (1983)
31. Iacovita, C., Rastei, M.V., Heinrich, B.W., Brumme, T., Kortus, J., Limot, L., Bucher, J.P.: Visualizing the Spin of Individual Cobalt-Phthalocyanine Molecules. Phys. Rev. Lett. **101**, 116602 (2008)
32. Wu, W., Harrison, N.M., Fisher A.J.: Electronic structure and exchange inter-actions in cobalt-phthalocyanine chains. Phys. Rev. B **88**, 024426 (2013)
33. Mandelbrot, B.B.: The Fractal Geometry of Nature; Chapter: 35, Gernal Tremas, and The Control Of Texture, pp. 319–325. W.H. Freeman and Company, New York (1983)

Chapter 7
Pattern of Sea-Shells Modelled by One-Dimensional Automata

Collision-Particles

Peter J. Plath, Ernst-Christoph Haß, and Jan K. Plath

Fig. 7.1 Sea shell *Cymbiola (Volutocorona) imperialis,* not uncommon occurrence in the Philippines (height 12 cm), The shell grows from top to bottom. The growth front is perpendicular to the direction of growth (*Photo* P. Plath), (compare: Documentaires alpha: les Coquillages) [1]

7.1 Introduction

The patterns on tropical mollusk shells of quite different spatial structure are extraordinarily rich in colors and shapes (see e.g. Fig. 7.1). There is a great fascination with the sight of them, which is not only related to the fact that these mollusk shells

P. J. Plath et al., *Imagery Synergetics*, Understanding Complex Systems,
https://doi.org/10.1007/978-3-030-95607-3_7

exert an exotic charm on us; rather they are—because these patterns, despite their seemingly arbitrary nature—subject to strict principles of their formation.

7.2 Historical Remarks

However, we cannot avoid briefly addressing the great cultural significance of the mussel shells, which goes back a long way into prehistoric times. This meaning is of course also based on the fascination with the pattern formation of these objects and their relative rarity.

Hoffman and Pike analyzed the artifacts of the Neanderthals discovered by the Portuguese paleontologist João Zilhão [2]: pierced and painted shells, which he discovered among other things in a cave in southern Spain in 2008 (see Fig. 7.2). They concluded that the processed shells were at least 115,000 years old [3, 4].

Shells made into jewelry chains were also suitable for ritual exchanges. When referred to as a kula or kula ring ritual exchange system with delayed reciprocity among the inhabitants of the Pacific Trobriand Islands, shell necklaces or shell bracelets are still used today [5]:

> These Melanesian islands are arranged almost in a circle, between them *soulava*, necklaces made of small red mussel plates, are exchanged clockwise. In the other direction, counterclockwise (in the mill direction), *mwali*, bracelets made from a white shell ring, are exchanged. The individual chains and maturities have a sacred character, each with its own orally transmitted story. All gifts must be exchanged after a while. [5]

In 2007, the sociologist R. Ziegler presented a simulation of this complex economic and ceremonial exchange system in his book "The Kula Ring of Bronislaw Malinowski" [6].

Fig. 7.2 Shell jewelry of the Neanderthals2; perforated and colored sea shells from the southern Spanish cave Cueva de los Aviones, which were used about 115,000 years ago

In a long-lasting social development process, the shell ring or the individual shell becomes an independent object that can be exchanged for other objects. The shell becomes money.

However, if instead the emphasis of exchange is placed on the sacred character, the conch shell can develop as an object for ritual use. In this way, it can be turned into a music or a signaling instrument.

As an object of science, the mussels and mollusks first appear as their fossils in Leibniz's geological treatises *Protogea* (Gottfried Wilhelm *Leibniz* (* 21. Juni/ 1. Juli 1646 in Leipzig; † 14. November 1716 in Hannover) [7]:

> Wenn wir nun der Überbleibsel des Meeres bey uns gedenken wollen, so müssen wir zuerst die *Seemuscheln* anführen, die hin und wieder in unseren Steinen sind. Es hat schon vorlängst der vortreffliche braunschweigische und hildesheimische Arzt Valerius Cordus von dem Argicola das meiste gehört, bey unseren Fossilien in den hannöverischen und hildesheimischen Steinbrüchen und Schachten und Kellern angemerkt, daß die *Meermuscheln* in unserer Gegend, auch bei Alfeld sehr häufig sind. ...
>
> Der Kammstein, steinern *Jacobsmuschel (ctenitits)* hat Spalten, und ist wirklich wie ein Kamm geformt. Insgemein ist er aschenfarbig, und wird in den Steinbrüchen, jenseits des S. Moritzberges gefunden. Agricol. Bey dem Geßner S. 165 ist ein hildesheimischer Kammstein in Form eines Walfischmauls."[7] (compare: Fig. 7.3b Scallop)

The *Protogea* was conceived between 1691 and 1693 as a contribution to the history of the *House of Hanover (German: Welfen)*, but it was not published during his lifetime. However, *Protogea* has made known by Eckhard [8] in 1719 and was translated into German by Scheid [7] in 1749.

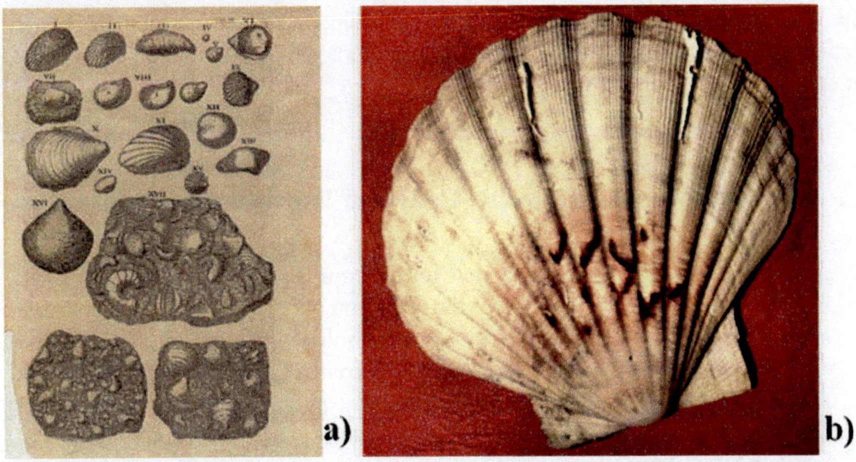

Fig. 7.3 **a** Illustration of the 1749 edition of *Protogaea* by Leibniz. File: Houghton GC6.L5316.749p—Leibniz, Svmmi polyhistorias, tab III.jpg [9]; **b** Scallop (German: Jakobsmuschel) (pecten maximus) Edible scallops are found in the Mediterranean and on the European Atlantic coast; (size: 14 cm; photo: P. Plath)

If you leaf through the wonderful, richly illustrated books about mussels and sea snails (mollusks) today, the names of Linnaeus, Lamarck and Gmelin appear very often as the scholars from the eighteenth century who first described these animals. How was that possible at that time?

Carl Nilsson Linnaeus (*May 23, 1707 in Råshult near Älmhult; †January 10, 1778 in Uppsala) was a Swedish scholar who worked with the binary nomenclature the basics of modern botanical and zoological taxonomy.

He owes his knowledge of the shape and appearance of seashells and mollusks to the possibility of viewing the collections of the Prussian princess and Swedish queen Luise Ulrike von Prussia (*July 24, 1720 in Berlin; † July 16, 1782 at Svartsjö Castle in Sweden) to study rare and exotic animals [10].

It was the French Revolution of 1789 that enabled the botanist and naturalist Jean-Baptiste de Lamarck (1744–1829) to become the founder of the "invertebrate". Wolfgang Lefèvre wrote a remarkable study on this [11].

By decision of the National Assembly of June 1793, a fundamental reorganization of the scientific institutions of France was carried out. The former "royal natural history cabinet" together with the former "royal Jardin des Plantes" and the former "royal manegerie" were combined to form the "Muséum national d'histoire naturelle".

Since Lamarck was among other things an outstanding collector and classifier of mussels, he received a professorship for the Linnaeus classes of insects and worms at this new national museum.

Lamarck had thus received a position that offered him the best conditions for his work. "For his descriptive, comparative and classifying studies of invertebrates, he would hardly have been able to find such excellent zoological collections anywhere else as at the Paris Muséum, although these holdings were considerably expanded during the Revolutionary Wars through confiscations of objects from natural history cabinets—especially in Holland [11]."

7.3 A Basic Model—Coupled Reaction Diffusion Equations

We were interested in developing models for the complex dynamics of the catalytic oxidation of carbon monoxide on supported catalysts containing palladium, when one of us got to know Hans Meinhardt's impressive results on the formation of pigment patterns on cochlea and mollusk shells at a conference in Leeds/England in 1989. What was astonishing for us was that the pigment patterns he showed were very similar to the reaction patterns of the catalyst in the oxidation of carbon monoxide.

Based on his biologically founded activator-inhibitor approach, H. Meinhardt had developed a continuous mathematical, spatially one-dimensional model—consisting of coupled non-linear differential equations—with which he could describe many of the known pigment patterns. His approach of the activator-inhibitor system of differential equations shows Eq. (7.1).

$$\begin{aligned}\frac{\partial a}{\partial t} &= s\left(\frac{a^2}{b} + b_a\right) - r_a a + D_a \frac{\partial^2 a}{\partial x^2} \\ \frac{\partial b}{\partial t} &= sa^2 - r_b + D_b \frac{\partial^2 b}{\partial x^2} + b_b\end{aligned} \tag{7.1}$$

This system describes a possible interaction between the concentrations of the autocatalytic activator $a(x)$ and its antagonist $b(x)$ at the point x in the one-dimensional space. sa^2/b is the production rate and $-r_a a$ the rate of removal, D_a and D_b are the respective diffusion coefficients of the activator and the inhibitor. b_a and b_b are the basic activator and inhibitor productions. With only minor variations of this basic system of equations, H. Meinhardt succeeds in simulating a considerable variety of pigment patterns in seashells.

Meinhardt's use of spatial one-dimensional reaction–diffusion equations goes back to the fundamental work of Alan M. Turing "The Chemical Basis of Morphogenesis" [12] in 1952:

> "It is suggested that a system of chemical substances, called morphogens, reacting together and diffusing through a tissue, is adequate to account for the main phenomena of morphogenesis. Such a system, although it may originally be quite homogeneous, may later develop a pattern or structure due to an instability of the homogeneous equilibrium, which is triggered off by random disturbances. Such reaction-diffusion systems are considered in some detail in the case of an isolated ring of cells, a mathematically convenient, though biologically unusual system. The investigation is chiefly concerned with the onset of instability. It is found that there are six essentially different forms which this may take." [12]

The great merit of Hans Meinhardt in simulating patterns of sea shells patterns using reaction–diffusion equations is undeniable [13]. Steven Wolfram's extensive work on "Universality and Complexity in Cellular Automata" [14] and the phantastic book "Cellular Automata machine" by Tommaso Toffoli and Norman Margolus [15] are also among the fundamental early works on cellular automata and opened up a true universe for modeling and simulations.

Figure 7.4 shows a section from the shell of *Olivia porphyria* (Linné) and its simulation by Hans Meinhardt. The shells of the mollusks only grow on their outer edge. We can therefore assume that the pigmentation takes place exclusively within this edge of growth. Therefore, the pattern formation should be describable with a "box model". This fact is also the reason why H. Meinhardt uses a spatial one-dimensional model. The mollusk shell grows by forming new edges—arranged parallel to each other, so to speak.

If a wave propagates in a one-dimensional system, the corresponding excitation moves along the one-dimensional space. If this space is mapped parallel to each other at the various time intervals, then the wave "runs" in the two-dimensional space–time structure created in this way on a straight line diagonally through this space, which can be clearly seen in Fig. 7.4.

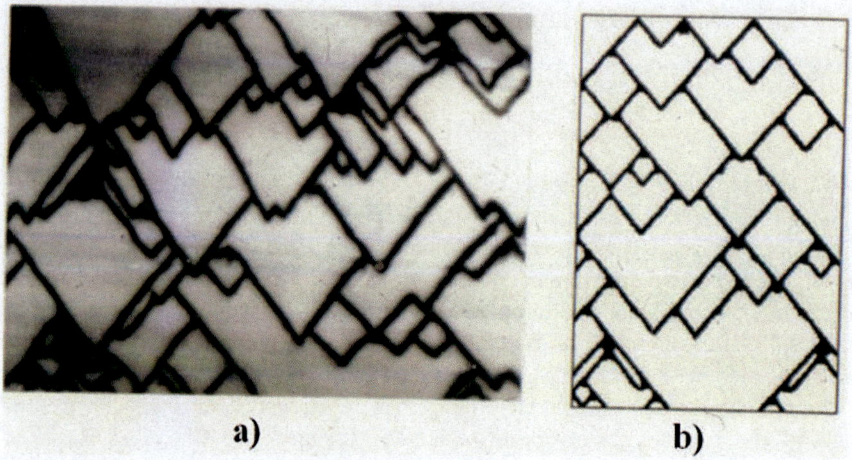

Fig. 7.4 **a** Section from the shell of *Olivia porphyria*, Linné; **b** simulation by Hans Meinhardt: In both cases the pattern grows from top to bottom. In other words, the arrow of time is directed downwards. The growth front, or location coordinate of the one-dimensional space, runs perpendicular to it from left to right. Travelling waves are observed, which are annihilated if they meet, and branching occurs in both cases (*Photo* Gift from Hans Meinhardt on the occasion of our joint discussion in Leeds)

7.4 Mapping a Texture onto Spatial Models of Sea Shells

A separate problem with the simulation of the pattern formation of and on sea shells is on the one hand the modeling of the three-dimensional structure of the shells and on the other hand the joining of these spatial structures with the pigment patterns. Ingo K. Hunsinger [16] and Frieder Nake [17] as well as Hans Meinhardt [13] dealt with this problem of modeling the three-dimensional geometry of the seashells and the mapping of texture maps onto the seashells almost simultaneously (see Fig. 7.5):

> Growing shells accumulate material forming whorls. New whorls <wrap up> atop older ones, producing spiral shapes. Often, an additional displacement can be observed along the direction of the spiral axis. This resulting trajectory is then called a *helico-spiral.* Modelling the shape of a snail incorporates the use of a *generating curve* that is swept along the helico-spiral, resulting in a snail-like surface. We employ parametric curves of the Bézier type [18]. In order to properly reveal the whorls and the aperture as well as to adhere to the gradual growth of shells, the generating curves must be rotated and scaled suitably. Illert applied the special coordinate system which, in differential geometry, is tied to the curve [19].

However, not all seashells have a smooth shape. The seashell *Cymbiola (Volutocorona) imperialis* (Fig. 7.1), for example, has horns that become larger and larger as the shell grows. It is really difficult to find simple mathematical functions for this, because they would also have to oscillate with increasing amplitude. As excellent as the simulations by Meinhardt and Hunsinger/Nake are, they are not originally derived from the growth of the animals. The geometric shape of the seashells, like the pigment pattern, would have to result from the different increase in cells in the

Fig. 7.5 **a** Mollusk shell *Olivia porhpyria* and **b** its simulation of the pigment pattern which is mapped onto the geometrically seashell shape (with friendly permission of Nake [17])

superimposed cell layers of the growth edge. In his book "A new kind of science" [20], which is very rich in new ideas, Stephen Wolfram uses discrete mathematical models to show possibilities of capturing even such complex growth structures, which are strongly reminiscent of rearing waves.

Certainly, one can assume a diffusion in a one-dimensional space and thereby disregard the cellular structure of the biological system. This is just an abstraction that finds its justification in success. However, this abstraction is not necessary.

A one-dimensional cellular automaton is certainly suitable to take into account the cellular structure of the biological system. We replace diffusion in continuous space by considering the neighborhood of cells. In other words, how a cell behaves in the next time $(t + 1)$ also depends on the state of its neighbors at time t. However, Plath, Plath and Schwietering [21] raised the basic question: "How can we construct a cellular automaton model, discrete in space and time, which is able to show the patterns being observed in the animals?".

To answer this question, they first asked: "What does diffusion mean in a cellular system?" Moreover, they asked whether random events are necessary in order to understand diffusion. Let us give a very first and rough answer to both questions. In the linear case, one does not need randomness to describe isotropic diffusion processes, because the linear space is isotropic by definition. So, diffusion becomes any kind of spatial averaging. For example, let us take Fick's second law (Eq. (7.2)):

$$\frac{\partial c}{\partial \tau} = D_c \frac{\partial^2 c}{\partial x^2} \tag{7.2}$$

As usual, one can rewrite the Laplacian term by the well-known spatial discretization (Eq. (7.3)):

$$\nabla^2 \boldsymbol{c} = \boldsymbol{c}(\boldsymbol{x}-1, \boldsymbol{\tau}) - 2\boldsymbol{c}(\boldsymbol{x}, \boldsymbol{\tau}) + \boldsymbol{c}(\boldsymbol{x}+1, \boldsymbol{\tau}) \tag{7.3}$$

So, Fick's law can be formulated by an iterative equation, where t is the discrete time and c the concentration. Let us call this discrete formulation of Fick's second law the *Laplacian diffusion* (Eq. (7.4)):

$$\begin{aligned} c(x, t+1) = c(x, t) + D_c[c(x-1, t) - 2c(x, t) + c(x+1, t)] \\ \text{with } \boldsymbol{x} \in \mathbb{Z}, \boldsymbol{t} \in \mathbb{N} \end{aligned} \tag{7.4}$$

Figure 7.6a shows an example of how this iterative equation works.

On the other hand, there is the well-known binomial smoothing procedure of a discrete function $Z(i, n)$ *with* $i \in \mathbb{N}$ (Eq. (7.5)):

$$\boldsymbol{Z}(\boldsymbol{i}, \boldsymbol{n}+1) = \frac{1}{4}[\boldsymbol{Z}(\boldsymbol{i}-1, \boldsymbol{n}) + 2\boldsymbol{Z}(\boldsymbol{i}, \boldsymbol{n}) + \boldsymbol{Z}(\boldsymbol{i}+1, \boldsymbol{n})], \boldsymbol{n} = 0, 1, 2, 3, \cdots \tag{7.5}$$

where n is the number of smoothing generation.

Interpreting this number n as discrete time t, this smoothing procedure formally represents a temporal averaging. One can easily rewrite this smoothing procedure in terms of the diffusion Eq. (7.6), replacing $Z(i, n)$ by $c(x, t)$:

$$\begin{aligned} c(x, t+1) &= \frac{1}{4}[c(x-1, t) + 2c(x, t) + c(x+1, t)] + \frac{2}{4}c(x, t) - \frac{2}{4}c(x, t) \\ &= c(x, t) + \frac{1}{4}[c(x-1, t) - 2c(x, t) + c(x+1, t)] \\ &= c(x, t) + D_c[c(x-1, t) - 2c(x, t) + c(x+1, t)] \end{aligned} \tag{7.6}$$

with the diffusion coefficient $D_c = \frac{1}{4}$. Let us call this special procedure the binomial diffusion (Fig. 7.6b). We can now state that, in the case of a linear cellular automaton, diffusion means any kind of a spatial averaging of the states of the cells over time.

In order to deal with natural numbers only, we define integral states $z(i, t)$ by means of the Gaussian-brackets notation $[u]$ going back to the temporal smoothing algorithm (Eq. (7.7)):

$$z(\boldsymbol{i}, \boldsymbol{t}+1) = \left[\hat{z}(\boldsymbol{i}, \boldsymbol{t})\right] \tag{7.7}$$

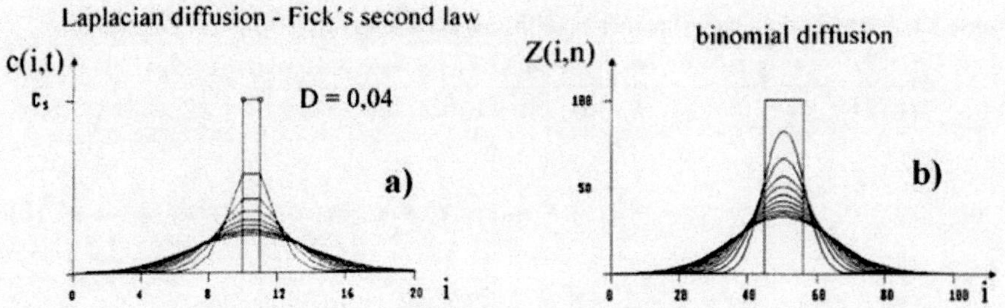

Fig. 7.6 **a** Graphs of the *Laplacian diffusion* of a sharp rectangular concentration profile at the beginning; **b** Graphs of the *binomial diffusion* of a sharp rectangular concentration profile at the beginning (Pictures were first published in Plath, Plath, Schwietering [21])

with

$$\hat{z}(i,t) = \frac{1}{4}(z(i-1,t) + 2z(i,t) + z(i+1,t)) \tag{7.8}$$

where $[u]$ is the largest natural number that is less or equal to u:

$$[u] \leq u \text{ and } [u] + 1 > u \tag{7.9}$$

This expression also represents a kind of diffusion, which we call *Gaussian diffusion* due to the Gaussian brackets notation (see Fig. 7.7b). In contrast to the binomial diffusion (Fig. 7.7a), however, it does not spread out the quantities all over the space. If we set a very sharp initial concentration profile, it does not run but stays in a restricted domain, although it shrinks over time.

Taking into account Eq. (7.7), one can easily express this iterative equation in the form of a transformation rule of a cellular automaton with the abbreviation Σ according to Eq. (7.10),

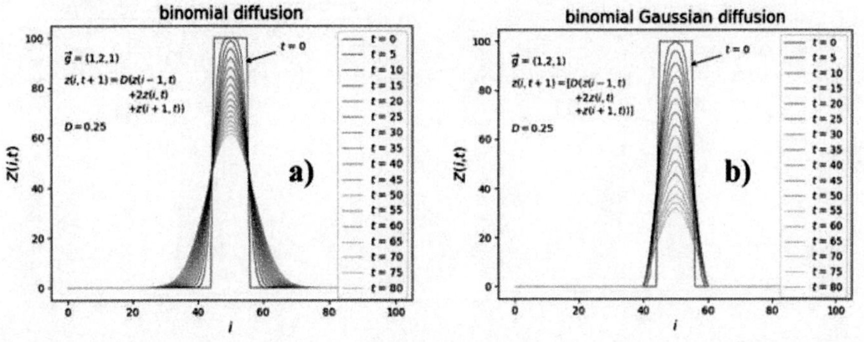

Fig. 7.7 Comparison between **a** binomial diffusion according Eq. (7.6) and **b** binomial Gaussian diffusion according Eq. (7.7)

Table 7.1 State rule for Gaussian binomial diffusion according Eq. (7.10)

Σ	0	1	2	3	4	5	6	7	8	9	10	11	12	13	...
$z(i, t+1)$	0	0	0	0	1	1	1	1	2	2	2	2	3	3	...

$$\boldsymbol{\Sigma \equiv g_{-1}z(i-1, t) + g_0 z(i, t) + g_{+1}z(i+1, t)} \tag{7.10}$$

with the weights $g_{-1} = g_{+1} = 1$ and $g_0 = 2$ as shown in Table 7.1:

Introducing an additive term b in the Eq. (7.7), $z(i, t+1) = \left[\hat{z}(i, t)\right]$, leads to the expression (7.11):

$$\boldsymbol{z_b(i, t+1) = \left[\hat{z}(i, t) + b\right]} \tag{7.11}$$

This iterative equation describes a spatially restricted diffusion as well, but the parameter b can now be interpreted chemically. For $b > 0$, one may think of a production and for $b < 0$ of a destruction, which is added to the diffusion. For example, if $b = 0.25$, diffusion will be stopped after some time by the production (Fig. 7.8b).

One obtains a stationary spatial distribution of states, $z(i, t+1) = \left[\hat{z}(i, t) + b\right]$, for all cells i. Let us call G the sum of weights of states of the neighboring cells (Eq. (7.12)):

$$\boldsymbol{G = \sum\nolimits_{k=-1}^{+1} g_k} \tag{7.12}$$

Increasing b, the stationary state is reached if b becomes larger or equal to $1/G$ (Eq. (7.13)):

$$\boldsymbol{\frac{1}{G} \leq b \leq 1} \tag{7.13}$$

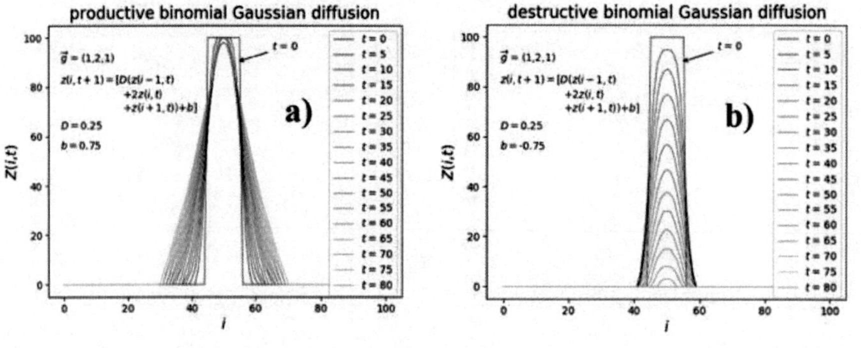

Fig. 7.8 Comparison between **a** productive binomial Gaussian diffusion ($\boldsymbol{b} > 0$) and **b** destructive binomial Gaussian diffusion ($\boldsymbol{b} < 0$)

Table 7.2 State rule for Gaussian binomial diffusion with $b = 1$

Σ	0	1	2	3	4	5	6	7	8	9	10	11	12	...
$z(i, t+1)$	0	1	1	1	2	2	2	2	3	3	3	3	4	...

Fig. 7.9 Temporal development of front-waves from two starting points concerning the automaton rule of Table 7.2 using cyclic boundary conditions. The front-waves realize the transition from $z(j, t) = 0$) (dark blue) to $z(j, t) = 1$ (turquoise)

Setting $b = 1$, it generally means a shift of the $z(i, t+1)$-row of Table 7.1 to the left by four digits (see Table 7.2).

The production now exceeds the diffusion and the states would grow anywhere, even if $z(i, t) = 0$.

Table 7.2 already represents a rule for a cellular automaton. If one sets the value $z(i, 0) = 1$ for the i-th cell at time $t = 0$, then it becomes the starting point of two front waves running in opposite directions (see Fig. 7.9).

The automaton of Table 7.2 realizes in the simplest way a one-dimensional, active, bistable medium in which an initial distribution spreads like A. S. Mikhailov describes in the introduction of his work Foundation of Synergies [22].

Introducing a lower limit L for opening the state growth of a cell (Eq. (7.14)),

$$z(\boldsymbol{i}, \boldsymbol{t}+1) = \begin{Bmatrix} 0; \text{if } z(\boldsymbol{i}, \boldsymbol{t}) = 0 \text{ and } \Sigma \leq L \\ \left[\hat{z}(\boldsymbol{i}, \boldsymbol{t}) + \boldsymbol{b}\right] \text{otherweise} \end{Bmatrix} \tag{7.14}$$

the states of those cells which have an appropriate neighborhood start to grow (see Fig. 7.10b). As a result, Gaussian diffusion is not spatially restricted almost anywhere.

If b equals minus one, $b = -1$, the future states become

$$z_{-1}(\boldsymbol{i}, \boldsymbol{t}+1) = \left[\hat{z}(\boldsymbol{i}, \boldsymbol{t}) - 1\right] \tag{7.15}$$

and the lower row of the transformation rule (Table 7.1) is shifted to the right.

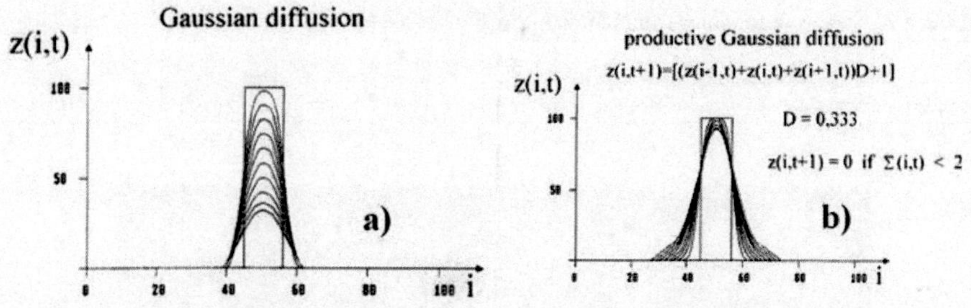

Fig. 7.10 **a** Graph of the spatially restricted Gaussian diffusion; **b** development of the productive Gaussian diffusion with the limitation of the state growth if $z(i, t) = 0$ and $\Sigma \leq L$

This value of b causes a destruction, which is added to the normal Gaussian diffusion.

7.5 The Vector Automaton Model

Both look-up tables (Tables 7.2 and 7.3) realize classical cellular automata. One can combine these automata with each other, constructing a new type of automata: cellular vector automata [23–26]. The advantage of such cellular vector automata is that one can select different rules within one expression by selecting different values of b.

The mathematical model for our simulation is a one-dimensional vector automaton [23]. This means, we take.

- **Topology**: a path graph P_l or a circular graph C_l with l vertices.
- **Cell**: each vertex is labelled by a vector in a two-dimensional concentration/ phase space, called *a cell*; the two components of the vector of the i-th cell at time $t \in \mathbb{N}$ are: the concentration $c(i, t) \in \mathbb{N}$ and the phase $p(i, t) \in \{0, 1\}$.
- **Neighborhood**: With respect to the temporal development of the cell a local neighborhood is defined consisting of the actual cell i and its adjacent vertices or cells, respectively.
- **Transformation rule**: Various transformation rules can be formulated which will transform both components of the vector from t to $t + 1$: $c(i, t) \rightarrow c(i, t + 1)$ and $p(i, t) \rightarrow p(i, t + 1)$, depending upon the neighboring cells.

Table 7.3 State rule for Gaussian binomial diffusion with $b = -1$

Σ	0	1	2	3	4	5	6	7	8	9	10	11	12	...
$Z(i, t + 1)$	0	0	0	0	0	0	0	0	1	1	1	1	2	...

These formal aspects define a one-dimensional cellular vector automaton. Let us now translate these ideas into a simplified model of the biological system under consideration.

If we now localize the state of the automaton at time $(t + 1)$ directly next to its state at time t, as is generally the case, then a two-dimensional spatiotemporal pattern results, which reproduces the movement in a one-dimensional space. The patterns created in this way are very similar to the pigment patterns of sea shells. However, the pigment patterns on the mussel shells are not about the movement in the one-dimensional space of the lip or the growth edge, but about each new growing edge, which consists of different, new biological cells, which, however, have emerged from their "mother cells". This means that the patterns on the snail shells are inheritance patterns.

Let us assume that the incorporation of pigments into a new cell, which occurs through cell division within an outer edge that is just being formed, is determined by:

- the concentration of pre-pigments, i.e., those substances from which the color pigments are synthesized in the mother cell (located within the "old" edge) and
- how high the concentration of these pre-pigments is in those cells that are adjacent to the mother cell.

Furthermore, it is certainly reasonable to assume that the cells can take over at least two different states of activity (phases) with regard to their ability to form pre-pigments: an active phase with $p = 0$ and a passive phase $p = 1$. This is their characteristic feature regarding the formation of pre-pigments during cell division. However, the transfer of the activity phase of the mother cell to the daughter cell should be also dependent on the activity phases of the cells neighboring the mother cell.

The behavior of the biological cells in these two phases should be characterized by a production ($p = 0, b > 0$) in the active state, and a destruction ($p = 1; b \leq 0$) in the passive state. So, the phases may be used to construct a switch between the different rules. But it should be an internal switch, the position of which should again depend on the sum Σ of states of the neighboring cells.

For example, if $p(i, t) = 0$, then one obtains for instance Table 7.4:

And if $p(i, t) = 1$, results Table 7.5.

Table 7.4 Phase rule for Gaussian binomial diffusion with $b = 1$ and $p(i, t) = 0$

Σ	0	1	2	3	4	5	6	7	8	9	10	11	12	...
$p(i, t + 1)$	0	0	0	0	0	0	0	0	1	1	1	1	1	...

Table 7.5 Phase rule for Gaussian binomial diffusion with $b = 0$ and $p(i, t) = 1$

Σ	0	1	2	3	4	5	6	7	8	9	10	11	12	...
$p(i, t + 1)$	0	0	0	0	1	1	1	1	1	1	1	1	1	...

Now, the state vector $\vec{z}\,(i,t)$ of a cell at time is represented by two different quantities: the integral concentration $z(i,t)$ and the phases $p(i,t)$ (Eq. (7.16)):

$$\vec{z}\,(i,t) = \begin{pmatrix} z(i,t) \\ p(i,t) \end{pmatrix} \tag{7.16}$$

The example shown in Tables 7.4 and 7.5 is well suited to show how the internal switch we require works: The production rule (Table 7.4) works up to $\sum \leq 7$, whereas the destruction rule (Table 7.5) governs the region for $\sum \geq 8$. However, if the cell is in the phase $p(i,t) = 1$ and the concentration sum Σ of the neighbours exceeds 3, $\sum \geq 4$, the destruction rule is responsible for the development of the vector state. Only if $\sum \leq 3$, the production rule starts to work again. This means that the cellular vector automaton displays a *hysteresis loop*.

Moreover, since the graph of our map $F : \vec{z}(i,t) \mapsto \vec{z}(i,t+1)$ is composed of two tent-like staircases, which are shifted against each other, we have a many-to-one map, which is essentially dissipative. Since the general transformation rule of the vector state $\vec{z}(i,t)$ realizes a dissipative map, the cellular vector automaton becomes a powerful tool for modelling the patterns of a natural system such as mussel and mollusk shells.

Of course, we are not limited to the binomial averaging, but can also use the more common arithmetic diffusion (see Fig. 7.11). In this way, we increase the influence of the neighboring cells $\vec{z}(i-1,t)$ and $\vec{z}(i+1,t)$ on the development of the i-th cell. If we use arithmetic averaging instead of binomial averaging, then Table 7.6 is realizing an excitable automaton system.

Starting with all cells being in the vector state $z(i,0)$ at $t = 0$ (Eq. (7.17)),

$$\vec{z}\,(i,0) = \begin{pmatrix} z(i,0) \\ p(i,0) \end{pmatrix} = \begin{pmatrix} 0 \\ 0 \end{pmatrix} \tag{7.17}$$

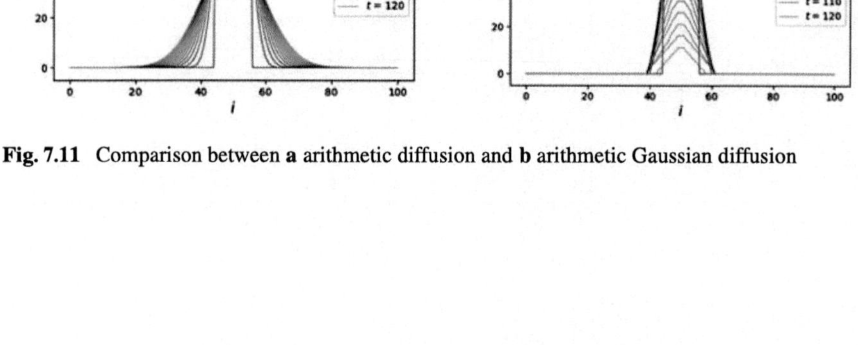

Fig. 7.11 Comparison between **a** arithmetic diffusion and **b** arithmetic Gaussian diffusion

Table 7.6 Excitable automaton system with arithmetic diffusion (1/6–8/4 rule)

Σ		0	1	2	3	4	5	6	7	8	9	10	11	12	...
$p(i,t)=0$	$z(i,t+1)$	0	1	1	2	2	2	3	3	3	4	4	4	5	...
$p(i,t)=1$	$z(i,t+1)$	0	0	0	0	0	0	1	1	1	2	2	2	3	...
$p(i,t)=0$	$p(i,t+1)$	0	0	0	0	0	0	0	0	**1**	**1**	**1**	**1**	**1**	...
$p(i,t)=1$	$p(i,t+1)$	0	0	0	0	**1**	**1**	**1**	**1**	**1**	**1**	**1**	**1**	**1**	...

and excluding the cells $j, k, l, \ldots$, which are in the vector state $z(i, 1)$ at $t = 0$ (Eq. (7.18)), for example,

$$\vec{z}(j,0) = \begin{pmatrix} z(j,0) \\ p(j,0) \end{pmatrix} = \begin{pmatrix} 1 \\ 0 \end{pmatrix} \tag{7.18}$$

one obtains *"chemical waves"* or *"reaction–diffusion waves"*, respectively, in this one-dimensional excitable cellular system, which annihilate if they meet each other and are extinguished at the edge of the automaton (see Fig. 7.12a, b). V. Krinsky called this kind of waves *"auto-waves"* [27] which are well known in Belousov-Zhabotinsky reaction system.

While, as the example of diffusion shows, it is quite clear what function the state-component $z(i, t)$ of the state vector $\vec{z}(i, t)$ has, this is not so easy to see for its phase-component $p(i, t)$. In order to be able to understand better the function of the phase component, the area of the active phase is enlarged: $p(i, t)) = 0$ for$\Sigma \leq 11$, i.e., we shift the transition from the active to the passive phase $p(i, t) = 0 \rightarrow p(i, t + 1) = 1$

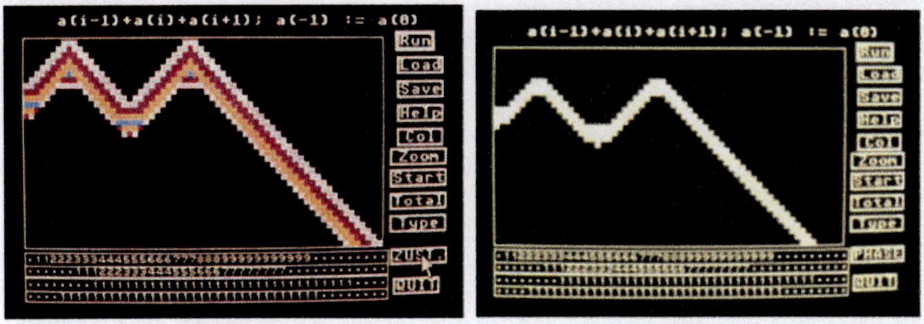

a) b)

Fig. 7.12 Space–time representation of reaction diffusion waves in a one-dimensional vector automaton concerning the rule of Table 7.6. The automaton represents an excitable system, which is excited by two cells in the state shown in Eq. (7.18). Time runs from top to bottom, while the cells are arranged horizontally forming a one-dimensional space. **a** The state-component $z(i, t)$ of the vector-state $\vec{z}(i, t)$ is shown; the waves annihilate each other. On the left side one can detect the annihilation of the wave on the edge of the automaton. **b** The photo shows the phase-component $p(i, t)$ of the vector state $\vec{z}(i, t)$. The white waves represent the phase-states with $p(i, t) = 1$ (Photos are taken from Atari computer by P. Plath, Sept. 1990) by four digits from $\Sigma = \Sigma_a = 8$ to $\Sigma = \Sigma_a = 12$ (compare Tables 7.6 and 7.7). Furthermore, the back-transformation from the passive to the active phase will now take place at $\Sigma = \Sigma_d = 3$ (Table 7.7) instead of $\Sigma = \Sigma_d = 4$ as shown in Table 7.6. An example for an automaton with this phase shift is depicted in Fig. 7.13.

Let us call the parameter $\Sigma = \Sigma_a$ the *activation parameter* for which holds $p(i,t) = 0 \rightarrow p(i,t+1) = 0$ for all $\Sigma < \Sigma_a$ and $p(i,t) = 0 \rightarrow p(i,t+1) = 1$ for $\Sigma \geq \Sigma_a$. In a similar way, we denote $\Sigma = \Sigma_d$ the *deactivation parameter,* which meets the conditions: $p(i,t) = 1 \rightarrow p(i,t+1) = 0$ for all $\Sigma < \Sigma_d$ and $p(i,t) = 1 \rightarrow p(i,t+1) = 1$ for $\Sigma \geq \Sigma_d$.

Doing so, there is no structural change detectable between the excitability of both the automata of Tables 7.6 and 7.7. Single solitary reaction–diffusion waves occur, which annihilate each other when they meet. This means, the parameters Σ_a and Σ_d span a space in which the system can be excited in a way that solitary waves are created.

Table 7.7 Excitable automaton system with arithmetic diffusion (1/6–12/3 rule)

Σ		0	1	2	3	4	5	6	7	8	9	10	11	12	…
$p(i,t) = 0$	$z(i,t+1)$	0	1	1	2	2	2	3	3	3	4	4	4	5	…
$p(i,t) = 1$	$z(i,t+1)$	0	0	0	0	0	0	1	1	1	2	2	2	3	…
$p(i,t) = 0$	$p(i,t+1)$	0	0	0	0	0	0	0	0	0	0	0	0	1	…
$p(i,t) = 1$	$p(i,t+1)$	0	0	0	1	1	1	1	1	1	1	1	1	1	…

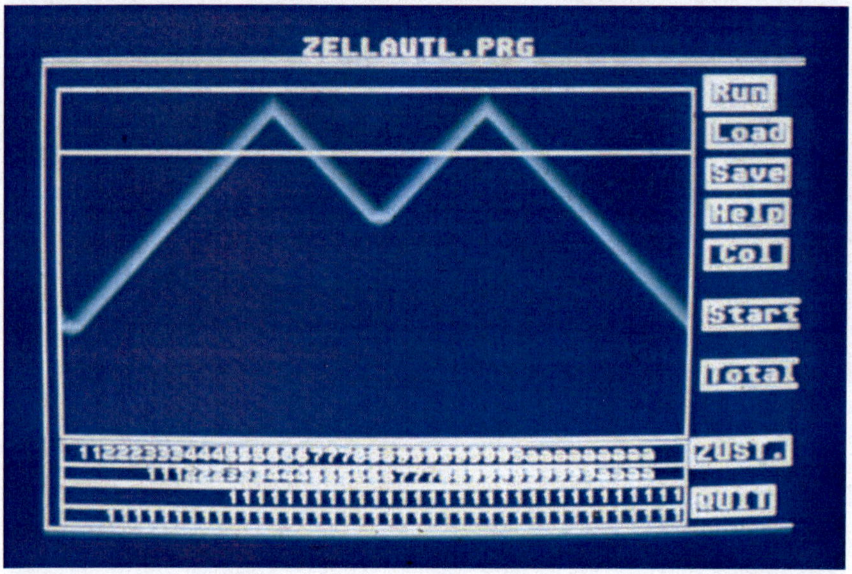

Fig. 7.13 Wave of the state component $z(i,t)$ in an excitable automaton system concerning Table 7.7, which is cyclically closed (Photo P. Plath)

However, if we change the parameters in a way, as it is shown in Table 7.8 choosing $\Sigma_a = 12$ but $\Sigma_d = 5$ and Table 7.9 with $\Sigma_a = 12$ and $\Sigma_d = 9$, the situation changes drastically. The automaton becomes *self-exciting*. In two-dimensional experiments like Belousov-Zhabotinsky reactions in a Petri-dish, these patterns are called *target pattern* (see Fig. 7.14).

The comparison of Figs. 7.13 and 7.14 shows that Σ_a and Σ_d become control parameters of our system. They control the change from the excitable system to the self-exciting system and control also the frequency of the sequence of waves as one can see comparing Fig. 7.14a, b. However, there is one major difference between the patterns in Fig. 7.14a, b. While in Fig. 7.14a all waves that meet cancel each other, as is also the case with chemical waves in the BZ reaction, slowdown of the wave velocities can be observed (Fig. 7.14b) by the waves that develop over time before

Table 7.8 Self-exciting automaton system (1/6–12/5 rule)

Σ		0	1	2	3	4	5	6	7	8	9	10	11	12	…
$p(i,t)=0$	$p(i,t)=0$	0	**1**	1	2	2	2	3	3	3	4	4	4	5	…
$p(i,t)=1$	$p(i,t)=1$	0	0	0	0	0	0	**1**	1	1	2	2	2	3	…
$p(i,t)=0$	$p(i,t)=0$	0	0	0	0	0	0	0	0	0	0	0	0	**1**	…
$p(i,t)=1$	$p(i,t)=1$	0	0	0	0	0	**1**	**1**	**1**	**1**	**1**	**1**	**1**	**1**	…

Table 7.9 Self-exciting automaton system (1/6–12/9 rule)

Σ		**0**	**1**	**2**	**3**	**4**	**5**	**6**	**7**	**8**	**9**	**10**	**11**	**12**	**…**
$p(i,t)=0$	$p(i,t)=0$	0	**1**	1	2	2	2	3	3	3	4	4	4	5	…
$p(i,t)=1$	$p(i,t)=1$	0	0	0	0	0	0	**1**	1	1	2	2	2	3	…
$p(i,t)=0$	$p(i,t)=0$	0	0	0	0	0	0	0	0	0	0	0	0	**1**	…
$p(i,t)=1$	$p(i,t)=1$	0	0	0	0	0	0	0	0	0	**1**	**1**	**1**	**1**	…

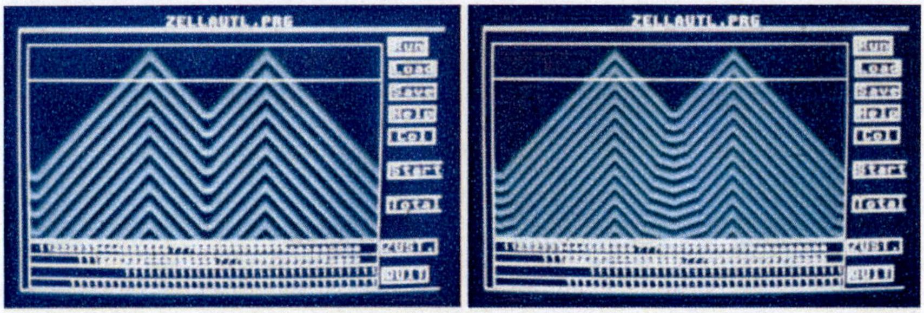

a) **b)**

Fig. 7.14 Self-exciting automaton systems with one-dimensional target patterns of travelling waves with time running from top to bottom and cyclic closure, **a** concerning Table 7.8 with $\mathbf{\Sigma}_a = 12$ and $\mathbf{\Sigma}_d = 5$ (1/6–12/5 rule); b concerning Table 7.9; $\mathbf{\Sigma}_a = 12$ and $\mathbf{\Sigma}_d = 9$ (1/6–12/9 rule) (Photos. P. Plath)

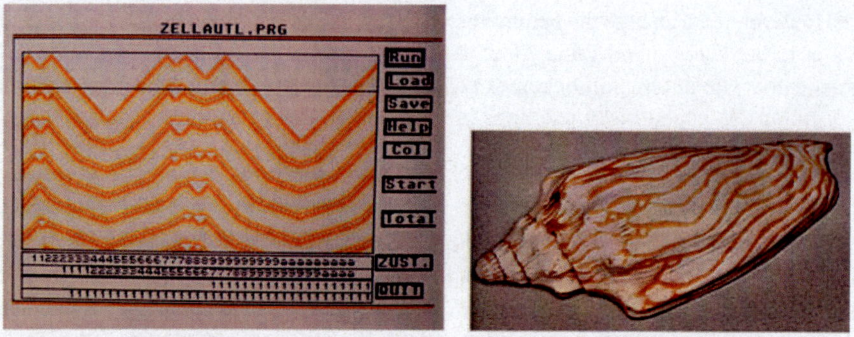

a) b)

Fig. 7.15 **a** Wavy patterns with "triangles" occur in the automaton "1/6–25/7" with $\Sigma_a = 25$ and $\Sigma_d = 7$ and circular boundary conditions (*Photo* P. Plath); **b** *Notovoluta perplicata*, Headley 1902 (Photo by P. Plath with friendly permission by Charles E. Tuttle, Company) [28]

they meet and extinguish each other. Obviously, the space of the control parameters is much more complicated than that it only consists of the areas of excitability and self-excitability.

If you enlarge the excitation area even further, e.g., to $\Sigma_a = 25$, and leave the deactivation almost unchanged, the behavior of the automaton becomes even more noticeable. More or less wavy wave trains arise perpendicular to the direction of propagation (Fig. 7.15a). But interestingly, "triangles" are formed within these waves if the number of original excitations is increased only slightly. Amazingly, one finds quite similar pattern formations among the specimens of marine snails, as the example of the sea shell *Notovoluta perplicata*, Headley 1902, shows (Fig. 7.15b).

It seems that we are on the right track to describe the patterns on the sea shells with the help of our vector automata. But we still need an important addition to our toolbox: the color of the patterns, because depending on the value of Σ, the color of the samples could change.

One has to define a further discrete function, which maps the vector components $z(i, t+1)$ onto a color: $z(i, t+1) \mapsto c(i, t+1)$ (see Fig. 7.16). In this way, a large variety of colored patterns can be produced, some of which may resemble the observed seashell patterns.

Figure 7.16 shows that the coloring of the states not only represents an aesthetic moment, but also provides further information. For example, the sudden breaks of the waves in Fig. 7.16a remain completely incomprehensible, while in Fig. 7.16b it becomes clear that they are due to the collision with remaining "stripes" parallel to the direction of growth.

Furthermore, if one regards carefully the famous fractal like patterns of *Oliva porphyria* L. (L.: Linné)) or *Cymbiola vespertilio (Aulicina)* (Fig. 7.17), one might get the impression that new Sierpinsky triangles occur during growing of the seashell even though no global disturbance of the growing front can be recognized. We consider these events to be rare, accidental occurrences, which, however, are no

Fig. 7.16 Purely deterministic pattern created by three exposed cells at the beginning with $(i_1, 0) = 1$, $z(i_2, 0) = 2$, and$z(i_3, 0) = 1$. **a** $c(i, t) \neq 0$, if $z(i, t) = 4$, in this case the colour black has been chosen; **b** $c(i, t) \neq 0$, if $z(i, t) = 4(black)$ or $z(i, t) = 3(grey)$

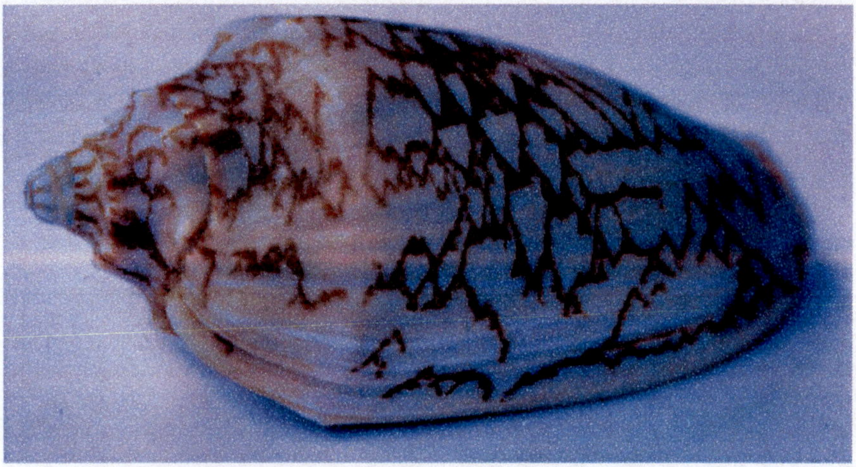

Fig. 7.17 *Cymbiola vespertilio (Aulicina)* (height 9,5 cm); (*Photo* P. Plath)

less necessary, such as mutations in living cells or organisms. For this reason, such improbable events might occur in each generation or time step of the automaton, respectively. So, "we ask for a random number in the range of 0 to 99. If this number is less than a chosen threshold n, one of the cells of the t's generation will be selected randomly [23]." The value of n is in the order of $n \sim 5$. This means that the corresponding probability is in the range of one-twentieth and thus, on average, the state $z(i, t+1)$ of a cell of the automaton is likely to be changed every twentieth time step. Furthermore, the deterministic value $z(i, t+1)$ of the chosen ith cell will randomly increased by one with the probability of one half.

In this context, it is interesting to look again at Fig. 7.4 in more detail. In the photograph (Fig. 7.4a) of *Olivia porphyria*, one can discover such rare events from which waves spontaneously emanate. But this is not the case in Hans Meinhardt's modeling (Fig. 7.4b). However, in both cases, the real seashell and its modelling based on reaction–diffusion differential equations, new waves are suddenly born from travelling waves. Sometimes one only discovers the incomplete approach of such a detachment of a new wave from an already running wave. The places where these new waves seem to arise spontaneously are, as far as this can be recognized, directly "below" previous events, such as the creation or annihilation of wave pairs. It is therefore reasonable to assume that "relics" of these previous events in the form of "invisible" stripes parallel to the direction of growth lead to this "spontaneous" emergence of new wave trains (compare Figs. 7.16 and 7.18).

Another detail in Fig. 7.4 is worth mentioning: not all waves cancel each other out when they meet. Some of them cross each other and at least in the modeling, dark "spots" can be seen both in the formation of the wave pairs and in the "crossing" of the wave trains, which certainly have nothing to do with a possible blurring of the images. When we model these patterns with vector automata, we observe this behavior in detail (Fig. 7.19).

The vector automaton in Fig. 7.19 shows solitary waves, which annihilate each other like autowaves (Krinsky [27]), but also those which cross each other with a delay of one cell like solitons. This raises the question of whether there are other types of solitary waves between the autowaves and the solitons, to which we will come back later. But for now, we still want to continue with the simple patterns which arise from the above discussed automata of the arithmetic Gaussian structure type governed by the 1/6–8/4 rule (compare Table 7.6 and see Fig. 7.20).

Fig. 7.18 Probabilistic automaton initialized, governed and colored by the rules of the automaton shown in Fig. 7.16, but with $n = 6$

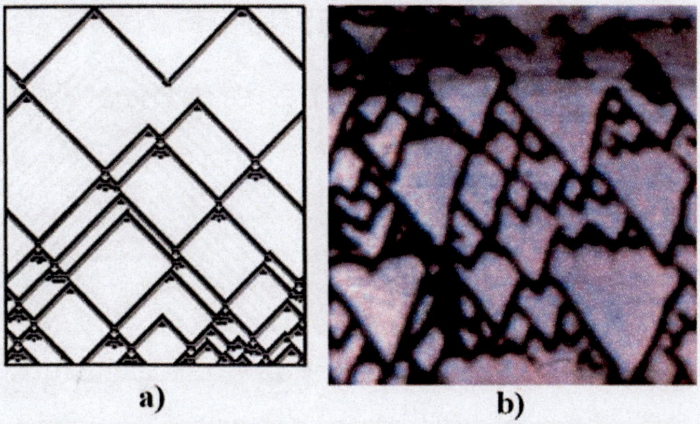

Fig. 7.19 **a** Probabilistic vector automaton created by two exposed cells at the beginning with: $z(i_1, 0) = 1$, and $z(i_2, 0) = 2$ and with $n = 6$. The circular boundary conditions are $z(l + 1, t) = z(1, t)$ and$z(0, t) = z(l, t)$. New, improbable events occur; waves annihilate or passes each other with a phase shift like solitons; **b** *Oliva porphyria* LINNAEUS (1758) with waves that annihilate or cross each other; enlarged section from the photo (with friendly permission; Grange Batelière–Paris (1969) p. 59, Fig. 102) [1]

Starting with only one cell j with $\vec{z}(j, 0) = \begin{pmatrix} z(j,0) \\ p(j,0) \end{pmatrix} = \begin{pmatrix} 1 \\ 0 \end{pmatrix}$ (Eq. (7.18)), while all the other cells $i \neq j$ are in the vector state $\vec{z}(i, 0) = \begin{pmatrix} z(i,0) \\ p(i,0) \end{pmatrix} = \begin{pmatrix} 0 \\ 0 \end{pmatrix}$ (Eq. (7.17)), the 1/6–8/5 rule creates a very interesting pattern: the result is a self-exciting automaton (compare 1/6–12/5 rule, see Table 7.8 and Fig. 7.14). The wave pairs behave as a typical auto-wave with respect to their mutual annihilation when they touch each other. However, with every further excitation, which occurs strictly periodically, the center changes into a fractal in one-dimensional spatial domain. This creates a spatiotemporal Sierpinsky-like fractal in the core of the propagating wave pattern (Fig. 7.20).

So far, we had not changed the basic structure of the automata. All vector automata which we considered so far were of the structure of the 1/6–*a*/*b* rule, which describes a diffusion based on the arithmetic mean. In practice, only the values $a = \Sigma_a$ and $b = \Sigma_b$ have been changed, which regulate the transitions from the active to the passive phase and vice versa. Let us see what a change in this diffusion does by simply playing a little with the rules of the automata.

The correspondence of the Sierpinsky-like pattern of the automaton (Fig. 7.21a) and the shell of the snail Cymbiolacca *wisemani B.* (1870) (Fig. 7.21b) is very surprising indeed. In both cases, we have white triangles of different sizes with dark dots in their baselines.

Figure 7.21a shows a very simple automaton based on Table 7.10, all cells of which are in their active state $p(i, t) = 0$ which cannot change. Only in the case that three neighbored cells are in the same state $z(i - 1, t) = z(i.t) = z(i + 1, t) = 4$,

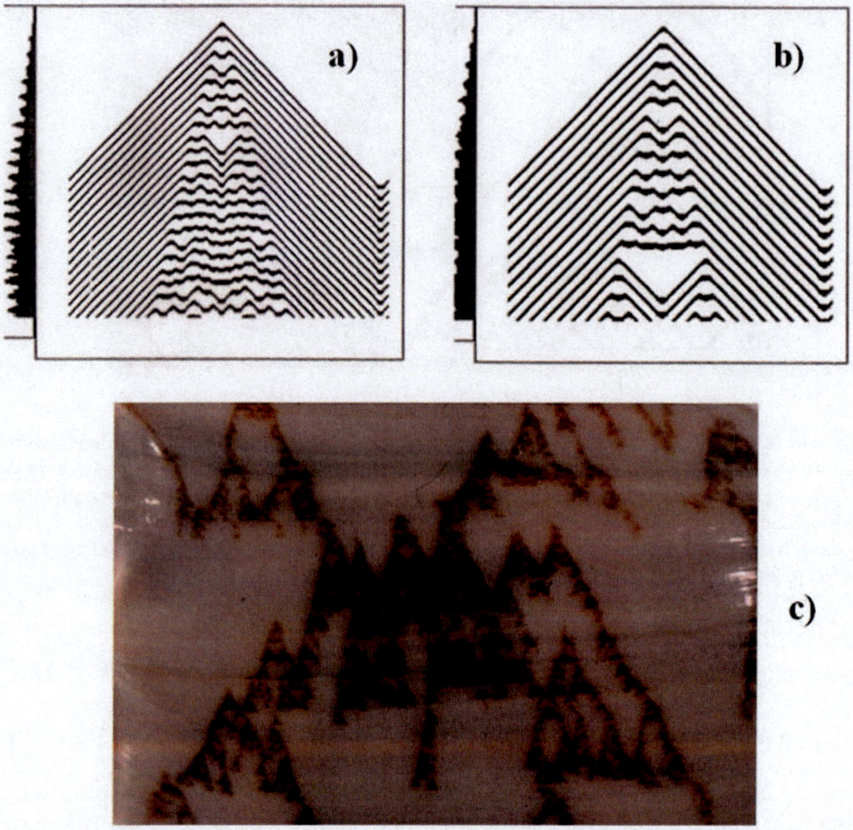

Fig. 7.20 **a** Phase pattern $p(i, t)$ which arise from the 1/6–8/5 rule; **b** and the 1/6–11/4 rule; time runs from top to bottom and space from left to right; **c** Volutidae; *Cymbiola nobilis nobilis*, Lightfoot (1786) (hight: 8 cm) (*Photo* P. Plath) showing a pattern which very is similar to the ends of Fig. 7.20a, b

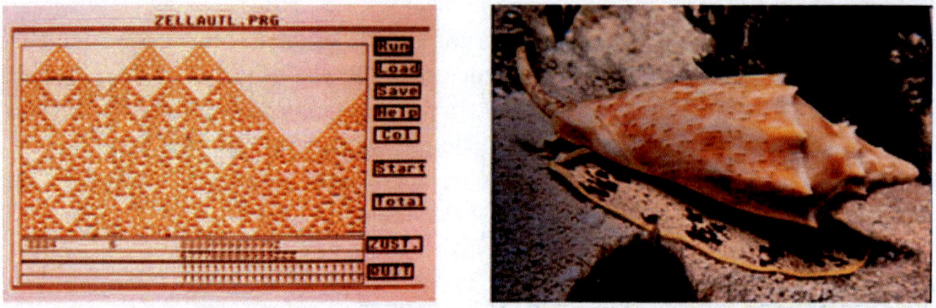

a) b)

Fig. 7.21 **a** Automaton with a spatial–temporal Sierpinsky-like fractal pattern; **b** *Cymbiolacca wisemani* Brazier (1870); in this photograph the shell is growing from bottom to top (Photo by P.Plath with friendly permission of Charles E. Tuttle Company INC [29])

Table 7.10 Automaton for a Sierpinsky-like fractal distribution

Σ		0	1	2	3	4	5	6	7	8	9	10	11	12	13	14
$p(i,t)=0$	$p(i,t)=0$	0	**2**	**2**	**2**	**4**	0	0	0	0	0	0	0	**5**	0	...
$p(i,t)=1$	$p(i,t)=1$	0	0	0	0	0	0	0	0	0	0	0	0	0	0	...
$p(i,t)=0$	$p(i,t)=0$	0	0	0	0	0	0	0	0	0	0	0	0	0	0	...
$p(i,t)=1$	$p(i,t)=1$	0	0	0	0	0	0	0	0	0	0	0	0	0	0	...

the sum $\Sigma = 12$ can be realized with the consequence that $z(i, t+1) = 5$. In this special case, the cells are colored dark brown. The numbers for higher values of Σ which can be seen in Fig. 7.21a are relicts of former automata used before which have no more effect on the automata of Table 7.10. They have no function on the current events like those parts of the genome, which remained as relicts from former developments.

This obvious comparison between the rules of our vector automata and the genetic code had also fascinated our colleague R. Neth when J. Schwietering and P. J. Plath presented it at the conference for Leukemia in Wilsede (1992) [30]. All blood cells are created by division and differentiation from stem cells that are located in the bone marrow. This also applies to blood cancer cells. The neighborhood rule encoded in the genome would therefore be responsible for cell differentiation, and in the case of mutations, also for the development of blood cancer cells.

Another exciting example are the pigment patterns of the shells like *Conus gloriamatis* CHEMNITZ (1777) and the related mollusk's shells of *Conus textile* LINNAEUS (1758) and *Conus aulicus* LINNAEUS (1758) [28].

Table 7.11 shows the rule table for vector automata which create soliton-like crossing waves, and Fig. 7.22 depicts some patterns of automata generated by this rule. Some examples of sea shells with patterns consisting of soliton-like crossings of waves can be seen in Fig. 7.23.

At the end of this section let us present the shell of the very tasty Sea Alomond (french: l' amande de mer) (Fig. 7.24a, b) which is very common at the Bay of Biscay (french: Golfe de Gascogne). The scientific name of this mussel is Glycyremis glycyremis LINNAEUS (1758). Playing around, we found a very curios automaton (Fig. 25a and b) the pattern of which is very reminiscent of these beautiful seashells.

Table 7.11 Vector automaton which creates soliton-like crossing waves

Σ		0	1	2	3	4	5	6	7	8	9	10	11
$p(i,t)=0$	$p(i,t)=0$	0	**1**	1	2	2	2	2	2	2	0	0	...
$p(i,t)=1$	$p(i,t)=1$	0	0	0	**1**	1	1	3	3	0	0	0	...
$p(i,t)=0$	$p(i,t)=0$	0	0	0	0	0	0	**1**	1	1	1	1	...
$p(i,t)=1$	$p(i,t)=1$	0	0	0	**1**	1	1	1	1	1	1	1	...

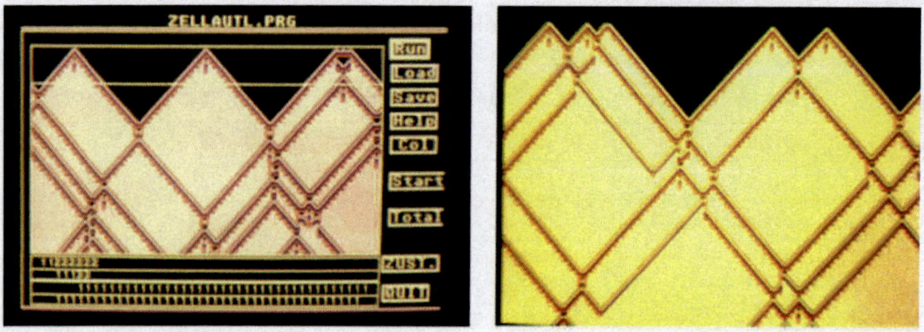

a) b)

Fig. 7.22 **a** Pattern of the automaton according to Table 7.11, starting from four accidently chosen points with $\overrightarrow{z}(j,0) = \binom{z(j,0)}{p(j,0)} = \binom{1}{0}$, while all the other cells are in the starting position $\overrightarrow{z}(i,0) = \binom{z(i,0)}{p(i,0)} = \binom{0}{0}$. The waves cross each other with a remarkable soliton-like phase shift.; **b** Pattern of a very similar automaton as shown in Fig. 7.22a), however, one can also detect waves which annihilate each other (both photos from P. Plath 1990)

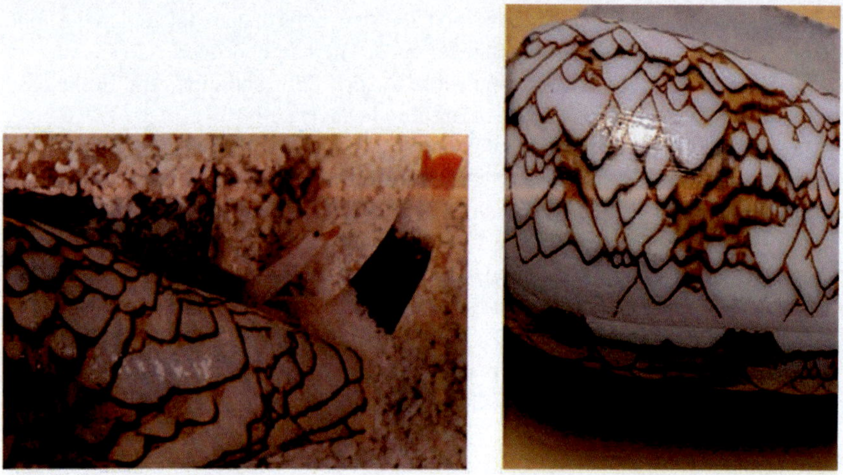

a) b)

Fig. 7.23 Soliton-like crossing of waves in pattern of seashells: **a** *Conus textile* LINNAEUS (1758) from the Great Barrier Reef (photo: K. Gillett) [31, 32]; **b** *Conus archiepiscopus* Hwass in Bruguière (1792) (found in Madagascar 2018 in Toliara) (*Photo* P. Plath)

7.6 Collision Patterns

In the early days of Synergetics V.I. Krinsky gave an overview of "Autowaves: Results, Problems Outlooks" at the Symposium on "Self-Organization—Autowaves and Structures Far from Equilibrium" in Pushino, USSR 1983 [27]. He wrote:

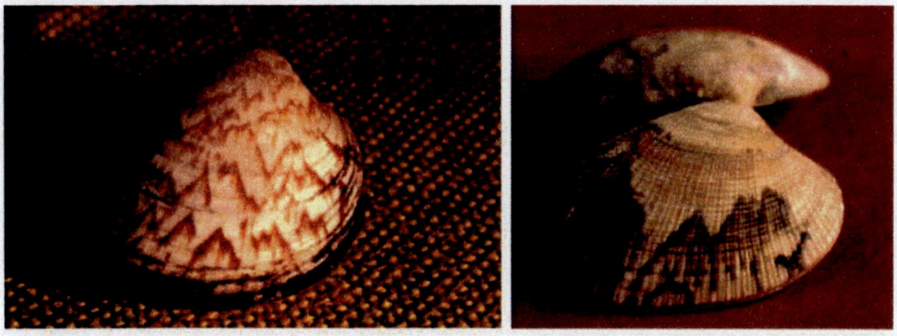

a) b)

Fig. 7.24 L 'Amande de mer or Sea Almond (*Photo* P. Plath 1990)

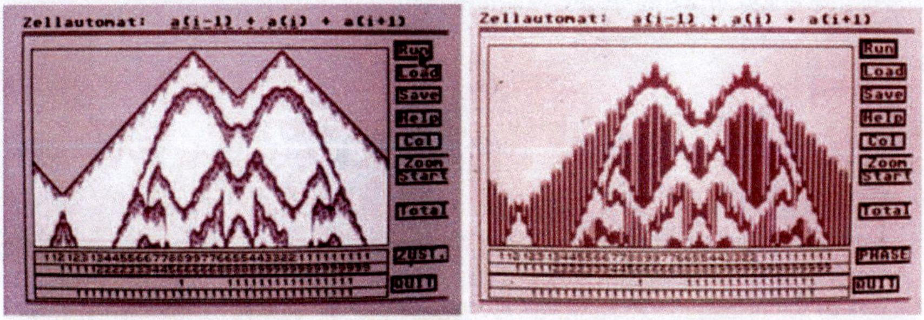

a) b)

Fig. 7.25 Automaton rule and **a** Pigment pattern $z(i, t)$ and **b** pattern of the corresponding phases $p(i, t)$ (Photos from Atari computer: P. Plath 1990)

> Autowaves differ fundamentally from waves in traditional (conservative) media (Table 7.12 in this article; the authors). They propagate at the expense of energy taken from an active media, and, therefore, cannot be considered as conservative systems. The shape and amplitude of autowaves remain constant during its propagation, whereas the amplitude of classical waves rapidly falls with the distance and the waveform is distorted by dispersion. In the case of autowaves, there is no reflection from either the medium boundaries or inhomogeneities.....

The term *autowaves*, which has become historical today, has been replaced by other terms and updated with regard to different areas. Today, instead, we often speak of *solitary waves* and in special cases of *solitons* of many different kinds depending on the differential equation whose solution they represent. These waves differ in their properties. However, its heuristic value is unaffected.

With the seashells as well as with our automata, however, waves also occur, which at the same time have the properties of "autowaves" as well as of "solitons". We are sure that in the differential equations with which Hans Meinhardt describes the pattern formation on the seashells, such hybrid situations also occur in his waves,

Table 7.12 Properties of waves and autowaves [27]

Properties	Waves	Autowaves
Conservation of energy	+	–
Conservation of amplitude and waveform	–	+
Reflection	+	–
Annihilation	–	+
Interference	+	–
Diffraction	+	+

see Fig. 7.4. We want to consider their behavior when they meet as a decisive criterion for differentiating between the different types of waves. Do the waves overlap (superposition principle) or cancel each other out (annihilation) or do they penetrate each other with a certain time delay (solitons)?

7.7 Stable Collision Particles

We consider waves as the spatial propagation of an excitation in particular, as the transmission of an excitation from cell to cell of the automaton. If two waves meet, a more or less large area of cells is formed in which excitations of the cells are generated, which can differ significantly from those in the waves themselves. We therefore regard this area of impact of the colliding waves as a new entity that can, for example, disintegrate instantaneously or only after a while or not at all. Thus, we assume that this area of impact is characterized by its stability. In an earlier work, we named this entity a *collision particle* [21]. We do not think that this is a misleading term, since we found a striking example for a stable entity, which emerges by the collision of waves of excitations on cells. In Fig. 7.26 the yellow line parallel to the direction of time represents such a stable collision particle. Besides the "yellow" particles, there are "red" particles as well, which are also created by the collision of the same waves. But they can furthermore be generated directly by a simple excitation of a cell, as a relic of the original excitation, so to speak, from which two oppositely leaching waves arise (Fig. 7.26). It should not go unmentioned that Stephen Wolfram, too, can create stripes using his rules No. 4, 12, 44, 76, 100 or 108 for one-dimensional automata parallel to the direction of growth starting from a simple cell [33]. However, these streaks are not the result either of colliding waves, nor of the decay of a starting particle.

In any case, the particles that are generated by the collision of the waves but also by the arbitrarily set initial excitations are surrounded by matching vector states in the neighboring cells. The collision-particles are very often stabilized to a certain extent by this simultaneously created environment. The original black-colored basic state $z(i, 0) = 0$ of the cells to be observed becomes dark red (Fig. 7.26) or yellow

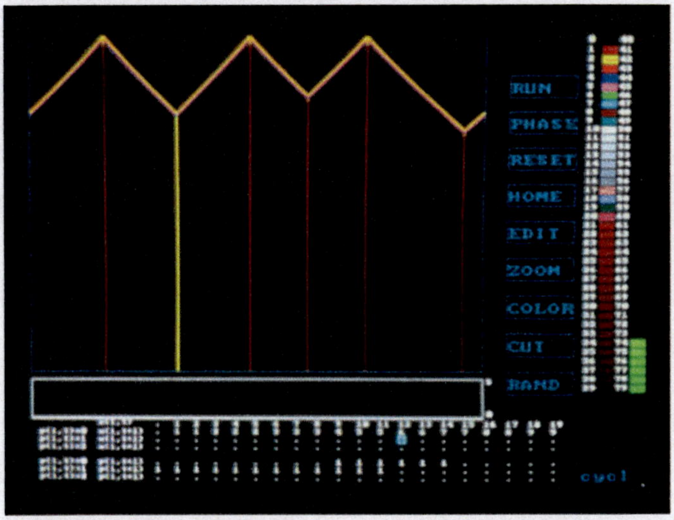

Fig. 7.26 A two-component vector automaton according to Table 7.13 that creates stable collision particles marked in yellow and red

Table 7.13 Two-component vector automaton which creates stripes by collision of waves

Σ		0	1	2	3	4	5	6	7	8	9	10	11	12	13	14
$p(i,t)=0$	$p(i,t)=0$	0	**1**	1	2	2	2	3	3	3	4	5	5	6	6	...
$p(i,t)=1$	$p(i,t)=1$	0	0	0	0	0	0	0	0	0	0	0	0	6	0	...
$p(i,t)=0$	$p(i,t)=0$	0	0	0	0	0	0	0	0	0	**1**	1	1	1	1	...
$p(i,t)=1$	$p(i,t)=1$	**1**	1	1	1	1	1	1	1	1	1	1	1	0	0	...

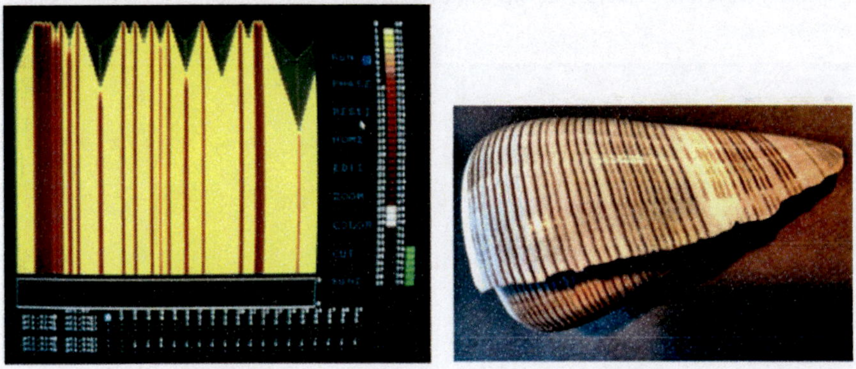

a) b)

Fig. 7.27 **a** An automaton that creates stripes parallel to direction of time according to Table 7.14; **b** *Conus (Cleobula) figulina* LINNAEUS (1758), which is characterized by a lot of fine stripes parallel to the direction of growth (*Photo* P. Plath)

Table 7.14 Vector automaton that creates stripes parallel to direction of time

Σ		0	1	2	3	4	5	6	7	8	9	10	11	12	13	14
$p(i,t)=0$	$p(i,t)=0$	3	3	3	4	4	4	5	5	5	6	6	6	7	7	...
$p(i,t)=1$	$p(i,t)=1$	0	0	0	0	0	0	**1**	1	1	2	2	2	3	3	...
$p(i,t)=0$	$p(i,t)=0$	0	0	0	0	**1**	0	0	0	0	0	0	0	0	0	...
$p(i,t)=1$	$p(i,t)=1$	0	0	**1**	1	1	1	1	1	1	1	1	1	1	1	...

or green (Fig. 7.27) according to the new corresponding state of the cells adjacent to the cells of the collision particle.

The coloring selected in Fig. 7.27a largely corresponds to the shell of the striped snail "Cepaea nemoralis", which is frequently encountered in Germany. The alternation of wide and narrow stripes in the automaton also corresponds to the pattern on these snail shells.

At this point, it is advisable to take a closer look at the structure of the collision particles using a few examples. Figure 7.28a shows the emergence and disintegration of the collision particle on the bases of a class of automata already discussed above in its basic structure, for example with the 1/6—12/3 rule.

The state profile $z(i,t)$ at an appropriate time t of such a generating wave is moving from right to left $(leftside\cdots,0,1,1,2,2,3,3,4,4,5,4,3,2,1,0,\cdots rightside)$. Moreover, this figure demonstrate also how the waves that arise annihilate themselves again. Their profile $z(i,t)$ is $(leftside\cdots,01,1,2,3,5,4,3,2,1,0,\cdots rightside)$ for waves moving from left to right.

A small variation in the activity structure of the 1/6 – 12 /3 rule cellular automaton (Fig. 7.28a) leads to the 1/6 – 12/2 rule automaton (Fig. 7.28b), which shows a completely different behavior of the colliding waves. The result is a larger, unstable

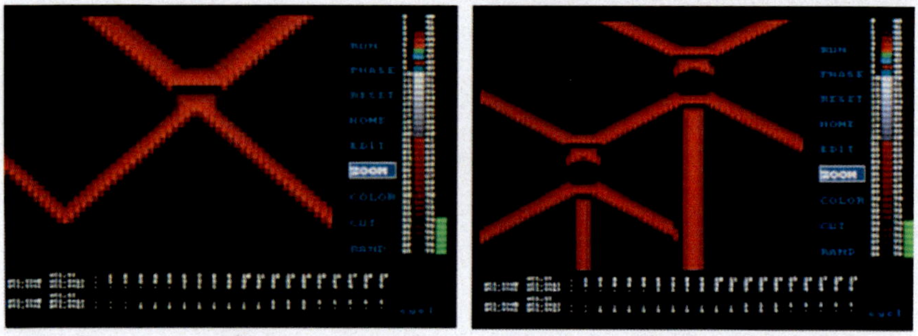

a) **b)**

Fig. 7.28 **a** 1/6—12/3 rule automaton.: while two waves are colliding, an instable "collision particle" is created that breaks down into two new waves, which differ from the waves that generate them. These new waves annihilate themselves when they meet their own kind; **b** 1/6—12 /2 rule automaton: the colliding waves form an instable "collision particle" which decomposes with a phase shift into two new waves and a stable particle that is localized at the collision space collision particle, which also splits into two diverging waves, but leaves a smaller, stable particle at the collision point. The space–time structure of this particle is parallel to the time axis (see Fig. 7.28b). Furthermore, it is surprising to see that waves of the type of these newly emerging waves unite with the stable particle, giving it an outer skin, so to speak.

It almost seems as if these stable collision particles act as attractors. But then do they have to be attractors with a fixed-point character? Could not oscillating structures also be possible? For sure, there is an analogy to John H. Convay's Game of Life developed in 1970 [34]. With special rules of this two-dimensional cellular automaton, one can create static as well as oscillating objects like beehive, blinker and glider. Meanwhile there exists a zoo of such inspiring cellular objects.

Figure 7.29 shows the space–time pattern of the cellular vector automaton according to Table 7.15, which answers exactly this question.

Let us regard at first the left part of both the Fig. 7.29a, b. When two waves collide, collision particles are created, which partially disintegrate again and emit pulsating

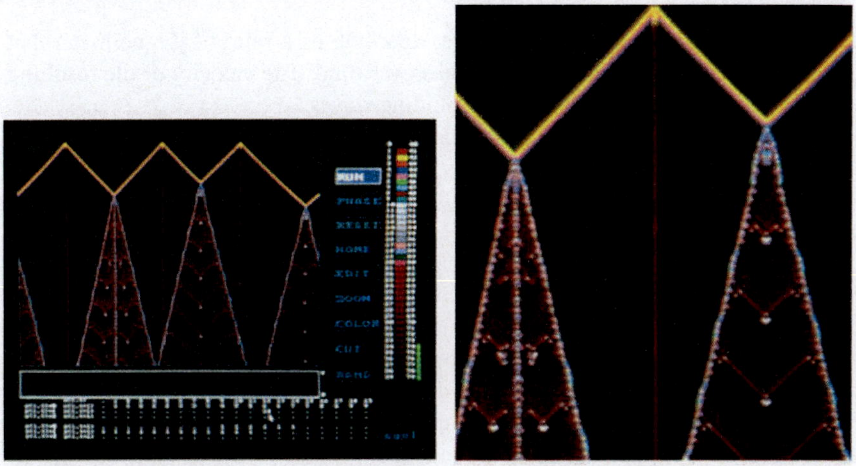

a) **b)**

Fig. 7.29 **a** Colliding waves, which create strings of pearls for waves and collision particles according to Table 7.15; **b** segment of Fig. 7.29a; the right pattern shows the occurrence of a periodic sequence of pearls which emerge as annihilation areas of colliding waves

Table 7.15 Automaton rule for collision pattern with string of pearls

Σ		0	1	2	3	4	5	6	7	8	9	10	11	12	13	14
$p(i,t)=0$	$p(i,t)=0$	0	1	1	2	2	2	3	3	3	4	5	5	6	6	...
$p(i,t)=1$	$p(i,t)=1$	0	0	0	0	0	0						0	**10**	0	...
$p(i,t)=0$	$p(i,t)=0$	0	0	0	0	0	0	0	0	0	**1**	1	1	1	1	...
$p(i,t)=1$	$p(i,t)=1$	**1**	1	1	1	1	1	1	1	1	1	1	1	0	0	...

a) b)

Fig. 7.30 **a** *Conus betulinus* Linnaeus, (1758); Periodic strings of brown dots are observable on this seashell (size ca. 7,5 cm) (Photo: P. Plath); **b** *Architectonica perspectiva* Linnaeus *(1758)*, (5.5 cm in diameter) (*Photo* P. Plath); strings of periodically structured pearls and strings of brown dots are observable

waves within themselves, but also leave a particle as a relic of the collision that remains localized, but constantly oscillates over time. The velocity of the resulting waves is about one third of that of the colliding waves.

These oscillating waves, but especially the remaining oscillating collision-particles, are very reminiscent of strings of pearls with pearls of different sizes in an arbitrary sequence. The diverging pulsating waves in turn generate waves that travel into the interior of the sector they enclose. If the waves generated in this way meet, they are annihilated in a remarkable collision particle. In Fig. 7.29a, b, these annihilation points light up like pearls, which form a periodic temporal system. Two examples of seashells with periodic pearls and strings are shown in Fig. 7.30.

The Sierpinsky triangle is a well-known example of the fascinating spatiotemporal fractal structure formation in one-dimensional cellular automata, as Stephen Wolfram's Rule No. 22 shows [35, 36]. But we have also found these structures in the course of the catalytic oxidation of carbon monoxide over time. Here, the conversion of CO to CO_2 at certain temperatures collapsed fractally to a considerable extent [37–39]. Variations of the Sierpinsky triangle can also be found frequently in the snail patterns. We have shown this at the beginning with the example of the seashell *Cymbiolacca wisemani* B. (1870) (Fig. 7.21a, b). In this section, it is important for us to show that such fractal patterns can also be the result of a collision of waves (see Fig. 7.31).

However, the same waves create stripes if they meet in different way, depending on the number of cells at time $t = 0$ being even or odd and the structure of the wave just before the collision is of appropriated manner (Fig. 7.32).

The original waves arise from excitations with a very complex structure. The excitation not only generates the two waves, but also two pairs of small particles, which are recognizable as red double stripes parallel to the direction of growth. When the growing Sierpinsky triangles hit these stripes, a chaotic shower is triggered

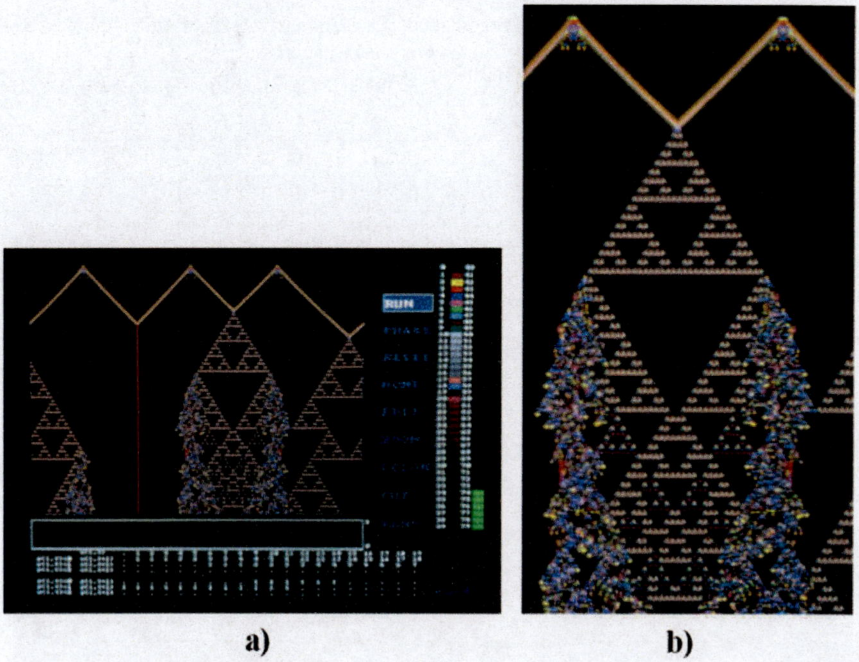

Fig. 7.31 **a** Collision of waves create a Sierpinsky pattern as well as stripes parallel to the direction of growth according to the rules of Table 7.16; **b** Partial enlargement of Fig. 7.31a: colorful showers arise when the Sierpinsky triangles meet the red lines originally created from the excitation at the beginning

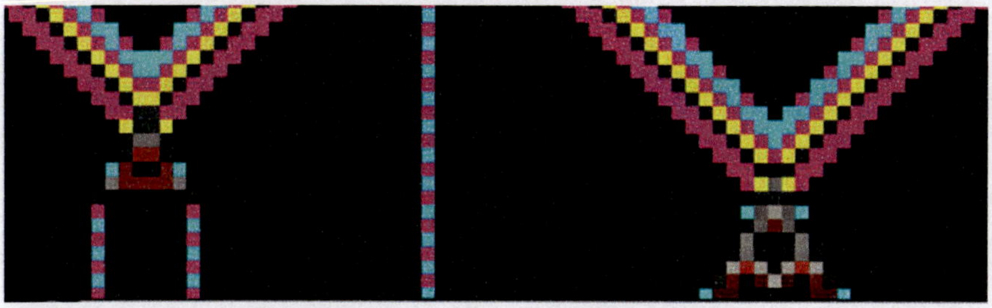

Fig. 7.32 Two waves that meet in different ways. This results in completely different collision particles. The corresponding automaton is described in Table 7.16

surprisingly (see Fig. 7.31). The complexity of the corresponding automaton is much higher than it is shown in Table 7.16. The arithmetic sum Σ of the neighboring cells in this automaton may go at least up to $\Sigma = 59$.

Again, the question arises whether such chaotic showers, which can also be found in the patterns of the seashells (compare Fig. 7.1), cannot also arise as the result of a collapsing impact particle. The cellular automaton in Table 7.17 (see Fig. 7.33) implements an example of this.

Table 7.16 First part of the automaton rule that creates Sierpinsky gaskets from colliding waves and stripes, which trigger showers if they hit the Sierpinsky pattern

Σ		Σ	1	2	3	4	5	6	7	8	9	10	11	12	13	14
$p(i,t)=0$	$p(i,t)=0$	0	**2**	1	1	2	2	3	3	4	4	5	8	12	12	...
$p(i,t)=1$	$p(i,t)=1$	0	0	0	0	0	0	**1**	1	1	2	2	3	3	4	...
$p(i,t)=0$	$p(i,t)=0$	0	0	0	0	0	0	0	0	0	**1**	1	1	1	1	...
$p(i,t)=1$	$p(i,t)=1$	**1**	1	1	1	1	1	1	1	1	1	1	1	0	0	...

Table 7.17 Automaton, the waves of which create chaotic showers if they met each other

Σ		0	1	2	3	4	5	6	7	8	9	10	11	12	13	14
$p(i,t)=0$	$p(i,t)=0$	0	**2**	**1**	**2**	**2**	**2**	**3**	**3**	**3**	**4**	**4**	**5**	**5**	**6**	...
$p(i,t)=1$	$p(i,t)=1$	0	0	0	0	0	0	0	0	0	0	0	0	**8**	0	...
$p(i,t)=0$	$p(i,t)=0$	0	0	0	0	0	0	0	0	0	**1**	1	1	1	1	...
$p(i,t)=1$	$p(i,t)=1$	**1**	1	1	1	1	1	1	1	1	1	1	1	0	0	...

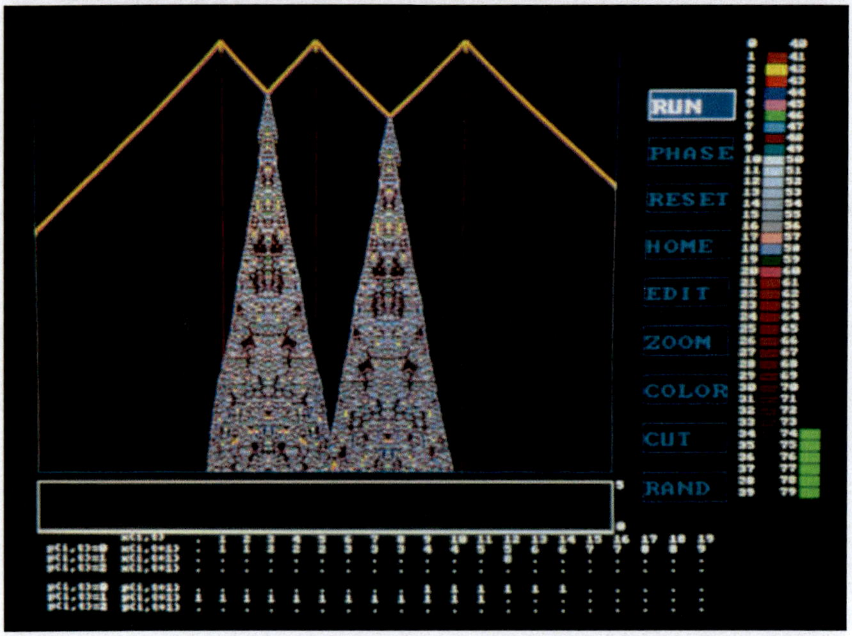

Fig. 7.33 Colorful chaotic showers as result of an impact of two solitary waves according to Table 7.17

7.8 Concluding Remarks

Mussel and snail shells are a very old cultural asset. The relative rarity of these objects and their fascinating shapes and pigment patterns let them to become jewelry and barter objects very early in human history. That did not change when them became objects in royal collections and later in scientific museums and in biological institutes, and is still the mainspring of great interest in them today.

Wonderful books about seashells in all seas were created and offered an illustrative overview of the diversity of this fascinating world. In the early days of synergetics, Hans Meinhardt impressed with his simulations of the patterns of seashells [13]. The reaction–diffusion equations he used were an excellent fit for the development of the theory of pattern formation in complex systems.

However, the question of what reality his equations described remained unanswered. How should a diffusion take place in the lip of the shells? Classical diffusion presupposes a continuum that is not created from one "time step" to another by new biological cells and therefore does not constitute a continuum. It remained a miracle how his wonderful simulations nevertheless described the reality of the pigment patterns on the seashells.

On the other hand, the cellular automata became a subject of mathematical research almost at the same time and were successfully used in the simulation of pattern formation processes. One of the many convincing results is the work of Stephen Wolfram [20], who also examined one-dimensional cellular automata and their possibilities. In our opinion, the cellular approach corresponds much more to the cell-biological process when the patterns in the seashells grow.

We assume that our rules, according to which the patterns develop from one time step to another, correspond to essential aspects of the genetic code that is inherited during cell division. In our opinion, this code says that cell division is a process in which cell communication between neighboring cells has a decisive influence. Hence, we consider these patterns to be *inheritance patterns.*

These neighborhood rules, which constitute the cellular automata, can be understood as a discrete variant of diffusion. In biological systems, where communication between individuals or individual cells is important, this neighborhood rules take on the role of diffusion, which works in continuous systems.

The typical classification of Krinsky [27], who distinguishes *autowaves* from other *solitary waves* and especially from waves to which the superposition principle applies, was of great heuristic value for us. Waves are also central to the patterns of inheritance. However, they are much richer in their characteristics than the classic waves. For this reason, it was important for us to highlight the typical characteristics of these new waves of inheritance. The concept of collision particles [21] was of crucial importance for the classification of the different types of waves.

References

1. Angeletti, S.: Documentaire Alpha: Les coquillages. Grange Batelière—Paris, p. 87n (1969)
2. Zilhãoa, J., Angelucci, D.E., Badal-García, E., d'Erricod, F., Daniel, F., Dayet, L., Douka, K., Higham, T.F.G., Martínez-Sánchez, M.J., Montes-Bernárdez, R., Murcia-Mascarós, S., Pérez-Sirvent, C., Roldán-García, C., Vanhaeren, M., Villaverde, V., Wood, R., Zapata, J.: Symbolic use of marine shells and mineral pigments by Iberian Neandertals. PNAS **107**(3), 1023–1028 (2010).https://doi.org/10.1073/pnas.0914088107.Accessed25Feb2021
3. Hoffmann, D.L., Angelucci, D.E., Villaverde, V., Zapata, J., Zilhão, J.: Symbolic use of marine shells and mineral pigments by Iberian Neandertals 115,000 years ago. Sci. Adv. **4**(2) eaar5255, 1–6 (2018). https://doi.org/10.1126/sciadv.aar5255
4. Plath, P.J., Haß, E.-C.: Construction of our reality and myths. Psychol. Behav. Sci. Int. J. **15**(2), 555909 (2020). https://doi.org/10.19090/PBSIJ.2019.10.555909
5. https://de.wikipedia.org/w/index.php?title=Kula_(Ritual)&oldid=199109160. Accessed 05 Mar 2021
6. Ziegler, R.: The Kula Ring of Bronislaw Malinowski, Bayrische Akademie der Wissenschaften, Philsophisch-Historische Klasse, Sitzungsberichte Jahrgang 2007, Heft 1, in Kommission Verlag C.H. Beck, München (2007)
7. Leibniz, G.W.: Protogea – oder Abhandlung von der ersten Gestalt der Erde und den Spuren der Historie in den Denkmaalen der Natur, aus dem Lateinischen ins Deutsche übersetzt von Scheid, Ch. L.; priviligierter Buchhändler: Johann Gottlieb Vierling, Leipzig und Hof, pp. 81–95 (1749); ETH-Bibliothek Zürich Shelf Mark: Rar 4300, Persistent Link: http://dx.doi.org/ https://doi.org/10.3931/e-rara-2875. Accessed 09 Mar 2021
8. Eckhart, J.G.: Beschreibung desjenigen, was bey Grabung des Herrenhäuser-Canals am Lein-Strome her Curiöses in der Erde gefunden worden, Neue Zeitungen von Gelehrten Sachen auf das Jahr 1719, Nr. 24, 185–192
9. https://en.wikipedia.org/wiki/Protogaea#/media/File:Houghton_GC6.L5316.749p_-_Liebniz,_Svmmi_polyhistorias,_tab_III.jpg. Accessed 15 Mar 2021
10. Linné, Cv: Museum Ludovicae, Ulricae reginae suecorum, gothorum, vandalorumque...: in quo animalia rariora, exotica, imprimis insecta & conchilia" describuntur & determinantur prodomi instar editum", ([Reprod.]) / a Carolo v. Linné. 1764, Source gallica.bnf.fr / Bibliothèque nationale de France
11. Lefèvre, W.: Jean Baptiste Lamarck (1744 – 1829)—ein Klassiker aus Mißverständnis?", Max-Planck-Institute for History of Sience, reprint 61 (1997) Berlin. https://www.mpiwg-berlin.mpg.de/sites/default/files/Preprints/P61.pdf. Accessed 17 Mar 2021
12. Turing, A.M.: Philos. Trans. R. Soc. Lond. Ser. B, Biol. Sci. **237**(641), 37–72 (1952)
13. Meinhardt, H.: The Algorithmic Beauty of Sea Shells—The Virtual Laboratory. Springer, Berlin (1995)
14. Wolfram, St.: Universality and Complexity in Cellular Automata. Physica **10D**, 1–35 (1984)
15. Toffoli, T., Margolus, N.: Cellular Automata Machines—A New Environment for Modelling. The MIT Press Cambridge, Massachusetts, London, England (1987)
16. Hunsinger, I.K.: Textursynthese nach Reaktions-diffusions-Modellen, Master's thesis, Studiengang Informatik, Universität Bremen (1994)
17. Hunsinger, I.K., Nake, F.: Textures on Seashell, Bericht Nr. 4/96, Fachbereich Mathematik und Informatik , Universität Bremen (1996)
18. Foley, J.D., Dam, van A., Feiner, St.K., Hughes, J.F.: Computer Graphics: Principle and Practice 2nd edn. Addison Wesley, Reading MA (1990)
19. Illert, C.: Formulation and solution of the classical seashell problem, II Nuovo Cimento 11 D, vol. 5, pp. 761–780 (1989)
20. Wolfram, St.: A new kind of science, Chapter: Growth of Plants and Animals. Wolfram Media, Inc., pp. 400–422 (2002)
21. Plath, P.J., Plath, J.K., Schwietering, J.: Collision Patterns on Mollusc Shells. Discret. Dyn. Nat. Soc. **1**, 57–76 (1997)

22. Mikhailov, A.S.: Foundation of Synergetics I—Distributed Active Systems, Springer Series in Synergetics, vol. 51, p. 24. Springer, Berlin (1990)
23. Plath, P.J., Schwietering, J.: In: Encarnação, J.L., Peitgen, H.-O., Sakas G., Englert, G. (eds.) Fractal Geometry and Computer Graphics. Springer, Berlin, pp. 162–172 (1992)
24. Gerhardt, M., Schuster, H., Tyson, J.J.: A cellular automation model of excitable media including curvature and dispersion. Science **247**, 1563–1566 (1990)
25. Gerhardt, M., Schuster, H., Tyson, J.J.: A cellular automaton model of excitable media: II. Curvature, dispersion, rotating waves and meandering waves. Physica D **46**, 392–415 (1990)
26. Schwietering, J., Plath, P.J.: Wachsende Muster. Wiss. und Fortschritt, vol. 42, pp. 73–75 (1992)
27. Krinsky, V.I.: Autowaves: Results, Problems, Outlooks. In: Krinsky, V.I. (eds.) Selforganization—Autowaves and Structures Far from Equilibrium. Springer Series in Synergetics, vol.28 (Series editor: H. Haken), Springer, Berlin, pp. 9–19 (1984)
28. Wilson, B.R., Gillett, K.: Australian Shells, Charles E. Tuttle Company, Rutland Vermont & Tokyo, Japan, p. 133 (1971), plate 88 Nr.3
29. Wilson, B.R., Gillett, K.: Australian Shells, Charles E. Tuttle Company, Rutland Vermont & Tokyo, Japan, p. 8 (1971), Fig. 3
30. Schwietering, J., Plath, P.J.: Frozen Pattern in Growing Systems. In: Neth, R., et al. (eds.) Haematology and Blood Transfusion", Vol 35, "Modern Trends in Human Leukemia IX, pp. 329–335. Springer, Berlin (1992)
31. Wilson B.R., Gillett, K.: Australian Shells: Illustrating and Describing 600 Species of Marine Gastropods Found in Australian Waters. Reed, Sydney, Melbourne, Wellington, Auckland (1971)
32. Gillett, K., McNeill, F.: The Great Barrier Reef and Adjacent Isles: a comprehensive survey for visitor, naturalist and photographer. Corel Press, Paddington Sidney Australia (1959)
33. Wolfram, S.: A New Kind of Science. Wolfram Media, pp. 54, 55 (2002)
34. https://de.wikipedia.org/wiki/Conways_Spiel_des_Lebens. Accessed 16 Apr 2021
35. Seifert, C.: Complex Systems Engineering—Zelluläre Automaten. University of Passau, WS 2017/2018, pp. 18 and 28, http://christinseifert.info/teaching/ws2017-cse/part-cellular-automatons-handout.pdf. Accessed 13 Apr 2021
36. Wolfram, S.: A New Kind of Science. Wolfram Media, p. 58 (2002)
37. Dress, A., Jaeger, N.I., Plath, P.J.: Zur Dynamik idealer Speicher—ein einfaches mathematisches Modell. Theoret. Chim. Acta (Berlin) **61**, 437–460 (1982)
38. Plath, P.J., Möller, K., Jaeger, N.I.: Cooperative effects in heterogeneous catalysis: part II: Analysis and modelling of the temperature dependence of the catalytic oxidation of CO on a palladium Al2 O3 -supported catalyst. J. Chem. Soc. Faraday Trans. I **84**(6):1751–1771
39. Plath P.J.: Modelling of heterogeneously catalysed reactions by cellular automata of dimension between one and two. In: Optimal Structures in Heterogeneous Reaction Systems, Springer Series in Synergetics, vol. 44 (editor. P.J. Plath, Series Edit. H. Haken), pp. 1–25. Springer, Berlin (1989)

Part III
Dissipative Structures

Chapter 8
Waves Which Move Uphill

Granular Gases, Fluids and Convective Solids of Quartz

Peter J. Plath, Ernst-Christoph Haß, and Sonja Sauerbrei

Fig. 8.1 Vibrated granular quartz (Rutile) layer (2 mm) at 66.4 Hz and an acceleration b ≈ 10.1 g. The wave is moving uphill the faraday heap which has been formed at the wall of the vessel (photo: P.J. Plath).

8.1 Introduction

Structure formation on surfaces of vertically vibrated fluids and granular media are well known phenomena since the fundamental investigations of Faraday [1] and Chladni [2]. At present, these phenomena are commonly called Faraday instabilities.

P. J. Plath et al., *Imagery Synergetics*, Understanding Complex Systems,
https://doi.org/10.1007/978-3-030-95607-3_8

In recent years, extensive research on granular media has been carried out in order to understand and simulate this kind of structure formation [3–19]. Many experiments and numerical calculations are performed for systems with comparatively large heights of the layer measured in units N of particle diameters d, i.e. $N = h/d$, where h is the height of the layer. The containers being used are cylindrical boxes with relatively small diameters of up to 5 cm. Under these conditions, granular convection is usually observed if $N \geq 10$ [16]. In these investigations, the movement of single particles within the granular material was of particular interest. Other studies were focused on the influence of the air on the formation of granular heaps [17]. The characteristic diameters of such heap patterns are in the range of some millimeters up to centimeters.

In a previous article, we reported briefly about our first results of vibrated granular quartz and rutile fillings (Fig. 8.1) [20]. The most striking observation was a strong change of the response function, which is correlated to the acceleration of the whole vibrating system in a rather short time. We suggested describing this phenomenon with a jerk caused by the non-linear, collective behaviour of the jumping particles which also leads to the formation of the Faraday hill. In the following, we discuss the observed patterns of vibrating quartz granulates and the corresponding acceleration curves depending on the frequency and amplitude of the driving force in much greater detail.

In our studies, we used a cylindrical container of a fairly large inner diameter and granular layers of low ratios $N \leq 2$ in the beginning of the experiments. We are interested on the one hand on the spatio-temporal development of the granular medium, which we recorded by camera photos and movies. On the other hand, we investigated the behaviour of the system as a whole—including the granular matter as well as the vertically oscillating container—by picking up the responding acceleration signal with an oscilloscope and analyzing its structure by a computer.

Surprisingly, starting with these thin granular layers, we observed structure formation in the layers as well as formation of granular hills building convective solids which show a characteristic length of about 5–10 cm and heights of 2–3 cm. Moreover, special patterns are observed on the surface of these Faraday hills, in particular "waves moving uphill". These structures are reflected by characteristic shapes of the corresponding response function of the acceleration signal in the oscilloscope display. The different patterns resulting by varying the vibration frequency and excitation amplitude are in detail discussed in the following.

We emphasize that our experimental investigations are directed towards the primary technical use of Faraday instabilities in granular media, wherefore our description is purely phenomenological but not based on brilliant theory using statistical physics as for example by Brilliantov and Pöschel [18, 19].

8.2 Experimental Setup

In order to describe the motion of the granular layer, we have carried out two kinds of investigations, i.e. measurement of the instantaneous acceleration following the driving force and visual inspections of the pattern formations by films and photographs. A sketch of the used experimental setup is provided in Fig. 8.2.

The container with a thin granular layer of quartz particles is vertically vibrated by means of a shaker system GW-V20/PA100E of *Data Physics (Deutschland) GmbH* which was extended by a head expander to carry a cylindrical box with a comparatively large diameter. The shaker in turn is fed by a *VOLTCRAFT*® frequency generator 8202, the periodic voltage signal U_d of which is the stimulating frequency f of the vibrated system. The height of its amplitude A and thus the strength of the driving force have been regulated by the power amplifier PA100E of the shaker system. In our experiments, we have varied the frequencies f in the range of 40–110 Hz and the amplitudes up to resulting peak accelerations a_{pk} of about 5 g (gravity $g = 9.807\ \mathrm{m\ s^{-2}}$).

The acceleration is measured with a shear-type accelerometer KS 76C-100 of *Metra Mess- und Frequenztechnik in Radebeul e.K.*, which is attached to the top of the container. Taking advantage of the piezoelectric effect, a low voltage signal is generated which is proportional to the acceleration of the whole vibrating system. This signal is pre-amplified by a conditioning module and picked up by a digital storage oscilloscope HM408 of ©*HAMEG GmbH*, where it is displayed as voltage curve $U_a(t)$. According to the manufacturer, the voltage sensitivity is the factor $B_{Ua} = 10.10\ \mathrm{mV/m\ s^{-2}} = 99.05\ \mathrm{mV/g}$. For comparison, also the signal U_d of the frequency generator is monitored by the oscilloscope display. In order to further processing the measured data, the oscilloscope is connected to a computer.

According to our test setup, we do not get our information about the vibrating system by relying on the amplitude A of the driving force, which we do not know

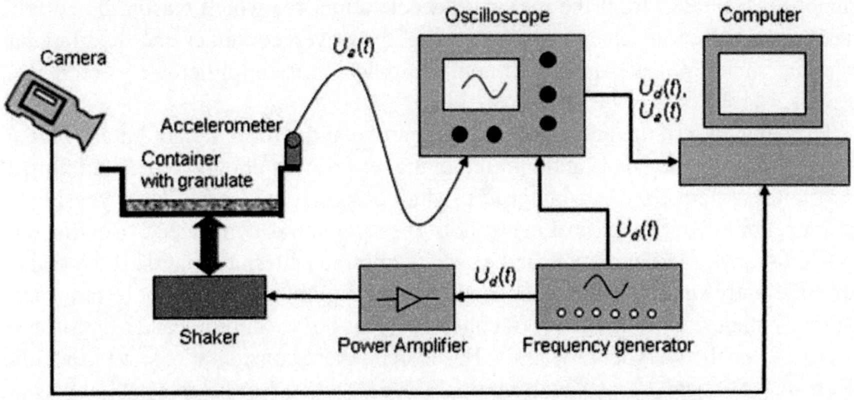

Fig. 8.2 Schematic representation of the experimental setup. A similar test arrangement was used by Pastor et al. [15, 16]

exactly, but rather by directly measuring the actual acceleration of the whole system, i.e. we keep strictly with the observed phenomena. Since the acceleration $a_{pk\text{-}pk}$ between the reversal points of a vibration is most important to explain the observed experimental phenomena, we applied the formula (8.1)

$$a_{pk-pk} = \frac{U_{a,\max} - U_{a,\min}\,[\text{mV}]}{B_{Ua}[\text{mV/g}]} = \frac{U_{a,\max} - U_{a,\min}}{99.05}[\text{g}], \tag{8.1}$$

where $U_{a,\max}$ and $U_{a,\min}$ are two succeeding maxima and minima of the voltage curve $U_a(\text{t})$, respectively. This correlates to the acceleration, which occurs within the time interval of a half period length $T/2$ of the oscillation. In order to be comparable with literature data of other authors (see e.g. [16] and references therein), we used a dimensionless quantity Γ according Eq. (8.2) to state (All subsequent equations must be increased by the value 2) the peak acceleration, where

$$\Gamma = \frac{a_{pk-pk}}{2g} = \frac{U_{a,\max} - U_{a,\min}}{2 * 99.05}. \tag{8.2}$$

The mean $\frac{U_{a,\max}-U_{a,\min}}{2}$ takes account for the experimental observation that the absolute values of $U_{a,\max}$ and $U_{a,\min}$ are different at higher amplitudes of the driving force.

The obtained acceleration data comprise the motion of the rigid container as well as the separate movement of the granular particles and thus allows insight into the behaviour of the vibration system as a whole. In this way it is possible to distinguish between the system response, on the one hand, if the granular matter follows rigidly the driving force, and on the other hand, if particles can loose from the bulk fulfilling a jump and falling back to the continuously vibrating container. In other words, the directly measured acceleration reflects the fact that the whole system behaves as non-linear weakly coupled oscillators. On the contrary, the amplitude of the driving force cannot be separated from the measured acceleration, for which reason the correct measure of both, the separate movements of the driven container and the granular medium, cannot be determined with our setup, which is simplified for its technical usage.

The behaviour of the granular matter, in particular the formation of Faraday hills, phase separation, hot spots and uphill moving waves, was photographed and filmed by a digital camera *Minolta* Dimage E50, which always provided sufficiently resolved pictures. For further analysis of these shots, the camera was connected to a computer.

Almost any granular matter can be used to observe pattern formation if the grains are sufficiently small. On the other hand, the single globules should not be too small to prevent undesirable side effects if cohesive forces due to humidity and electrostatic forces exceed the weight of the grain. Based upon these considerations, we used thin layers of 100 g quartz grains with a particle size d from 0.6 to 1.2 mm enabling us to observe Faraday instabilities; the density of the quartz crystals is $\rho_{SiO_2} \approx 2.7$ g/cm^3. It should be pointed out that the quartz grains being used are no spherical particles or balls, respectively.

The container is a fairly large cylindrical aluminium box weighting about 1.7 kg with an inner diameter of about 18 cm, so that its diameter is much larger than that of a single quartz globule. The height h of the layer in the beginning of the experiments is about 1.5–2 mm and thus we start with low ratios $N = h/d \leq 2$. The total weight of container and granulate (1.8 kg) allows a free vibration path between maximal and minimal elongation of the shaker.

8.3 Results

8.3.1 Phase Separation—Convective Solid

Starting with a vibration frequency of 50 Hz with very small excitation amplitudes, the thin layer of granular bulk follows exactly the driving force with only a small phase shift of the response signal as compared to the excitation signal. Accordingly, no structure formation of the granular medium is observable in case of tiny excitations. In Fig. 8.3 a photo of the reposing quartz layer (a) and the respective oscilloscope display (b) is shown.

Increasing the amplitude of the driving force such that it just exceeds the threshold for structure formation in the granular medium, the layer very slowly begins to contract forming a Faraday hill close to a partial region of the cyclic vessel wall.

For a better understanding it should be mentioned that we use the term "Faraday hill" for those relative high and extended granular convective patterns which are formed close to the walls of the vessel. We herewith refer to the work of Garcimartín

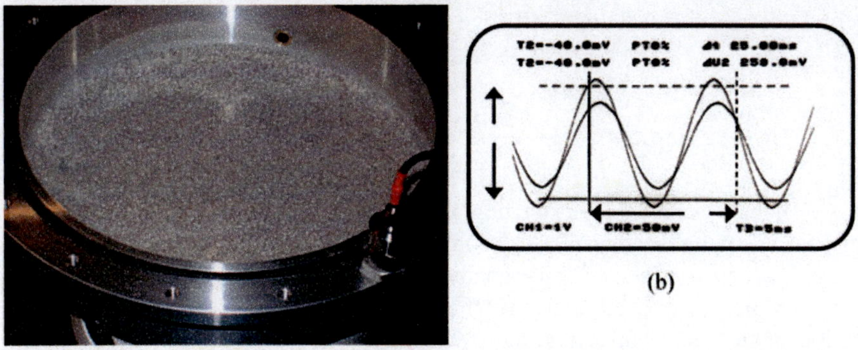

(b)

(a)

Fig. 8.3 Vertically vibrated thin quartz layer, driven by a frequency of 50 Hz and tiny excitation amplitude with resulting $a_{pk\text{-}pk} \approx 1.9$ g and $\Gamma \approx 0.95$ just below the threshold for structure formation ($\Gamma \geq 1$). **a** Photo of the reposing quartz layer, i.e. the whole button of the container is equally covered with a thin layer of h < 2 mm; **b** corresponding oscilloscope display of the excitation function $U_d = U_1$ and the response function $U_a = U_2$

et al. [16] who showed that the properties of the wall will not influence essentially the granular convective dynamic of the layer as a whole. This dynamic is caused by the air-grain interaction induced convection mechanism [21] of the vibrated granular media as documented extraordinarily by Akiyama et al. [22, 23]. Within this work we will not report on the well known small Faraday heaps far from the boundaries of the container which are discussed everywhere.

If the system is reaching its stationary state, almost no single particles remain at the flat bottom of the vertically shaken container. On the contrary, all particles are now collected in the Faraday hill which peaks a height of about 2.5–3 cm. The formation of these Faraday hills starting from a layer with $N < 2$ could either been caused by a not completely plain container bottom or friction with the wall or a combination of both. There is the interesting work of Miao et al. [24] on formation and transport of a sand heap in an inclined and vertically vibrated container showing that very fine quartz sand with grains of 0.15–0.20 mm in diameter forms a heap which moves towards the wall of the container. This effect is very similar to our observation of forming Faraday hills.

Almost all particles of this Faraday hill seem to remain in a fixed position, i.e. only a few globules tumble down the relatively steep slope of the hill. However, if a particle reaches the foot of the hill at the bottom of the vessel, it will immediately attracted and caught by the hill. A photo of this situation and the corresponding oscilloscope display is shown in Fig. 8.4a, b.

As a consequence, we observe a clear phase separation between the hill—which acts as a convective solid—and the almost empty bottom of the container. Only in the direct vicinity of the foot of the hill very few particles form a diluted (highly evacuated) two-dimensional granular gas. This situation is very similar to a sublimation state, i.e. a solid with a low vapour pressure.

At the oscilloscope display this situation is characterized by a shoulder at the periodic response signal on its left edge (see Fig. 8.4b).

8.3.2 Phase Separation—Convective Solid—Liquid—Gas

If one slightly increases the excitation amplitude of the driving force, the particles on the slope of the Faraday hill start to fall down more and more rapidly. In addition, avalanches can be observed and—finally by reaching a stationary state—the whole slope is rolling down continuously. In this stationary state, also a "granular liquid" at the foot of the Faraday hill can be observed (see Fig. 8.5). Both phases, the convective solid and the granular liquid, are separated by a sharp phase-boundary (red coloured dotted line in Fig. 8.5a). Beyond this granular liquid, a third phase of fast jumping particles emerges, which is behaving as a "granular gas". This phase is spreading over the remaining part of the container at its plain bottom, such that the density of its particles diminishes almost exponentially with increasing distance from the sharp phase boundary to the granular liquid (magenta coloured dotted line in Fig. 8.5a).

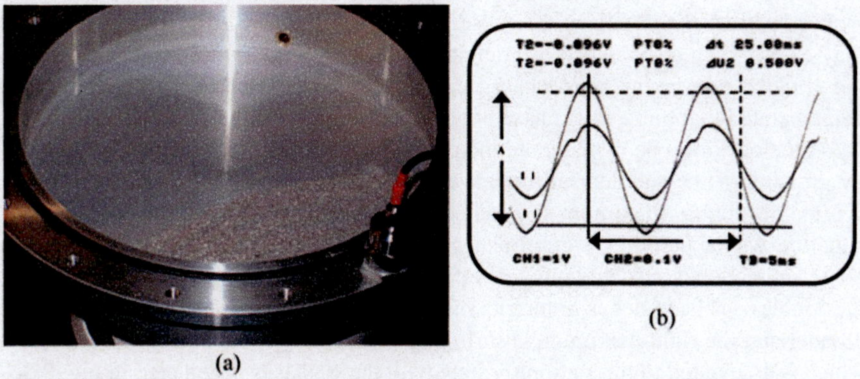

Fig. 8.4 Vertically vibrated thin quartz granulate, driven by a frequency of 50 Hz and excitation amplitude just over the threshold for structure formation with resulting $a_{pk\text{-}pk} \approx 2.7$ g and $\Gamma \approx 1.35$. **a** Photo of the Faraday hill together with the diluted granular gas; **b** corresponding oscilloscope display of the excitation function $U_d = U_1$ and the response function $U_a = U_2$. The response curve shows a shoulder at the left side (see also [5, 16])

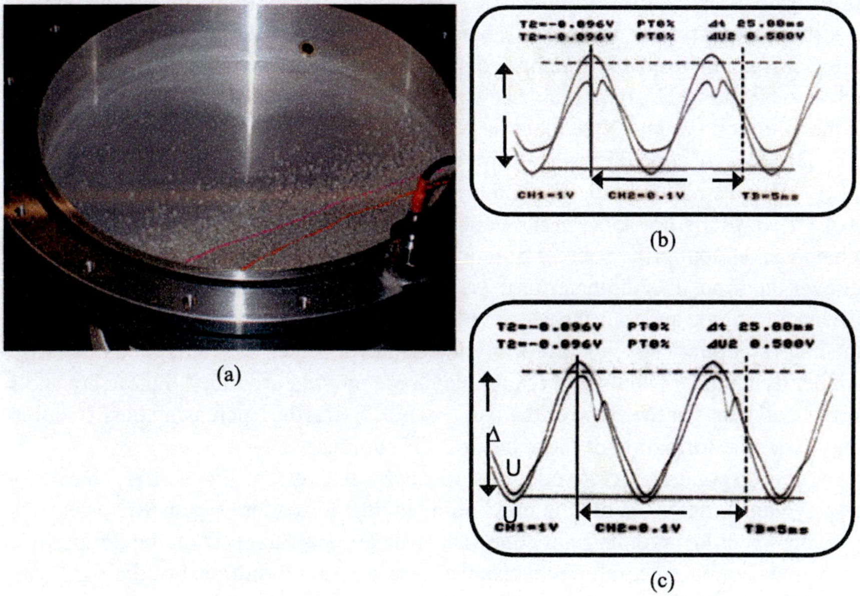

Fig. 8.5 Vertically vibrated thin granulate of quartz, driven by a frequency of 50 Hz and higher excitation amplitudes. **a** Photo of the stationary state of three coexisting phases, i.e. convective solid (the Faraday hill), granular liquid and granular gas (the phase boundaries are marked with red and magenta coloured dotted lines, respectively); **b** oscilloscope display of the excitation function $U_d = U_1$ and the response function $U_a = U_2$ corresponding to the intermediate situation of permanently rolling down particles, but no significant phase separation ($a_{pk\text{-}pk} \approx 3.4$ g and $\Gamma \approx 1.7$); **c** oscilloscope display at stationary state with three phases ($a_{pk\text{-}pk} \approx 4.8$ g and $\Gamma \approx 2.4$)

For brilliant theoreticians this observation might not be surprising, since they may expect such coexistence of granular liquid or granular gaseous phases beside the granular convective solid phase, but we proofed this idea experimentally and demonstrated that there exist phase boundaries between all these three phases.

At the oscilloscope display, the increase of the excitation amplitude is reflected by growing of the shoulder at the left edge of the response signal until an almost symmetric double maximum is reached (see Fig. 8.5b). This corresponds to the situation where particles are permanently falling down the Faraday hill, but the phase separation is just beginning. If one further increases the excitation amplitude, the double peak becomes asymmetric, such that the left maximum growths higher and broader and the right maximum is shrinking significantly. The oscilloscope display which is associated to the stationary state with three phases is depicted in Fig. 8.5c.

8.3.3 Hot Spots as Sputtering Sources in Convective Solids

Shifting up the excitation amplitude (acceleration) at higher frequencies ($\geq$60 Hz), at first structure formation and phase separation proceeds in the same way as in the case of lower frequencies, which we have discussed above. But working with frequencies between 60 and 70 Hz, remarkable new phenomena arise at even higher accelerations of the vibration system. Note that this is just the frequency range where Garcimartín et al. [16] have observed Faraday heaps using larger values of particle diameter units ($N \geq 10$) and small cylindrical containers. In our experiments using a frequency of 66.4 Hz we found hot spots in the Faraday hill which are developing in time. They seem to act as sputtering sources of much smaller quartz grains which eject particles all over the time. The diameter of these sources is about 1–2 cm. Depending of the driving amplitude and thus the acceleration of the vibrating system, there exist on the one hand stationary states with locally fixed hot spots (see Fig. 8.6a) and otherwise smaller dynamic hot spots (0.5–1 cm) which are moving around. All these hot spots form small heaps at the slope of the Faraday hill. By further increasing the excitation amplitude, the formation of these hot spots is suppressed.

The corresponding picture on the oscilloscope display (see Fig. 8.6b) is characterized by vanishing of the double maximum leaving a broader asymmetric peak with two small shoulders on its right edge. The broader peak arises from the left peak of the former double maximum whereas the shoulders are a rudiment of the right one. This statement is the outcome of stepwise careful variations of the driving frequency in a lot of experiments.

8.3.4 Convective Solids—Waves Moving Uphill

If one boosts the excitation amplitude at the same frequency (66.4 Hz) to a very high level, which is significantly beyond the situation where hot spots occur, a new

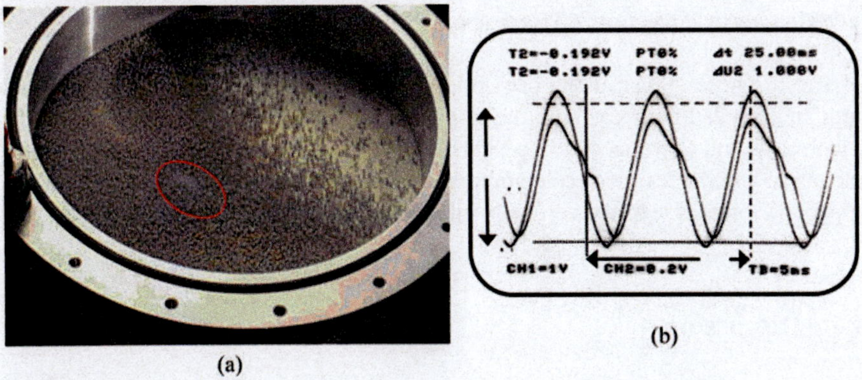

Fig. 8.6 Vertically vibrated thin quartz granulate, driven by a frequency of 66.4 Hz and a high excitation amplitude with resulting $a_{\text{pk-pk}} \approx 8.4$ g and $\Gamma \approx 4.2$. **a** Photo of the granular convection solid with a hot spot; **b** corresponding oscilloscope display of the excitation function $U_d = U_1$ and the response function $U_a = U_2$

stationary state is reached which is characterized by uphill moving waves. These waves originate periodically at the phase boundary between the convective solid and the granular fluid or even the granular gas and move from the foot of the Faraday hill up to the crest of the hill where they vanish. The lifetime of such a moving wave and hence the periodic length of the wave migration process is about 8 s. A snapshot of this situation together with the respective oscilloscope display is shown in Fig. 8.7a, b. The observed particle flow on the surfaces of the hill and a moving wave is discussed in more detail in Sect. 8.5. It is remarkable that in our experimental setup the occurrence of both, hot spots and uphill moving waves, are only observed

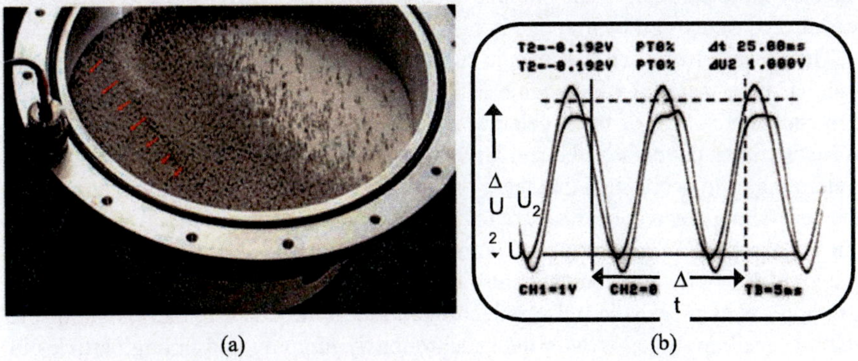

Fig. 8.7 Vertically vibrated thin quartz granulate, driven by a frequency of 66.4 Hz and a very high excitation amplitude with resulting $a_{\text{pk-pk}} \approx 10$ g and $\Gamma \approx 5$. **a** Photo of the granular convection solid with a wave moving uphill; **b** corresponding oscilloscope display of the excitation function $U_d = U_1$ and the response function $U_a = U_2$

in the frequency range from 60 Hz to about 70–80 Hz, but never at frequencies higher than 80 Hz.

The response signal at the oscilloscope display shows a broad periodic peak with an extended flat maximum the slope of which is varying periodically slightly downwards and upwards (see Fig. 8.7b). This may be due to the fact that at large excitation amplitudes the acceleration is limited at the maximum expansion on the upper and lower side of the vertically vibrating system.

8.4 Discussion

8.4.1 Collective Behaviour of the Granular System

It is our opinion that the cooperative movement, i.e. the *jerk*, of many particles in a granular medium is responsible for both, the spatio-temporal structure formation of the vertically vibrating bulk of quartz globules as well as the corresponding shapes of the acceleration response function at the oscilloscope display.

The granulate is neither a rigid body nor a smooth and flexible system of loosely connected particles, which are linked via hinges or springs or movable parts, respectively, but rather a conglomeration of small grains which are coupled by collisions and the associated impulse transmissions. It can also be considered as a system of inelastic clashing spherules which are freely movable in a stream of air and can interact with adhesive forces [9].

If the system is accelerated sufficiently strong against gravity of earth, after a temporary decrease of gravitational acceleration single particles can move from the granular bed and jump upwards. This results into a low-pressure at the remaining bulk whereby more particles of the granular medium are drawn up. A similar explanation was already suggested by Faraday in his pioneering work [1].

The downward falling particles in turn are striking onto the flexible amorphous granular base material which again leads to an impact on the whole bulk. This in turn causes a change of the acceleration of the vibrating system. At certain values of acceleration, cooperative behaviour by a large number of particles of the granular system seems to occur such that these particles are jumping upwards synchronously and also falling down concertedly onto the underlying granular material. This impact can be observed as an abrupt alteration of acceleration, i.e. as a *jerk*. Since the system is driven by a periodical sinusoidal force, the sudden change of acceleration also appears as a periodic pulse reflection—the jerk. In a sense, the system of the granular medium together with the synchronously jumping and falling particles in the air-filled space can be compared to an air-cushioned residuum.

The strength and the time of the collective jumps of many particles are obviously depending on the acceleration caused by the sinusoidal excitation amplitude. In the case of only a few jumps of single particles at low acceleration of the driving force, only a shoulder at the left edge of the sinusoidal response curve is observed. This corresponds exactly to the situation where the Faraday hill begins to emerge from the bulk which is a smooth thin layer ($N \leq 2$) in the beginning of the experiments. Below this threshold, i.e. the formation of a shoulder, the granular medium only follows the sinusoidal excitation of the driving force with a small time difference.

If one gets more and larger collective jumps at somewhat higher accelerations, the impact and thus the retroaction of the air-filled residuum, i.e. the underlying convective bulk, also plays a major role. In this situation, one can observe a strong movement of single grains which are rolling down on the surface of the Faraday hill. Furthermore, a multiphase system emerges consisting of the granular "convective solid", a "granular liquid" and a "granular gas", respectively (see Fig. 8.5a, c). This state is reflected at the oscilloscope display by appearance of a double maximum in the response curve of the acceleration.

Hot spots in form of sputtering sources as well as waves which move uphill occur only if a large number of particles are fulfilling high jumps collectively. But one should keep in mind that the upstroke of the vibrating system is limited for technical reasons such that the excitation is no longer purely sinusoidal, but occurs with truncated elongation at its extreme points.

8.4.2 Response Behaviour of the System as a Whole

As described in the experimental setup (see Sect. 8.2), we monitored the voltage signal $U_a(t)$ of the measured acceleration on an oscilloscope display and subsequently recorded and evaluated it on a computer. It should be mentioned that the accelerometer does neither measure the time-dependent path, i.e. the displacement $z(t)$ of the vibrating system from the rest position, nor the current velocity $v(t) = \frac{\mathrm{d}z(t)}{\mathrm{d}t}$, but rather the acceleration $a(t) = \frac{\mathrm{d}^2 z(t)}{\mathrm{d}t^2}$ of the driven system at time t.

As described in detail in Sect. 8.2, the oscilloscope serves to control the periodic driving signal which is stimulating the shaker. This signal is generated by an external frequency generator and simultaneously transferred to the oscilloscope where the periodic driving voltage $U_d(t)$ is displayed as well as to the power amplifier of the vibratory system. The amplification magnitude can roughly be ascertained from the position of the adjustment button of the power amplifier. On the other hand—as described above, the response behaviour of the whole vibrating system allows determining exactly the acceleration which results from the periodic excitation. In this way, one gets a reliable measure of the acceleration by evaluating the voltage curve $U_a(t)$ of the response signal. In order to understand the influence of the vibration behaviour of the empty container and to separate it from the excitation of the whole system containing the bulk, we have also studied the acceleration of the vessel without any granular material at all frequencies and excitation amplitudes in comparison with the other experiments.

Starting from the time series of the voltage curves $U_d(t)$ and $U_a(t)$, which are displayed at the oscilloscope and recorded on a computer, we suggest in the following a mathematical description which does not refer to the motion of a single particle, but to the moving granulate as a whole. The granular system is stimulated by a sinusoidal driver function $U_d(t)$ according to Eq. (8.3) (see Fig. 8.8),

$$U_d(t) = U_{d,0} + U_{d,1} \sin(\omega_{d,1} t) = U_{d,0} + U_{d,1} \sin\left(\frac{2\pi}{T_{d,1}} t\right) \tag{8.3}$$

where $U_{d,0}$ represents the zero excitation amplitude at a constant voltage offset assigned to a starting time $t = 0$ of a period, $U_{d,1}$ the maximum amplitude, $\omega_{d,1}$ the angular frequency and $T_{d,1}$ the duration of an excitation period, respectively.

In the case of driving functions $U_d(t)$ with small excitation amplitudes, the system responses also with a sinusoidal voltage function $U_a(t)$ as shown in Eq. (8.4) (see Fig. 8.8) which is proportional to the measured acceleration. This function is only slightly phase-shifted as compared to the excitation function $U_d(t)$:

$$U_a(t) = U_{a,0} + U_{a,1} \sin(\omega_{a,1} t + \varphi_{a,1}) = U_{a,0} + U_{a,1} \sin\left(\frac{2\pi}{T_{a,1}} t + \varphi_{a,1}\right) \tag{8.4}$$

with $U_{a,0}$ = excitation amplitude at $t = 0$, $U_{a,1}$ = maximum amplitude, $\omega_{a,1}$ = angular frequency, $T_{a,1}$ = duration of a period of the acceleration cycle and $\phi_{a,1}$ = phase shift, respectively. As expected, in all experiments applies $\omega_{a,1} \approx \omega_{d,1}$ or $T_{a,1} \approx T_{d,1}$, respectively.

If the amplitude of the driving function $U_d(t)$ exceeds the threshold, at which particles begin to jump simultaneously, we extend Eq. (8.4) by a series of periodic

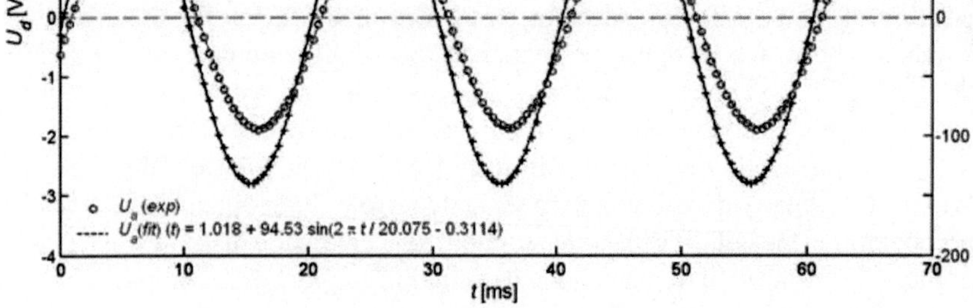

Fig. 8.8 Experimental and fitted driving functions $U_d(t)$ and response functions $U_a(t)$ in the case of no moving particles, i.e. the system follows exactly the driving force. There is only a slight phase shift between $U_d(t)$ and $U_a(t)$ (see also Fig. 8.3b)

Gaussian functions such that our approximation for the voltage curve $U_a(t)$ of the acceleration takes the form of Eq. (8.5) (see Fig. 8.9):

$$\begin{aligned} U_a(t) &= U_{a,0} + U_{a,1}\sin\left(\omega_{a,1}t + \varphi_{a,1}\right) + \sum_{i=2}^{n}\left\{U_{a,i}\exp\left[-\alpha_{a,i}\sin^2\left(\omega_{a,i}t + \varphi_{a,i}\right)\right]\right\} \\ &= U_{a,0} + U_{a,1}\sin\left(\frac{2\pi}{T_{a,1}}t + \varphi_{a,1}\right) + \sum_{i=2}^{n}\left\{U_{a,i}\exp\left[-\alpha_{a,i}\sin^2\left(\frac{2\pi}{T_{a,i}}t + \varphi_{a,i}\right)\right]\right\} \end{aligned} \tag{8.5}$$

where $U_{a,i}$ means the maximal amplitude and $\alpha_{a,i}$ a dimensionless ansatz coefficient related to the half-width of the i-th Gaussian function. The parameters $\omega_{a,i}$ and $T_{a,i}$, respectively, describe the periodicity of the i-th Gaussian function and the variable $\phi_{a,i}$ its phase shift relative to the driving function $U_d(t)$. We like to emphasize that the single terms (8.6) of periodic Gaussian functions,

$$\exp\left[-\alpha_{a,i}\sin^2\left(\omega_{a,i}t + \varphi_{a,i}\right)\right] \text{ and } \exp\left[-\alpha_{a,i}\sin^2\left(\frac{2\pi}{T_{a,i}}t + \varphi_{a,i}\right)\right], \text{ respectively,} \tag{8.6}$$

are taking in account the effect of the jerk which we discuss more detailed in the following section.

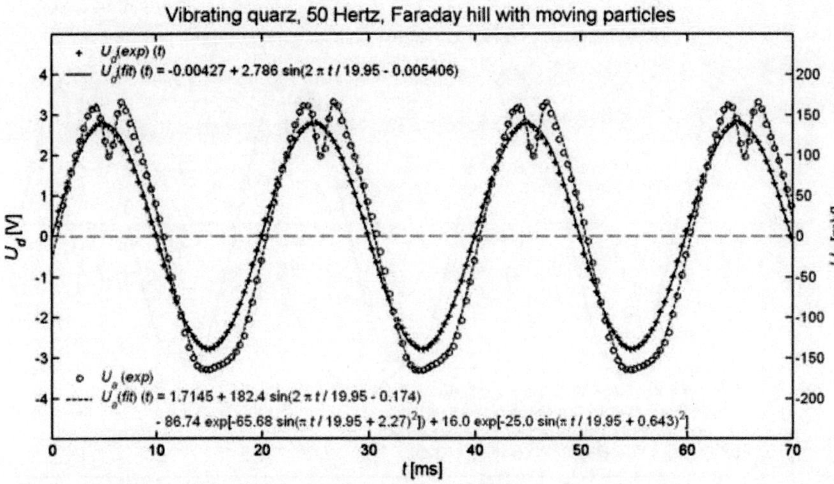

Fig. 8.9 Experimental and fitted driving functions $U_d(t)$ and response functions $U_a(t)$ in the case of simultaneously jumping particles. A pronounced double peak on the maxima and a broadening of the minima of function $U_a(t)$ can be observed (see also Fig. 8.5b)

In our approximations of Eq. (8.5), we have used at most 2 periodic Gaussian functions ($n = 2, 3$, see e.g. Fig. 8.10). The resulting fits show also that $\omega_{a,i} = 1/2\omega_{a,1} = 1/2\omega_{d,1}$ and $T_{a,i} = 2\,T_{a,1} = 2\,T_{d,1}$, respectively, due to the addition theorems of trigonometric functions. The phase shifts which are obtained from the approximations are in the range of $0 \leq \varphi_{a,i} \leq \pi$ and act generally in such a way that the periodic Gaussian functions are almost not overlapping. This leads to the conclusion that during one oscillation period the granular system experiences several jerks of probably unequal intensity which occur at different times within the period.

As an alternative to the periodic Gaussian functions (8.6), one can also use series of terms given by Eqs. (8.7) and (8.8), respectively, in order to approximate the impact of the jerk at higher excitation amplitudes:

$$U_{a,i}\left[\sin\left(\omega_{a,i}t + \varphi_{a,i}\right)\right]^{2\beta_{a,i}}, \quad U_{a,i}\left[\sin\left(\frac{2\pi}{T_{a,i}}t + \varphi_{a,i}\right)\right]^{2\beta_{a,i}} \tag{8.7}$$

$$\exp\left[-\gamma_{a,i}\cos\left(\omega_{a,i}t + \varphi_{a,i}\right)\right], \quad \exp\left[-\gamma_{a,i}\cos\left(\frac{2\pi}{T_{a,i}}t + \varphi_{a,i}\right)\right] \tag{8.8}$$

In the case of the exponentially sinusoidal approach (Eq. 8.7), one obtains suitable fits of the voltage curve $U_a(t)$ of the acceleration only if $2\beta_{a,i} > 100$, and—as a consequence—this approximation was not pursued any more. On the other hand, the approach of Eq. (8.8) can be converted into Eq. (8.6) due to the addition theorems of angular functions; but because of the better interpretability of Gaussian functions we have exclusively worked with Eq. (8.6).

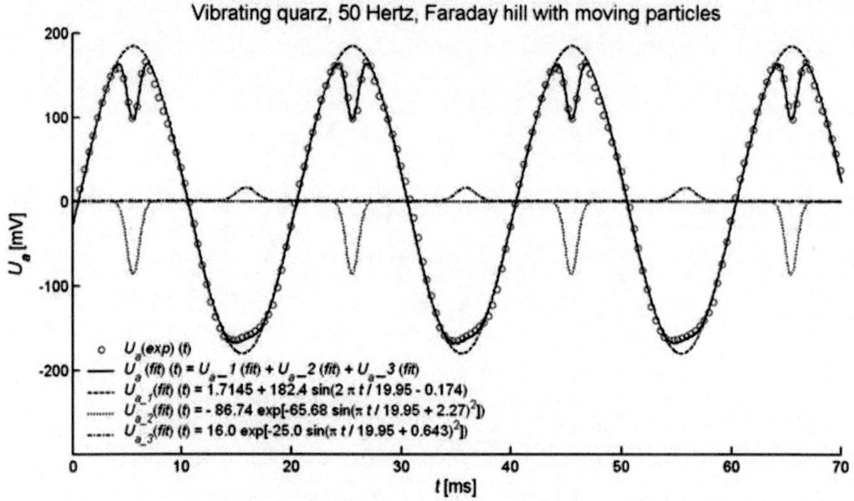

Fig. 8.10 Experimental und fitted response functions $U_a(t)$ in the case of simultaneously jumping particles and the split of $U_a(t)$ into its constituent functions $U_{a\_1}(t)$, $U_{a\_2}(t)$ and $U_{a\_3}(t)$, respectively

8.4.3 *Excitation and Jerk of the Vertically Vibrating System Containing Granular Quartz*

Since the whole vibrating system is stimulated by a sinusoidal driving function, its response function is also a sinusoidal curve proportional to the acceleration, and hence its temporal alteration e_j is proportional to the third derivative of the displacement $z(t)$ of the vibrating system from the rest position with respect to time (Eq. 8.9):

$$e_j \sim \frac{\mathrm{d}\,U_a(t)}{\mathrm{d}t} \sim \frac{\mathrm{d}^3 z(t)}{\mathrm{d}t^3} \tag{8.9}$$

The third derivative of change of position in time or—equivalently—the derivative of the acceleration with respect to time is usually denoted by the term *jerk* (German: *Ruck*). In the common usage of language, however, the notion *jerk* describes the *sudden movement or change* of acceleration. We prefer to adapt this definition of the term *jerk* and suggest denoting the change of acceleration in time more generally as *excitation* (German: *Erregung*). Based upon this interpretation, the *time-dependent alteration of acceleration* e_j—observed at the oscilloscope display—*is* in the case of small driving amplitudes *a sinusoidal excitation* which gets *superimposed by a jerk* if the amplitude of the driving function increases.

Based upon these considerations, the approximation approach of Eq. (8.5) was chosen, which resulted in excellent fits ($R^2 > 0.99$), such that all subtleties of the response function $U_a(t)$ could be reproduced. According to the time-dependent derivative of the acceleration, also the derivatives of these approximation functions in time describe the *excitation* and the *jerk*, respectively, in the sense explained above (see Fig. 8.11). Referring to Eq. (8.5), these two aspects of the response function $U_a(t)$ can be separated on a time scale. On the one hand, the *excitation function*, given by Eq. (8.10),

$$\frac{\mathrm{d}\,U_{a,1}\sin\left(\omega_{a,1}t + \varphi_{a,1}\right)}{\mathrm{d}t} \tag{8.10}$$

corresponds to the time-dependent change of acceleration of the whole vibrating system which occurs on a long time scale. On the other side, the different *jerk functions* (Eq. 8.11),

$$\begin{aligned} \frac{\mathrm{d}\,U_{a,i}\exp\left[-\alpha_{a,i}\sin^2\left(\omega_{a,i}t + \varphi_{a,i}\right)\right]}{\mathrm{d}t} &= \frac{\mathrm{d}\,U_{a,i}\exp\left[-\alpha_{a,i}\sin^2\left(\frac{\omega_{a,1}t}{2}t + \varphi_{a,i}\right)\right]}{\mathrm{d}t} \\ \frac{\mathrm{d}\,U_{a,i}\exp\left[-\alpha_{a,i}\sin^2\left(\frac{2\pi}{T_{a,i}}t + \varphi_{a,i}\right)\right]}{\mathrm{d}t} &= \frac{\mathrm{d}\,U_{a,i}\exp\left[-\alpha_{a,i}\sin^2\left(\frac{\pi}{T_{a,1}}t + \varphi_{a,i}\right)\right]}{\mathrm{d}t} \end{aligned} \tag{8.11}$$

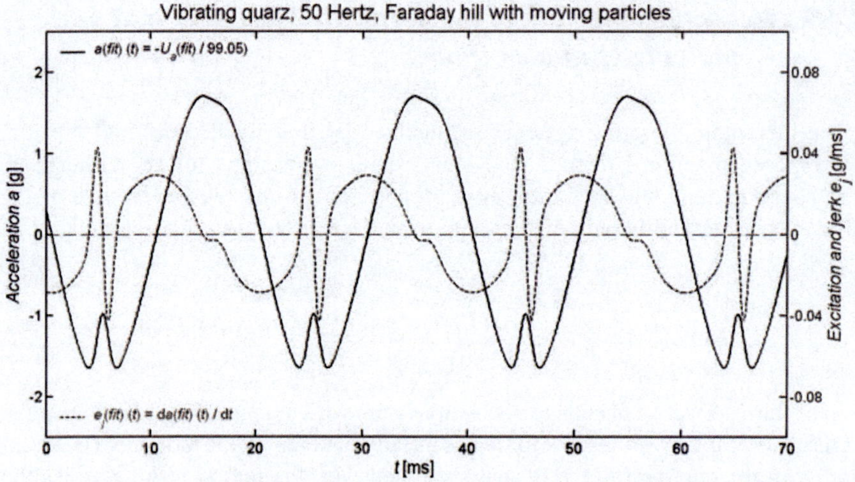

Fig. 8.11 Deduced acceleration function a(t) and its derivation $e_j(t)$ describing excitation and jerk in the case of simultaneously jumping particles. It should be noted that the function a(t) is of opposite sign as compared to the function $U_d(t)$ of the driving force and the displacement z(t). In this diagram the acceleration function a(t) was intentionally mirrored by the time axis as compared to Figs. 8.5b, 8.9 and 8.10. The reason is that the acceleration is proportional to the back driving force of the oscillation system and thus of opposite sign with respect to the stimulating force and the elongation z. Hence, if one plots $U_d(t)$ and z(t) in positive direction of the vertical coordinate axis (not shown here for simplicity), a(t) must correctly be drawn in the opposite—i.e. negative—direction

take place on a rather short time scale as compared to the period length T. This can be demonstrated by calculating half-widths $t\_half_{a,i}$ of $U_{a,i}(t)$ $(i \geq 2)$ according to Eq. (8.12):

$$t\_half_{a,i} = \frac{2T_{a,i}}{\pi} \arcsin\left(\sqrt{\frac{\ln 2}{\alpha_{a,i}}}\right) \tag{8.12}$$

In the case of the fitted jerk functions shown in Figs. 8.9 and 8.10, one obtains $t\_half_{a,2} = 1.31$ ms and $t\_half_{a,3} = 2.125$ ms, respectively, which is significantly below the period length $T = 19.95$ ms. As a consequence, the slope of the derivation function e_j of the acceleration changes almost suddenly within a very short time interval. *This is the jerk!*

8.4.4 *Power Spectrum of Response Function $U_a(t)$*

For the sake of completeness, the power spectrum of the discussed response function $U_a(t)$ (Figs. 8.9 and 8.10) was generated by fast Fourier transform (FFT). Figure 8.12

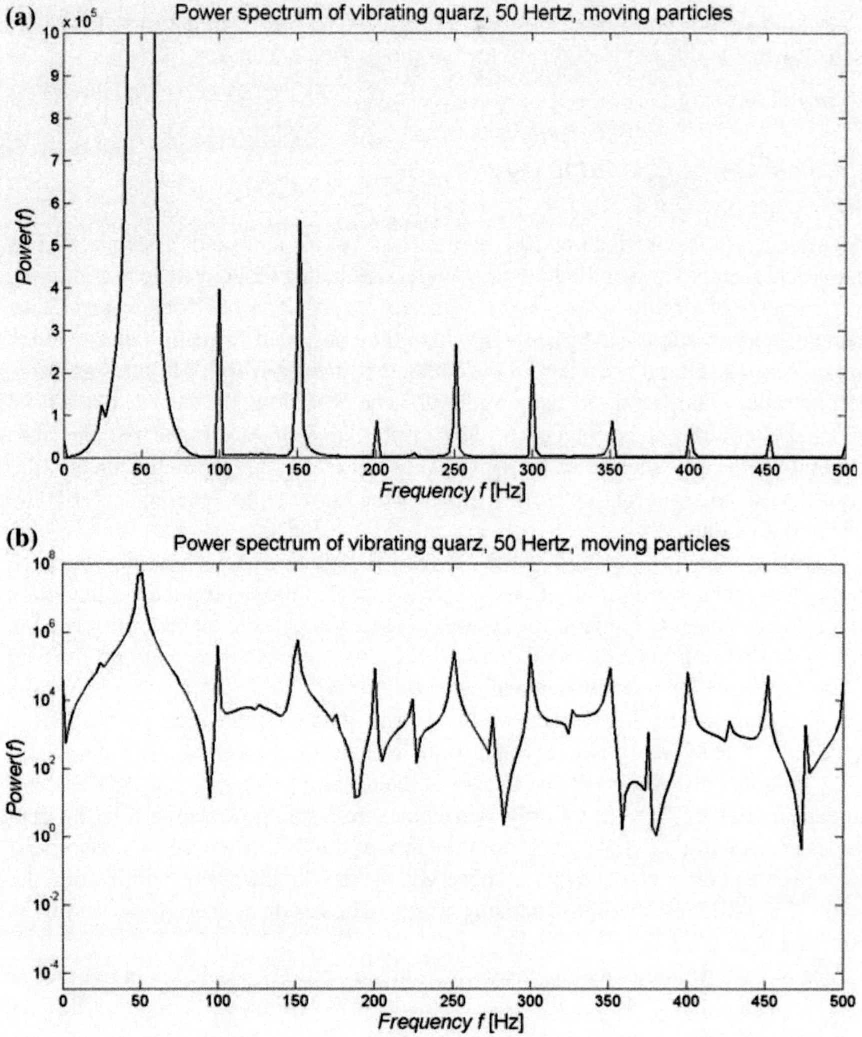

Fig. 8.12 Power spectrum of the response functions $U_a(t)$ in the case of simultaneously jumping particles. **a** Representation in Cartesian coordinates; **b** representation in semi-logarithmic coordinates

shows this spectrum in Cartesian (a) and semi-logarithmic (b) representation, respectively. This power spectrum exhibits in addition to the main frequency 50 Hz of the driving force a sub-harmonic frequency of about 25 Hz with much lower magnitude. This corresponds to a slight modulation of the amplitudes within two subsequent periods. Beside this sub-harmonic, no further significant frequency is observable in Fig. 8.12. The subtleties, which we have found by decomposition of the function

$U_a(t)$ into a pure sinusoidal term and a sum of periodic Gaussian terms is only reflected by the different heights of the harmonic frequencies.

8.4.5 *Flow of Granular Quartz Beats*

Gerner et al. [17] described in 2007 the interplay of air and sand. They showed by numerical simulation that the flow of granular media in a Faraday heap is connected with the flow of air through this heap. Some years ago, Laroche, Douady, and Fauve [5] executed excellent experiments in which they observed formation of a Faraday hill close to the lateral boundary of the container dimension 100 * 12 mm^2 as stable configuration. They reported also on an avalanche flow along the surface of the heap which they assumed to be compensated by an internal circulation of the particles from the bottom towards the top of the heap. Akiyama et al. [22] presented spectacular snapshots of this internal convective flow of the particles in heaps of a vertically vibrated granulate.

Laroche et al. [5] stated, that the convective regime is not observed when the granular layer is too thin, i.e. if $N = \frac{h}{d}$ is too small. The experiments which they reported have been carried out for $10 < N < 140$ and particle diameter dispersions of 0.63 – 0.80 mm. For $N \approx 45$ at the starting point of the experiment, they observed a hill, the height between top and valley of which was about 30 mm. This is just the height of the Faraday hills which we have investigated. But in contrast, we started with $N \leq 2$. However, we observed also the formation of a convective solid which is the Faraday hill. We found that in case of the driving frequency $f = 66.4$ Hz and an acceleration of $\Gamma \approx 5$ not only avalanches are running down the hill, but also waves are moving uphill, starting from the foot of the hill, where $N \approx 1$. By visual inspection and recording movies we observed the flow of the grains within both, the body of the hill and the uphill moving waves. The result of these observations is sketched in Fig. 8.13.

The sketched flow of grains inside the Faraday hill (Fig. 8.13) is based on the ideas of Miao et al. [24] and the measurements of Akiyama on the convex flow of grains in vertically vibrated granular beds [23].

In our experiments, the hill is always formed close to the border of the container as earlier observed by Laroche et al. [5]. A steady downhill flow of beats is created on top of the hill as well as on top of the wave moving uphill. At the line in which the surface of the moving wave meets the slope of the hill, the beats which move on both the surfaces vanish into the bulk of the convective solid.

Caballero and Melo [25] observed uphill running droplets of fine silica particles of 0.1 mm in diameter on convex surfaces of curvature radius $R_c = 57$ mm vibrating vertically at $f = 20$ Hz. The surface on which the droplets are running belongs to the container. In contrast, the uphill moving waves we observed are formed on the slope of the Faraday hill but not on a specially curved ground of the container.

In case of a somewhat lower amplitude of the sinusoidal driving force and hence the acceleration ($\Gamma \approx 4.2$), we observed the formation of sputtering sources on the

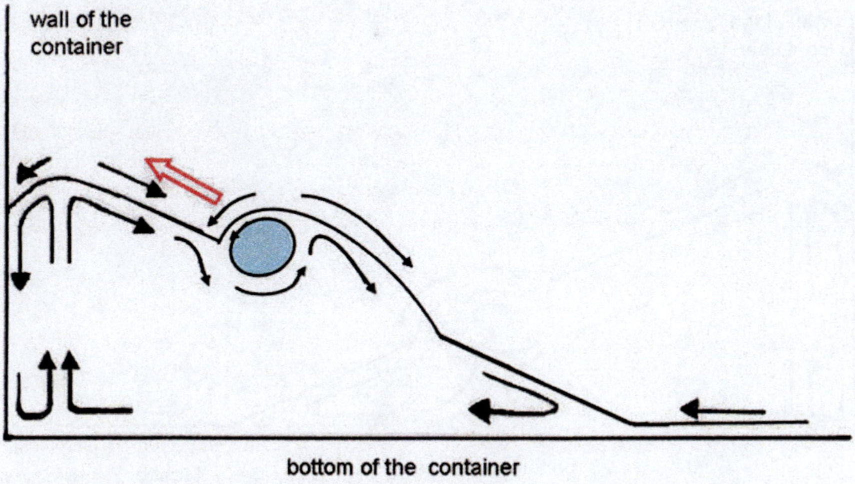

Fig. 8.13 Schematic representation of the observed particle flow and the moving wave on the Faraday hill (for comparison see also Fig. 8.7a, c)

slope of the Faraday hill. If they are located close to the foot of the hill, they remain stable in space and time, whereas they walk around and vanish after some minutes, if they are created far away from the foot of the hill. The resulting flow of the quartz beats on the surface of the hill and within the sputtering sources which we observed visually and by recording films leads us to the proposal for the granular convection as shown in Fig. 8.14.

In the centre of the small circular source heap, the beats are sputtering to the surface and flow down in radial direction away from the centre. In order to understand this flow, we suggest that there exists a toroidal vortex beneath the surface of the source and hence an underlying saddle point (see Fig. 8.14).

8.5 Summary and Outlook

It was our interest to study the behaviour of thin granular layers in an enlarged container under vertical vibrations which might be useful for technical purpose. Furthermore, our container should be large enough to observe granular pattern formation almost undisturbed by its boundaries. Because of the lack of accuracy in the inclination of the vertically driven container as compared to the apparatus of de Bruyn et al. [26], the large Faraday hills move incidentally to an arbitrary higher part of the inclined container wall. A similar effect was already observed in case of the studies carried out by Laroche et al. [5].

In contrast to other authors, we started the experiments with small ratios $N = \frac{h}{d} \leq 2$ of almost equally distributed grain particles. As a result, the symmetry breaking at

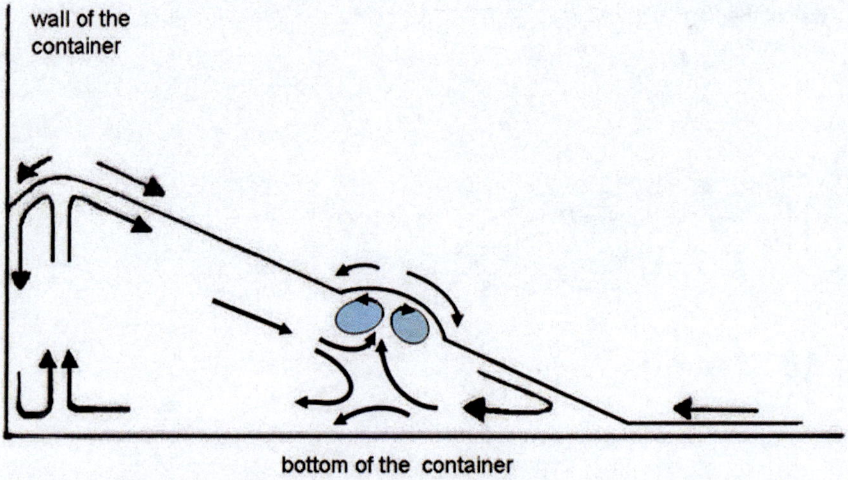

Fig. 8.14 Schematic representation of the experimentally observed surface flow of beats in the case of sources on slopes. The arrows symbolize the flow of the grains on the surfaces as well as in the body of the convective solid (see for comparison also Fig. 8.6a)

higher amplitudes ($\Gamma \geq 1$) enabled us to detect a variety of phenomena which have not yet been carefully investigated, i.e.

- "phase-separations" between the "granular convective solid", the "granular liquid" and the "granular gas", respectively, and the sublimation of the "convective solid" without avalanches,
- "granular waves" which move uphill, starting at the foot of the Faraday hill, and
- stable sputtering sources located close to the phase separation line at the foot of the Faraday hill.

The described experiments on inert quartz particles serve us as a starting point to obtain the basic physical properties of vertically shaken systems in an air flow with Faraday instabilities.

It was our aim to use an experimental setup with technical relevance like mining, fine chemicals and pharmaceutical processes.

In 1997 Linz [27] wrote:

> "On the macromechanical level, however, the collective dynamics of the grains is often comparably simple, although often very surprising since we have not reached an intuitive understanding yet. Facing the fact that theories from first principles are not analytically manageable at present, it is nevertheless worthwhile to study how much we can learn about granular systems without knowing the micromechanical details of the problem."

We agree with the statement of Linz. Technical development is not depending on deep theoretical understanding of all the problems which are connected with the implementation of a new technology. Therefore, the intuitive understanding of the processes which are inherent in the technical procedures are of great importance. Our work should serve this purpose.

In a future work, we intend to apply the results to carry out tribochemical reactions [28] in vertically vibrated reactors under atmospheric pressure and low temperatures (room temperature or only slightly above).

Acknowledgements We are indebted to Markus Eiswirth and Alexander Mikhailov for useful support and stimulating discussions and the mechanical factory of the Fritz-Haber-Institut der Max-Planck-Gesellschaft for the manufacturing the container.

References

1. Faraday, M.: Philos. Trans. R. Soc. London **52**, 299 (1831)
2. Chladni, E.F.F.: Die Akustik. Breitkopf und Härtel, Leipzig (1802)
3. Savage, S.B.: J. Fluid. Mech. **194**, 457 (1988)
4. Evesque, P., Rajchenbach, J.: Phys. Rev. Lett. **62**, 44 (1989)
5. Laroche, C., Douady, S., Fauve, S.: J. Phys. (France) **50**, 699 (1989)
6. Clémont, E., Duran, J., Rajchenbach, J.: Phys. Rev. Lett. **69**, 1189 (1992)
7. Pak, H.K., van Doorn, E., Behringer, R.P.: Phys. Rev. Lett. **74**, 4643 (1995)
8. Melo, F., Umbanhowar, P.B., Swinney, H.L.: Phys. Rev. Lett. **75**, 3838 (1995)
9. Jaeger, H.M., Nagel, S.R., Behringer, R.P.: Rev. Mod. Phys. **68**, 1259 (1996)
10. Ramirez, R., Risso, D., Cordero, P.: Phys. Rev. Lett. **85**, 1230 (2000)
11. Ristow, G.H.: Pattern Formation in Granular Materials. Springer, Berlin (2000)
12. Garcimartín, A., Maza, D., Ilquimiche, J.L., Zuriguel, I.: Phys. Rev. E **65**, 031303 (2002)
13. Behringer, R.P., van Doorn, E., Hartley, R.R., Pak, H.K.: Granul. Matter **4**, 9 (2002)
14. Risso, D., Soto, R., Godoy, S., Cordero, P.: Phys. Rev. E **72**, 011305 (2005)
15. Pastor, J.M., Maza, D., Zuriguel, I., Garcimartín, A., Boudet, J.-F.: Phys. D **232**, 128 (2007)
16. Garcimartín, A., Pastor, J.M., Arévalo, R., Maza, D.: Eur. Phys. J. Spec. Top. **146**, 331 (2007)
17. van Gerner, H.J., van der Hoef, M.A., van der Meer, D., van der Weele, K.: Phys. Rev. E. **76**, 051305 (2007)
18. Brilliantov, N.V., Albers, N., Spahn, F., Pöschel, T.: Phys. Rev. E **76**, 051302 (2007)
19. Brilliantov, N.V., Pöschel, T., Kranz, W.T., Zippelius, A.: Phys. Rev. Lett. **98**, 128001 (2007)
20. Haß, E.C., Plath, P.J.: VFI Der Versuchs- und Forschungsingenieur. **42**(1), 42 (2009)
21. Thomas, B., Squires, A.M.: Phys. Rev. Lett. **81**, 574 (1998)
22. Akiyama, T., Shinmura, K., Murakawa, S., Aoki, K.M.: Granular Matter **3**, 177 (2001)
23. Akiyama, T., Aoki, K.M., Yamamoto, K., Yoshikawa, T.: Granular Matter **1**, 15 (1998)
24. Miao, G., Huang, K., Yun, Y., Zhang, P., Chen, W., Wei, R.: Phys. Rev. E. **74**, 021304 (2006)
25. Caballero, L., Melo, F.: Phys. Rev. Lett. **93**, 258001 (2004)
26. de Bruyn, J.R., Lewis, B.C., Shattuck, M.D., Swinney, H.L.: Phys. Rev. E **63**, 041305 (2001)
27. Linz, S.J.: Lect. Notes Phys. **484**, 306–318 (1997)
28. Thiessen, P.-A., Meyer, K., Heinicke, G.: Grundlagen der Tribochemie. Akademie-Verlag, Berlin (1967)

Chapter 9
Dissipative Sculpturing of Beige Jasper of the Eastern Desert of Egypt

Hartmut Linde

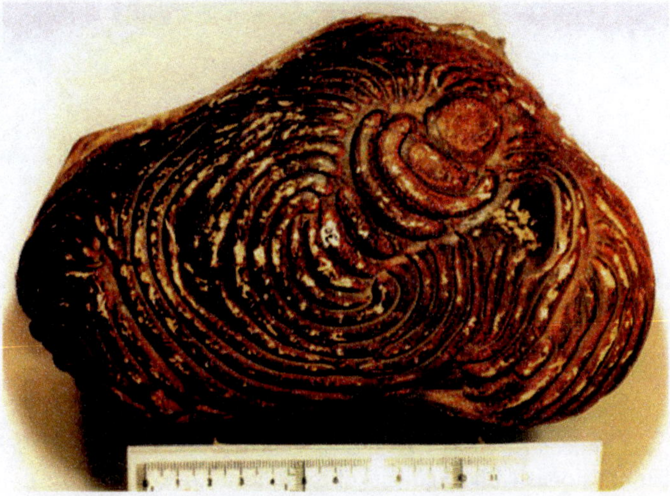

Fig. 9.1 We found also a stone, at which by chance and disturbance a banana-like center was created

9.1 Introduction

H. and G. Linde found these stones first gathered by Bedouins in their camp and later with help of them directly at the spot of their origin, a chalk-like hill (between the 26th and 28th parallel of latitude and the 33th and 34th degree of longitude) in the eastern mountain-desert of Egypt (Fig. 9.1). Figure 9.2 shows the biggest found

P. J. Plath et al., *Imagery Synergetics*, Understanding Complex Systems,
https://doi.org/10.1007/978-3-030-95607-3_9

Fig. 9.2 Prof. Hartmut Linde, sitting as a human sale on a sculptured stone with a fragment of a relief (see arrow) of four concentric rings with a "wavelength" of 5 upto 8 cm

stone with H.L. as scale (and the chalk-like hill as background) with a fragment of a relief of four concentric rings with a "wavelength" of 5 upto 8 cm (See also in Fig. 9.3 other big stones with fragments of the relief).

The surface of the hill is covered with many of these stones, of which the bigger ones are mostly broken. At cliffs these stones are embedded in the walls of the chalk-like ground. A part of the stones shows the relief in view. Stones laying at the surface of the hill show at their top a crackled brown desert lacquer of Fe/Mn-compounds at their top, the underside is beige like freshly broken stones or if they are taken immediately out of the wall.

The analysis of the chalk yields 93.3% $CaCO_3$ and 6.7% of clay. We identified different small marine organisms (including silica-algae) and nearby a fossil coral reef, thus we claim the chalk-like hill for an uplifted fossil marine sediment—after geologic studies—of the Nummulite-formation of the lower Eocene. The as jasper identified nodules have a micro-crystal quartzite-like structure with narrow packed quartz-grains of 0.005 mm and are scarcely polluted by clay or chalk [1, 2]. After the far-reaching destroying of this formation, the nodules of jasper are destroyed and

Fig. 9.3 Another stone with sculptured patterns of quite different sizes: on the upper side wavelengths of cm scale; in the middle and lower part with in mm scale

fragmented by thermal and sand-born erosion and cover now also areas of the gravel-desert (Sserir) in the desert [2]. These fragments exhibit very seldom such small traces of the relief in view, that we could identify them only afterwards only with the knowledge of the intact stones. Such fragments—we found afterwards also in the desert—are available in the mineral collection—misunderstood as "Nilkiesel"—of the "Mineralogische Sammlung des Naturkundemuseums der Humboldt Universität zu Berlin"3 but without any knowledge of the intact relief and its origin. We did not found such intact stones with reliefs in other mineral collections. Likewise we found neither their description nor a relevant theoretical background to these phenomena in the literature [3–6]. Thus we were encouraged to select (and to buy) a well-aimed series from the most simple up to the most complex stones (already gathered in the camp of the Bedouins and offered to visitors). We did this for the reconstruction of a physico-chemical process in the far past finished at casual intermediate states by external influences.

9.2 Stones, Successions, Concretions and Reliefs Over a Broad Range of Sizes

The smallest stones found are often spherical (with diameters in the range 0.5–2 cm) without surface relief. Figure 9.4 shows e.g. examples of a small, unbroken stone and a broken one. Yet, inside the last exhibits at the sectional view one concentric ring with the outer surface, discriminated by its colour relative to the background material. This shows evidence of a superposed layer due to growth in two steps

Fig. 9.4 Smallest spherical stones with surface reliefs

forming a succession. (Larger stones of 15 cm can indicate until eight internal levels!) Further, we see three steps of two or three united nodules as concretions, of which a dumbbell-shaped one is superimposed around the waist with a flat layer.

One broken part of another bigger dumbbell is also superimposed around the broken waist but with a relief consisting of one and a half concentric ring (a second similar ring-system at its backside and signs of over-layered inner levels at the broken side exist but are here not shown. (The other broken part of the dumb-bell is not available). Such favoured over-layering of the concave waist even with the fine-sculpture (reliefs with elliptical rings and spirals at one stone) are also shown in Fig. 9.5a, b).

Larger stones found (in the range of 10 cm diameter) have mostly disc-shaped or parabolic, egg-shaped forms. Figure 9.6a–c shows three sides of a large concretion of two egg-shaped parts. Figure 9.6a looks like Bismarck with a smooth and bald neck and with a destroyed armor. At its backside (Fig. 9.6b) we see the continuation of the bulges from the first side, coming from a central disc with disturbed central ring-bulges, visible from the side at Fig. 9.6c. The second bald part of the head shows a smooth basic surface of this part and at least three sources of irregular bulges.

Another big stone (Fig. 9.7) shows—as expected when cut—in its sectional view elliptic rather than circular rings inside. Such stones are often both, successions and concretions, here connected with three parts. Also as units looking stones are often composed of several parts unified by a common layer, as one can see at the sectional view.

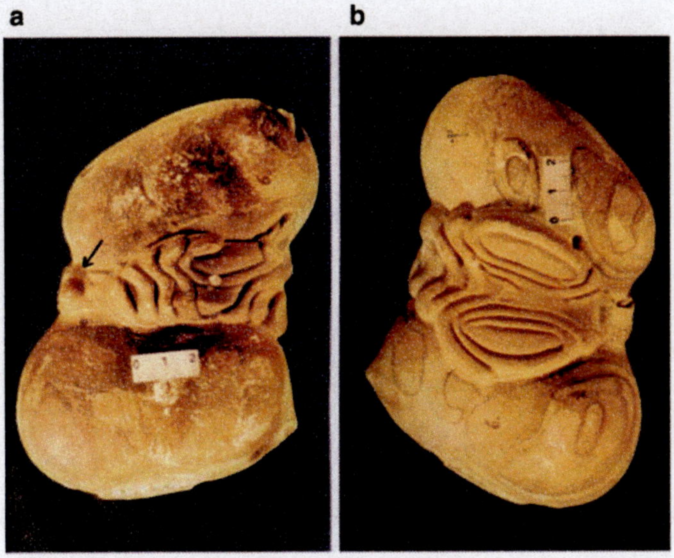

Fig. 9.5 **a, b** Another larger dumbbell is also placed around the broken waist, but with a relief consisting of one and a half concentric ring

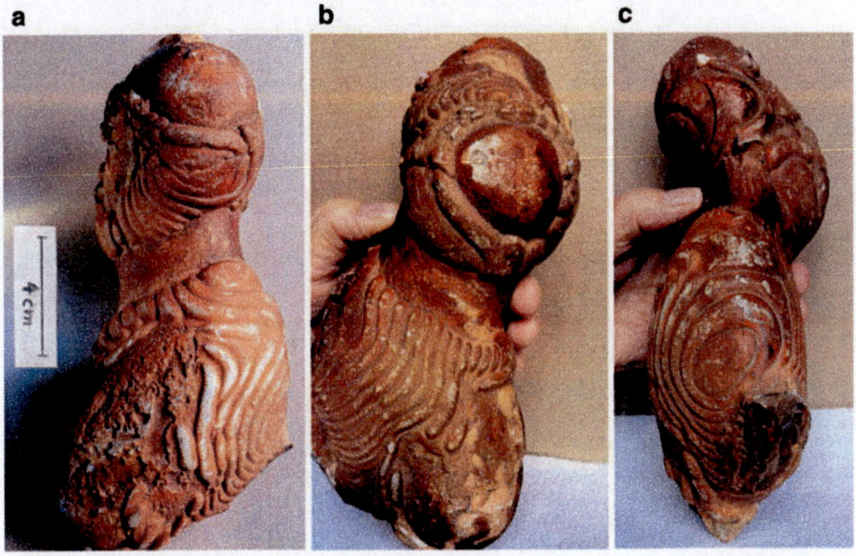

Fig. 9.6 **a, b c** Larger stones found (in the range of 10 cm diameter) have mostly disc-shaped or parabolic, egg-shaped forms

Fig. 9.7 Also stones, looking as units, are often composed of several parts unified by a common layer, as it is to seen at its sectional view

9.2.1 An Imaginable Basis Relief

In relatively large stones (10 cm diameter or higher) a layer with a relief may be cover just one or several areas not all regular although illustrating a common origin as we shall discuss below. We have found surface reliefs with concentric rings as regularly formed bulges around a central disk. Figure 9.5b shows also two elliptical central discs, which created elliptical bulges and an elliptical spiral. (At this stone, we see further the traces of strongly eroded reliefs as a proof that too late created reliefs could not harden before the final stop of the physical processes by external geological reason.) Figure 9.8a shows a stone covered with regular concentric rings at the one side, while the other side shows according to Fig. 9.8b regular spiral-shaped bulges both around a central disc. The spiral occurs, when the central disc is disturbed accidently by a step sideways of the surround. Figure 9.8b imposes also additionally by two nucleating half-spheres discussed later.

The concentric ring-shaped relief around a central disc impresses as the imaginable basic form of the relief. Figure 9.9 shows a stone with the smallest "wavelength" with 0.08 cm and a second stone with a wavelength of 1.5 cm. The largest wavelength found was about 8 cm (Fig. 9.2) at the largest found stone in 80 cm diameter range. The radius of the central disc is about two times larger than the wavelength in first order approximation, and these values, radius and wavelength, are related to the thickness of this layer.

On occasion when a stone is disc-shaped or paraboloid, the relief layer is mostly localized with the central disc at the less convex sides. At dumb-bell shaped stones the over-layering—with or without reliefs—prefers the waist between the both spheres

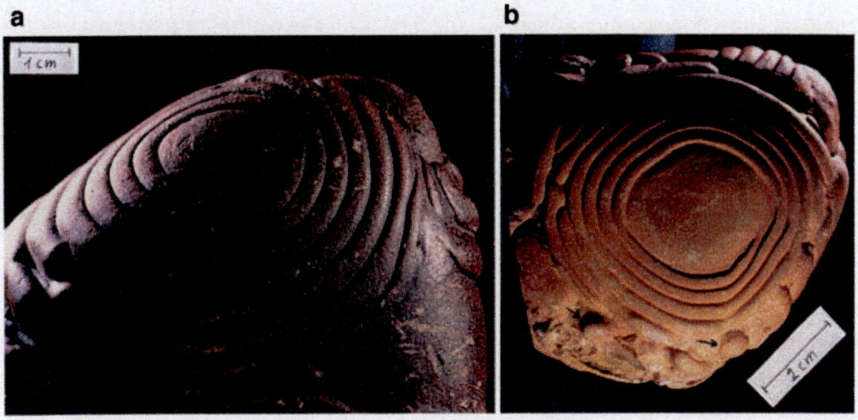

Fig. 9.8 The same stone with quite different structures: **a** concentric rings on the foreside, but **b** a spiral on its rare

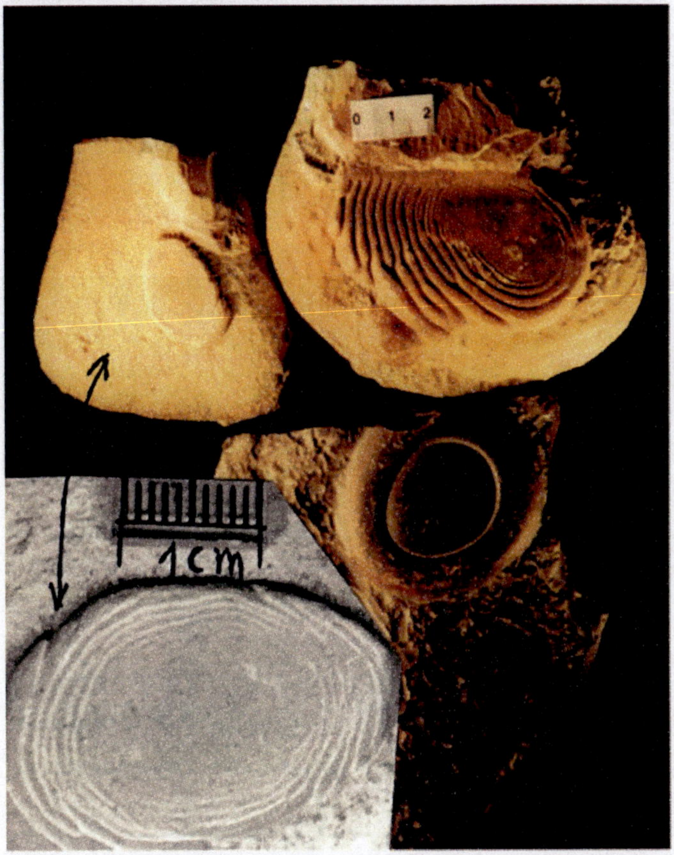

Fig. 9.9 Concentric rings on a very small scale; partially with period doubling (right above)

that is the most concave line (Figs. 9.4 and 9.5a, b). (The central disks appear preferably at the most concave spots by heterogeneous nucleation at the hardened surfaces, as later discussed).

9.2.2 The Variability of the Relief by Deformations, Dislocations and Bounds

Often the central disks are deformed, e.g. to elliptical discs, then also surrounded by elliptical bulges. There are also bounds, which make the discs un-complete so that the bulges are also un-complete and cover only one side of the disc. Figure 9.10 shows an overview about some concentric ring-bulges and spirals-bulges for comparison. However, a dislocation at the periphery of the disc is connected with both left- and right-handed spirals and even with two-armed ones.

The spirals can be regular (Figs. 9.8b and 9.11) but also deformed around an elliptical disc (Fig. 9.5a) or as wrinkled spiral (Fig. 9.12) restricted by bounds. See in Fig. 9.12 also two banana-like nucleation plots in the groove between bulges at the sculptured surface. (Usually nucleation spots are half-spherical, as visible in Fig. 9.8b). That can be caused by the additional over layering over an already sculptured layer (with a modification of the sculpture, see the concluding remarks). There are also simple side step dislocations of the bulges, see e.g. both sides of a stone in Fig. 9.13b. The bulges can stepwise surround bounds, or penetrate gaps in them

Fig. 9.10 Overview on some concentric ring-bulges and spirals-bulges for comparison

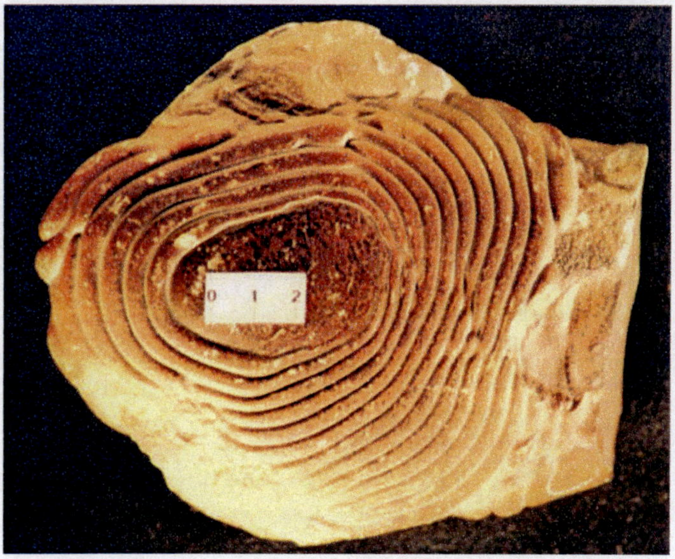

Fig. 9.11 Regular spirals

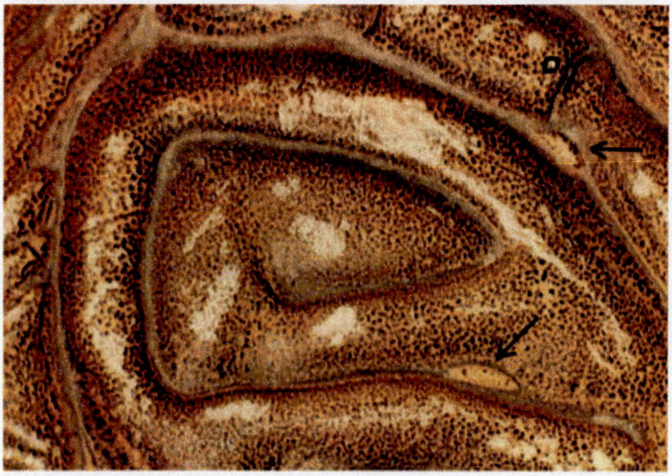

Fig. 9.12 Wrinkled spiral with two banana-like nucleation plots in the groove between bulges (have a look at the arrows)

and can even be unified afterwards when before separated by a bound (Fig. 9.14). When more than one relief area appears at a stone, no interference or overlapping is recognizable when they touch each other (Fig. 9.15). Many peripheral bulges are irregularly covering large parts of the stone, showing a lot of the mentioned

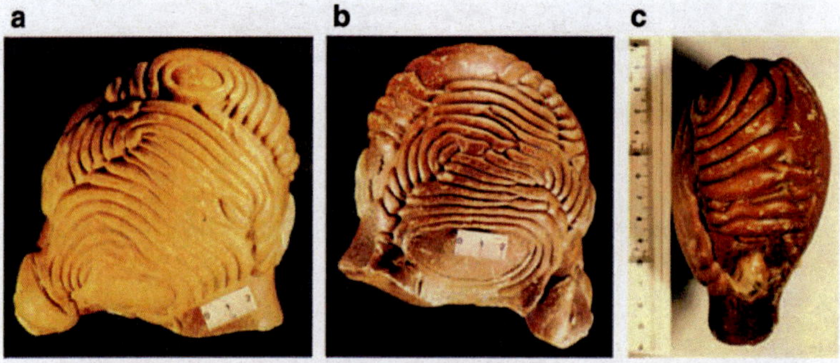

Fig. 9.13 Simple side step dislocations of the bulges. **a** Front side, **b** back side, **c** right side of (**a**)

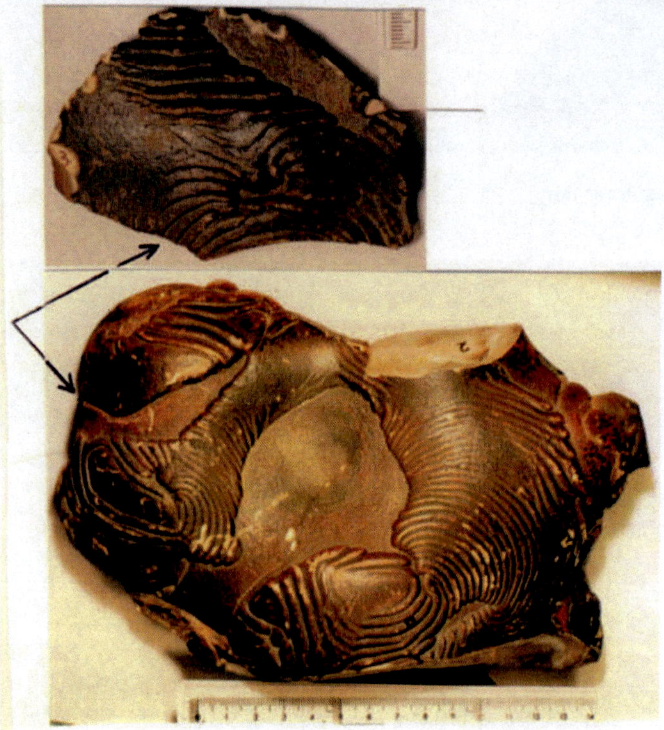

Fig. 9.14 Separated bulges can unify

Fig. 9.15 Bulges do not interfere

dislocations. There are also boundary-lines as well between the reliefs coming from different leading centers as even inside of reliefs coming from only one disc.

9.3 Growth of an Amorphous Body When Diffusion, Geometry and Physical-Chemical-Thermodynamic Conditions Play Together

It seems to be plausible that silica-gel accumulates as precursors of jasper had originated at first by local homogeneous nucleation in the chalk-like sediment when formed at the ground of the shallow sea covering North Africa in the early Eocene. The necessary supersaturated solution of $Si(OH)_4$ in this sediment can be produced by simultaneously sedimented silica algae because of their highly solubility and metastable SiO_2-skeletons as un-equilibrium crystals. This source of soluble $Si(OH)_4$ can be maintained over long time intervals.

Then for the overall process the porosity of the amorphous silica-gel, internal diffusion and accumulation of $Si(OH)_4$ are expected to play a crucial role in the growth of the stone found without to shrink. The chalk particles do not participate as nucleation points they are separated from the silica-nodules and are not bounded by silica bridges. Rather the homogeneous and heterogeneous nucleation seems to be a local spontaneous phase separation of the silica-gel from the aqueous chalk-like sediment with an inter-phase between them. This local accumulation of $Si(OH)_4$ forms at first growing spherical accumulates and increases locally in the gel the accumulation pressure and helps overcoming the expected yield shear-stress of the expandable chalk-like sediment.

On this way, only at critical values of shear stress the nodule tends to keep growing. Neighbouring nodules when touching each other during growing can be backed together to units called concretions (see Figs. 9.5a, b, 9.6a–c and 9.7). Soon after an insulated nucleation point a nucleus starts to grow in the bulk of the sediment, centripetal diffusion helps spherical growth of the small silica nodules. This process can be seen approximately as the growth of a spherical Newtonian drop (easily deformable silica gel with negligible Bingham-behaviour [7]) in extended sediment with Bingham-behaviour. The latter corresponds to a continuous medium in which besides the viscosity η and the shear-stress τ and also a critical shear-stress τ_B gradient of the so-called Bingham-behaviour is important for the possibility of an elastic deformation or of a flow [8]. When the applied shear-stress remains below the critical value τ_B, there is only elastic deformation following Hooke‘s law, but once this limit is exceeded, the Bingham body flows—in first order approximation—as a Newtonian liquid. Note also that at variance with the Newtonian flow, the flow of a Bingham body in a tube does not follow a parabolic velocity profile but rather exhibits a flat front with very sharp decline to the zero velocity at the wall. Accordingly, the shear-stress at solid boundaries is quite high. This behaviour influences just as the Bingham-flow at other solid surfaces and prefers a similar more easy expansion along solid walls, when there the pressure is coming from an at this wall sitting nucleus of an expanding second phase. (Possibly, the maximal shear-stress near the solid surface—later additionally increased at corners and grooves and working at molecular-smooth surfaces of the hardened accumulate—can explain the favoured wall-led and groove-led expansions.)

9.3.1 The Grow-Stop Dynamic Due to Decreasing Shear Stress with Increasing Radius of the Nodule

For a growing nodule, the shear stress in the surrounding quasi-solid sediment is relatively high for lower radii. This phenomenon—confirmed by model-calculation of the shear stress [9, 10]—is also well known to occur at the tip of a needle or when using knifes both with very small radii at the tip or blade cutting butter or clay, two Bingham-like materials. (It needs a much higher pressure to deform these materials with a tennis ball). Thus depending on these shapes we may locally be above or below the critical radius respectively yield stress and therefore the growth of the nodule is possible or not. Thus, the form-constant growth of spherical nodules stops very soon when the critical radius is reached and the shear stress remains below the critical value. Remember, that the accumulation-pressure is limited! Other-sides, an instability during growing leads mostly to the mentioned eggs and discs (and finally also to a layer) because a growing disc can preserve at their circumference an under-critical radius.

9.3.2 *The Diffusion-Controlled Accumulation Pressure Leads also to the Grow-Stop Dynamic, Which Limits the Volume Increase of the Silica Body*

As noted earlier [11, 12], diffusion controls the supply and eventual accumulation of $Si(OH)_4$, which increases for instance the spherical volume V_n, and, consequently, the outer spherical surface of the silica nodules, A_n. As time proceeds, pressure evolves proportional to the relative volume increase

$$\frac{dP_{acc}}{dt} = \frac{dV}{V}dt \tag{9.1}$$

An estimate can be obtained if we take into consideration in the diffusion law

$$\frac{dn}{dt} = -DA\frac{dc}{dx} \tag{9.2}$$

the relative volume increase of one gmol $SiOH_4$ by multiplication with V_{gmol}/V. Then we have:

$$\frac{dP_{acc}}{dt} = \left(\frac{dnV_{gmol}}{V_n}\right)dt = \left(\frac{dV}{V_n}dt\right) = -DV_{gmol}\left(\frac{dc}{dr}\right)\left(\frac{A_n}{V_n}\right) \tag{9.3}$$

(r is the radius of the spherical accumulate). Assuming that the in the time scale considered diffusion coefficient D, the concentration-gradient dc/dr and V_{gmol} remain constant, then the rate of pressure increase is proportional to the ratio of surface to the volume in a nodule, thus it scales

$$\frac{A_n}{V_n} = \frac{4\pi r^2}{\frac{3}{4}\pi r^3} \sim \frac{1}{r} \tag{9.4}$$

Thus dP_{acc}/t is decreasing if the spherical nodule gets bigger. That is on principle in similar form also valid for other shapes of the growing nodules and even for growing cover layers with and without the reliefs. This is a second mechanism of stop in the growth of the nodule, when the accumulation pressure P_{acc} becomes to short. As mentioned before, the very un-ripe accumulate exhibits already during its slow growing a steady solidification by internal diffusion and accumulation of $Si(OH)_4$, what is a competing process against the external diffusion and accumulation. The complete volume V_n of the freshly formed nodule works therefore as additional sink for the diffusion of $Si(OH)_4$.

Thus the above diffusion controlled stop of growth is additionally supported.

Furthermore, this internal diffusion and accumulation is maintained after stop of growing of the nodule until the next solid intermediate phase of silicate is reached. Thus the restoring of a sufficient high concentration gradient of $Si(OH)_4$ necessary for a further accumulation is prevented and delayed. The resulting long enough time of rest enables other-sides an effective solidification process of the attained intermediate state of the nodule as precondition for the following local heterogeneous nucleation and for the facilitated expansions along a solid surface.

These considerations are sustained by the following findings: In some cases with several relief patches at the same surface we can recognize that one relief is completely preserved while nearby another relief is very much eroded, see Fig. 9.5b.

It is obvious that the eroded relief is the last developed relief, which has had no time enough for complete solidification because of the final stop of internal and external accumulation due to the final stop of diffusion by external geological reason. (The nearly same "wavelengths" and thickness of the ring-bulges show, that this final stop is not caused by a higher Bingham yield or viscosity of the sediment, because this would likely change the wavelength!).

This diffusion-controlled accumulation pressure limits not only the radius respectively volume of the spherical primary accumulate body—but also the radius (and volume) of other shaped bodies—and even of nuclei created at the surface of the basic body. It influenced also the complete or un-complete over-layering, the last responsible for the development of the relief.

9.3.3 *The Role of Heterogeneous Nucleation with Respect to the Creation of Local Leading Centers for Further Growing by Over-Layering and Fine-Sculpturing*

It is not sure, whether the already mentioned first point-like nucleation is a homogeneous or heterogeneous one. More important is, that with the occurring of already solidificated silica nodules as intermediates, local heterogeneous nucleation of the accumulate (the less condensed phase) are possible and favoured at these higher condensed and not more deformable surfaces. This is well known from thermodynamic criteria, which favour in addition the most concave spots of this surface for nucleation. This enables the locally selected restart of expansion—with falling favour—at grooves, concave corners, and plane surfaces, because here at more concave surfaces the volume of the initial nucleus and therefore the nucleation work can be minimized [13]. We have found evidence of such behaviour e.g. by studying the localization of the leading centers of the reliefs at the surface of the solid silica bodies. At Fig. 9.5a, b, we recognize in addition an incomplete over-layering of the waist of the dumb-bell. In addition, a leading center is stopped in the early state of growing—at one side in the groove at Fig. 9.5a (see the arrow)—because of external influences. In the photos of Fig. 9.8b, such two nuclei restarting in over-layering at different spots, are visible, marked by the arrows. Generally, we see at the most disc-shaped stones that the central discs as the leading centers for concentric rings and spirals of the sculptures localized near the flattest and most concave sides.

9.3.4 Concentric Spheres Inside of Spherical Silica-Stones by Over-Layering Due to Facilitated and Therefore Preferential Tangential Slide-Way Expansion Along the Solidificated Surface of Their Precursor Bodies, in Short: Wall-Led Expansion

Let us have, e.g., a perfect spherical intermediate solidificated by internal accumulation after the growth has been stopped plus an adequate residence-time until the supersaturated concentration of $Si(OH)_4$ is restored. As there is a supersaturated medium and heterogeneous nucleation is possible this may well occur at any point of the sphere. However, as a half-spherical nodule expands along solid walls then soon or later, as noted above, the shear-stress decreases as the radius of these half-sphere increases, the shear stress finally takes on values below the yield threshold τ_B at the calotte thus stopping there locally the growth of the nodule. (This growth process is finished, when due to the diffusion $dP_{acc} \sim 1/r$ and $dP_{acc} \sim 1/V_n$, we reach locally or for the whole volume a value below the yield stress.) However, the remaining maximal shear stress at the corner of the nodule with the basic wall enables a remaining tangential expansion along the wall. This process is called "tangential wall-led expansion".

Note, that the stop of the further expanding at the calotte is ruled by τ_B. A low τ_B: has to lead to a large thickness of over-layering and wavelength, respectively, and vice versa, whereby the velocity of over-layering, which is anyhow very slow because of the diffusion, is here seemingly without influence.

The over-layering can continue until the surface of the initial spherical nodule is completely covered and the growth stops because the condition for the facilitated wall-led expansion is not more given. That is an additional stop mechanism.

This can provide after repeated solidifications followed by heterogeneous nucleation—multiple layer structures, one kind of the so-called successions. We have found stones with up to eight layers! Only small ones can be concentrically over-layered by regular spheres, because the above-mentioned instability leads also to discs and ellipsoids consisting of several covering layers (compare Fig. 9.7). These over-layering are not very regular because of inhomogeneous conditions.

An over-layering can also be seen in Fig. 9.16a.

There exist also typical successions looking like discs, which show additionally a relatively thicker layer—like a unique very thick ring-bulge—around the peripheral border (Fig. 9.16a). (From Fig. 9.16b it can be seen that the backside of this stone

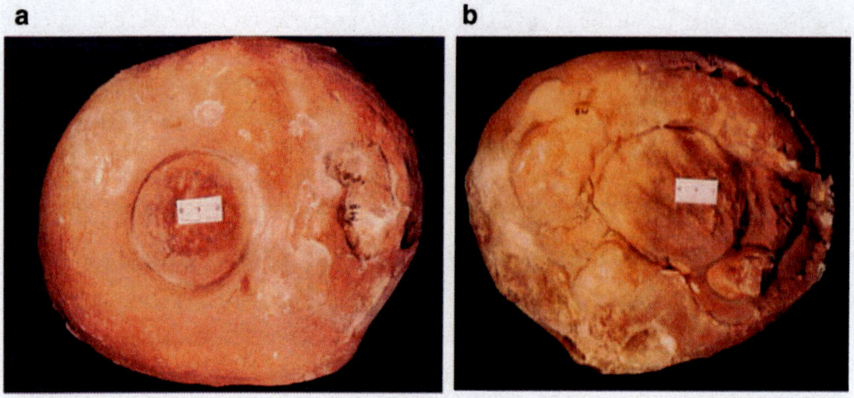

Fig. 9.16 **a** Typical consequences that look like discs, forming a relative thick layer—like a single, very thick ring bulge—around the circumference, **b** (right one) back of the sculpture (**a**)

is a large concretion backed together from at least 6 parts.) The common periphery ring-bulge is the extreme result of the fact, that reliefs covering whole discs show relatively larger wavelengths of the bulges at this peripheral border, which is more convex. Though this border is not favoured for nucleation, the over-layering from a nearby occurring nucleation seems to be facilitated and only stopped at a larger thickness. The reason may be the maintaining of a constant boundary-angle of about 120° between bulge and the basic wall, which influences the radius of the growing accumulate surface. It can be shown, that the very convex radius of the border induces a relative smaller radius of the spherical surface of the accumulate nucleus during growing under the influence of this constant boundary angle of about 120°, which is estimated between nuclei and the basic surface, between bulges and the basic surface and even between the touching ring-bulged themselves. See also Fig. 9.14a–c, showing the different wavelengths and thicknesses of the relief at the flat and at the convex sides of the same stone.

9.3.5 Concentric Ring-Bulges Around the "Central Disc" Due to Repeated Local Over-Layering by Facilitated Slide-Way Expansion Along the Grooves at the Border of the Central Disc and Equally at the Border of the Consecutively Step-Wise Developed Ring-Bulges, in Short: Groove-Led Expansion

As above mentioned, the premature stop of such over-layering due to $P_{acc} \sim 1/r$ respectively $P_{acc} \sim 1/V$ leads to an disc-like layer (central disc!) sitting at the surface which is also subjected by the following solidification and later heterogeneous nucleation. Now, the heterogeneous nucleation is preferred in the groove around the—in the regular case circular—central disc, because this is the most concave line.

The nucleus find here two solid walls linked each other also with an angle of about 120°. That leads to a stronger shear stress along this groove than along an only flat surface. Thus, the so-called "tangential groove-led expansion" is favoured against the tangential expansion led only from one wall. Consequently, this groove-led expansion is restricted to a near region of the groove and forms there a ring-shaped bulge around the central disk. Only in seldom cases—obviously of disturbances at the grooves—the area-covering over-layering according to the tangential—of only one—wall led expansion was observed too, instead of this kind of forming a ring bulge.

The unifying by coagulation of the two fresh fronts of the bulges (at the side opposite to the nucleation spot at the central disk) complete a ring-bulge and this kind of facilitated expansion is also stopped. This is again an additional growth-stop mechanism, as mentioned also before with respect to the complete over-layering of a nodule. Thus, the immediate further expansion is prevented and after solidification of the ring-bulge and following favoured nucleation at its external groove, the next ring-bulge can be developed. In an analogical way as multiple concentric layers, also numerous concentric ring-bulges around the central disc can by formed at the surface.

The forming of the central disc with the concentric ring-bulges initiates the developing of the basic relief at the surface. Note, that the formation of these typical concretions and successions and of the basic relief rules by the same principles.

9.3.6 Possible Reasons for the Increase of Variability of Concretions and Successions and Especially of the Complex Relief (Dissipative Sculpture)

It plays obviously a crucial role for the variability of the relief, whether touching fronts of bulges—either during surrounding the central disk or when they came from different central discs—are able to coagulate or not. There are—deduced from topological images of the reliefs—three possibilities:

1. Both fronts are liquid-like fresh and coagulate to a unified form. The circumstantial evidence for this is the existence of completely regular concentric ring-bulges and of two-times over-layered spherical nodules, both without traces of the unification.
2. One front e.g. of a bulge is already stopped and aged and the other colliding bulge is fresh enough and follows the wrinkled groove by a side-step dislocation, see Fig. 9.14b.
3. Both fronts e.g. of colliding bulges are just before they are forced to a stop by the diffusion controlled dynamic. We expect in this case not a change of wavelength

but a deceleration of the growing velocity leading to an already premature aging and solidifying of the surface.

That can prevent an easy coagulation and can lead to an incompletely unified thinner spot of the bulge or to a complete stop of both fronts with a sharp grooved boundary in between. Such examples are bodily available shown e.g. in Fig. 9.14a–c.

4. A break or shearing of the expanding front of the central disc is responsible for the origination of spiral-shaped bulges and the two possible directions of the shearing break correspond to the two directions of the spirals (Figs. 9.8b, 9.10 and 9.12).

   That can be caused by local differences of the expansion velocity possibly by very local differences of the yield value and/or the viscosity of the Bingham-body. But if these material properties are only small and wide spread, they can be discussed as reasons for the deformation of the—under homogeneous conditions—regular circular central discs with circular concentric bulges. (They are often elliptically or un-regularly deformed).
5. Other-sides can very sharp local material properties can lead to the bounds as not penetrable boundary lines for the expanding elements of the relief. Typical examples show Figs. 9.14a–c and 9.16.
6. Simultaneous nucleation—a random process preferred in the most concave places—can help develop reliefs that are more complex. This process also takes place on extensive surfaces such as grooves along bulges. Together with premature stops of expanding elements of the relief, these processes can also contribute to the development of complex reliefs (Figs. 9.14, 9.15 and 9.16).
7. Note, that a regular ring-bulge system contains very defined convex backs of the bulges and sharp grooves between them (also with an touching angle of about 120 °C). The slope of the backs looking into the direction of the central disc is a little bit steeper than the slop looking to the periphery. This can help to understand the origin of disturbed and un-complete sculptures.

   These dislocations together with the concentric rings and the different kinds of spirals can show—at larger stones—a higher variability of the relief than of fingerprints.

9.3.7 Concluding Remarks

In our search of the literature, we have found neither evidence of the sculptured silica stones descript in this report, nor a theory that could be used to explain the sculpturing phenomena we have discussed here. However in the Catalogue by A. Seilacher [14, 15] there is an interesting photo concerning a

> --huge block (length 184 cm) of Upper Jurassic limestone fallen into a clay-filled fissure (see Fig. 9.17). This is surprising, because karst fissures are results of limestone dissolution, while the micritis crust bearing these ornaments originated certainly by precipitations. Even though the origin of sculptures is still problematic, they clearly belong to the family (morphospace) of zebra-patters found in very disparate situations.--

That is really very far from a try of explanation of this fascinating sculpture.

Here we are immediately forced to compare the origin of the sculptures of this stone with the principles developed for the understanding of our stones from the eastern desert of Egypt.

Maybe this—in German: "Gedankenexperiment"—intellectual game can motivate the search for the physical and chemical conditions of the origination of this fascinating sculpture in view. (Every theory is in the beginning an intellectual game like also in our try to understand the sculptures originated in Egypt and we cannot avoid the following speculations).

This stone (Fig. 9.17) contains obviously at the complete—also as flat assumed surface—the basic sculpture as reported here of silica-concretions. However, this sculpture is only at one end of about 1/3 of the surface of the stone visible recognizable with the two typical central disks surrounded by concentric rings. The middle part of the big stone is additionally over-layered at the second time finally the other end of the stone is again—at the third-time—over-layered both with a very similar but typically modified sculpture.

These modified sculptures can be explained—with our intellectual speculation corresponding to our above problem—by a starting with banana-like nucleation-germs in the bended grooves of the foregoing sculptures. These germs are growing for a short time by favoured expansion along the groove originating than the banana-like shape. Than follows at the backside of the banana—and further at the backside of the horse-shoe-like bended bulges new nucleation, which form finally a heard-like sculpture, when the open ends of bended bulges are unified. That is the result of the same groove-lead lateral expansion as explained also in our above problem. (The grooves around the half-concentric bulges have a dominant influence on the expansion in comparison with the grooves of the already sculptured basic surface).

It was a help for this speculation, that we found in our stones two Banana-like nucleation-points at already with the basic-sculpture covered surface (Fig. 9.13, see the arrows). We checked also a long series of other photos of gathered sculptured stones (which we did not buy) and fund no signs of a secondary over-layering. There was in the far past obviously no enough time for creating the modified sculpture by secondary over-layering. However, we found also a stone, at which by chance and disturbance a banana-like center—as part of the first overlaying—was created for the following developing, see Fig. 9.18. We recognize at least the horseshoe open bended (half-concentric) bulges.

By the way, there exist also concentric-ring-sculptures observed at SiO_2 and Cornelian Chalcedon. Here we would now not exaggerate the speculation.

Fig. 9.17 The stone that gives rise to our speculation. It contains obviously at the complete—also as flat assumed surface—the basic sculpture as reported here of silica-concretions

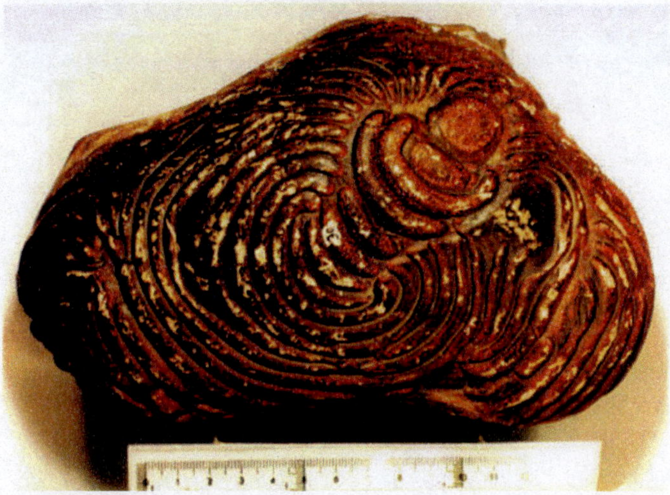

Fig. 9.18 A stone, at which by chance and disturbance a banana-like center—as part of the first overlaying—was created

Appendix

Finally, it is worth to speak about a simple experiment, which shows, that the so-called nonlinear behaviour of a Bingham-body is able to form a dissipative sculpture by its deformation. You take a knife with a not dentate but sharp side and push it into vertical direction one or two mm deep into the flat surface of not to warm butter or margarine or clay. Those are Bingham-bodies because of the a little bit stable packing of small droplets or particles. Then move the knife sideward along the horizontal surface of this body. That results in a wavy system of this material, which climb upwards at the blade. The usually visible dislocations depend on different wavelengths caused by different deep pushing of the knife into the paste. That is a proof for a two-dimensional periodic flow-stop behaviour, which deforms the Bingham-body and moves them along the slide even against the gravity (see Fig. 9.19a, b).

Very surprising is a rough analogy to our problem:

The pressure—working by the moving knife—forms at first a bulge, but the growing of the bulge stops at a definite size of the bulge. That corresponds to the stop-dynamic of the silica-accumulate in the Bingham-sediment because of the not over a critical value increasing pressure. Then the bulge moves along the blade surface even against the gravity. That corresponds to the favoured wall-led expansion of the silica-accumulate. Simultaneously follows further inflations of bulges and their moving along the blade, respectively wall, and that in repetition.

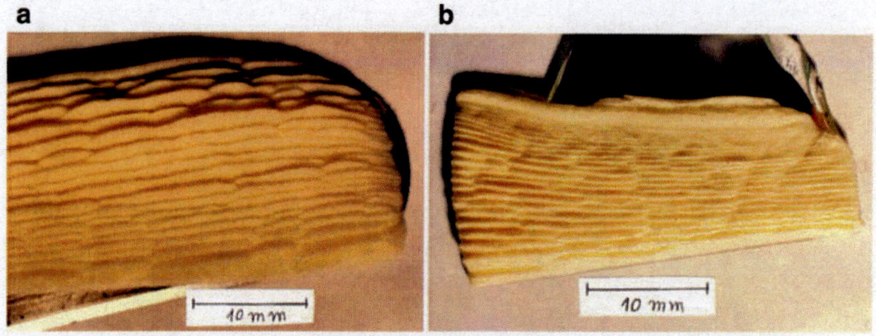

Fig. 9.19 **a**, **b** Butter proof for a two-dimensional periodic flow-stop behaviour, which deforms the Bingham-body and moves them along the slide even against the gravity

References

1. Kühne, K.: Institut für Silikatforschung, Humboldt-Universität Berlin, personal communication (1998)
2. Bautsch, H.B.: Mineralogische Sammlung Naturkunde-Museum, Humboldt-Universität Berlin, personal communication (1998)
3. Bauer, M.: Edelsteinkunde, pp. 568–593. Tauschnitz, Leipzig (1909)
4. Landmesser, M.: Mobility by metastability: silica transport and accumulation at low temperatures. Chem. Erde **55**, 149–176 (1995)
5. Landmesser, M.: extra Lapis 7, Cristian Weise Verlag München (1994), pp. 49–80
6. Landmesser, M.: Mobility by metastability: in sedimentary and agate petrology: applications. Chemie der Erde **58**, 1–22 (1998). Gustav Fischer Verlag Jena
7. Bingham-fluid: Special, non-Newtonian fluids named after Eugene Cook Bingham are called Bingham fluids. Their dynamic viscosity is a function of the shear rate (or shear gradient) and leads to a linear flow behaviour; https://www.de.wikipedia.org/wiki/Bingham-Fluid. Accessed 8 Aug 2020
8. Gibbs, J.W.: Thermodynamische Studien. Verlag Wilhelm Engelmann, Leipzig (1892)
9. Grassmann, P.: Physikalische Grundlagen der Chemie-Ingenieurtechnik, 158, 276,294,796,800, Aarau/Frankfurt a. M. (1961)
10. Diener, G.: Expansion of a hollow sphere in the environment with Bingham-behaviour. Technical University Dresden, personal communication
11. Linde, H., Linde, G., Velarde,M.G.: Morphogenese von Kiesel-Kongretionen mit komplexen Dissipativen Skulpturen, Non-equilibrium processes and dissipative structures in geoscience, (Selbstorganisation, Jahrbuch für Komplexität in den Natur-, Sozial- und Geisteswissenschaften), vol. 11, pp.141–185 (2000). Dunker & Humblot, Berlin
12. Linde, H., Linde, G., Velarde, M.G.: Esculturas disipativas, Circulos y espirales en rocas del desierto, Investigation y ciencia, Edition espaniola de Scientific American. Marzo (2008)
13. Rusanov, A.I.: Phasengleichgewichte und Grenzflächenerscheinungen. Akademie-Verlag (1978)
14. Seilacher, A.: Fossil Art, Catalogue by A. Seilacher, An Exhibition of the Geologisches Institut Universität Tübingen (Germany), The Royal Tyrell Museum of Palaeontology, Drumheller, Canada (1977)
15. Seilacher, A.: Self-organizing mechanism in morphogenesis and evolution, in constructional morphology and evolution. In: Schmidt-Kittler, N., Vogel, K. (Eds.), pp. 151–271. Springer-Verlag, Berlin (1991)

Chapter 10
Complex Dissipative Structures Mainly at Liquid/Liquid and Liquid/Gas Interfaces

Examples

Hartmut Linde, Kerstin Eckert, and Karin Schwarzenberger

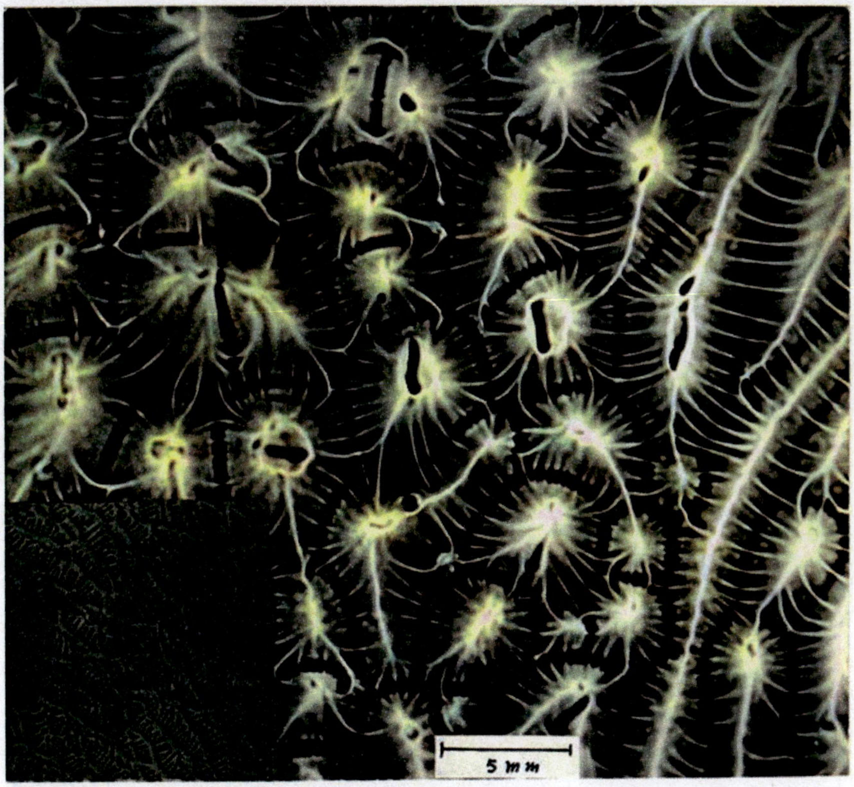

Fig. 10.1 An impression of collective behavior due to wave-like relaxation oscillation

Figures and captions inserted by E. C. Haß according to templates of H. Linde.

P. J. Plath et al., *Imagery Synergetics*, Understanding Complex Systems,
https://doi.org/10.1007/978-3-030-95607-3_10

10.1 Introduction

Well-known—and often to see in the daily weather forecast—are the chaotic macroscopic flow-systems of air and sea-water in one's own land and occasional also about the whole world. Both flow systems, the first of air and the second of water, are coupled mutually to a partially ordered and also chaotic disordered spatial and temporal system (see e.g. Fig. 10.1, demonstrating collective behavior). The physical reason for these dissipative structuring's is the energy-input by the sunshine resulting into temperature differences of the air- and water-phase. Even the solid crust of the earth respectively their parts are subjected to breaks, uniting's and migrations, in all cases with a large time scale. In the last case the driving force is the heat-production deep in the earth with heat-transfer to the surface, resulting in dissipative structures of flows in the melted mineral area. Lastly, we are confronted with Rayleigh-Instability under geometric inhomogeneous conditions. (In the last-mentioned case we can roughly make out systems of convection-cells similar to Fig. 10.4 but with such roll cells in only one liquid phase and of course driven by buoyancy forces because of a vertical gradient of temperature.)

At first, we show here partially similar but also more complex—however small-scaled—hydrodynamic dissipative structures at liquid/liquid and liquid/gas inter-phases under the driving force of interfacial tension differences by mass- or heat-transfer. (In the report about the dissipative sculpturing of jasper—the deformation of the plastic Bingham-sediment by microscopic breaks is a little bit similar to the macroscopic breaks during the deformation of the solid crust of the earth also by hydrodynamic forces).

10.2 Dissipative Structures by Heat- and Mass-Transfer Through Liquid/Liquid and Liquid/Gas Inter-Phases Driven by Self-Organized Differences of the Surface-Tension

Let us start with the Bénard experiment [1], reported in 1900. Figure 10.2 (mass-transfer from 5% butanol + cyclohexanol/water, 4 s after start) shows such Bénard cells (respectively roll cells of first order (RCI) by mass-transfer of surface-active agents from the higher viscose liquid phase into a lower viscose phase—or by heat-transfer from a liquid into air). The heating of the fluid lowers also its surface-tension $\frac{d\sigma}{dc}$, which is together with the concentration-gradient $\frac{d\sigma}{c} \cdot \frac{dc}{dx}$ the driving force for the interfacial-convection of Marangoni-instability [2].

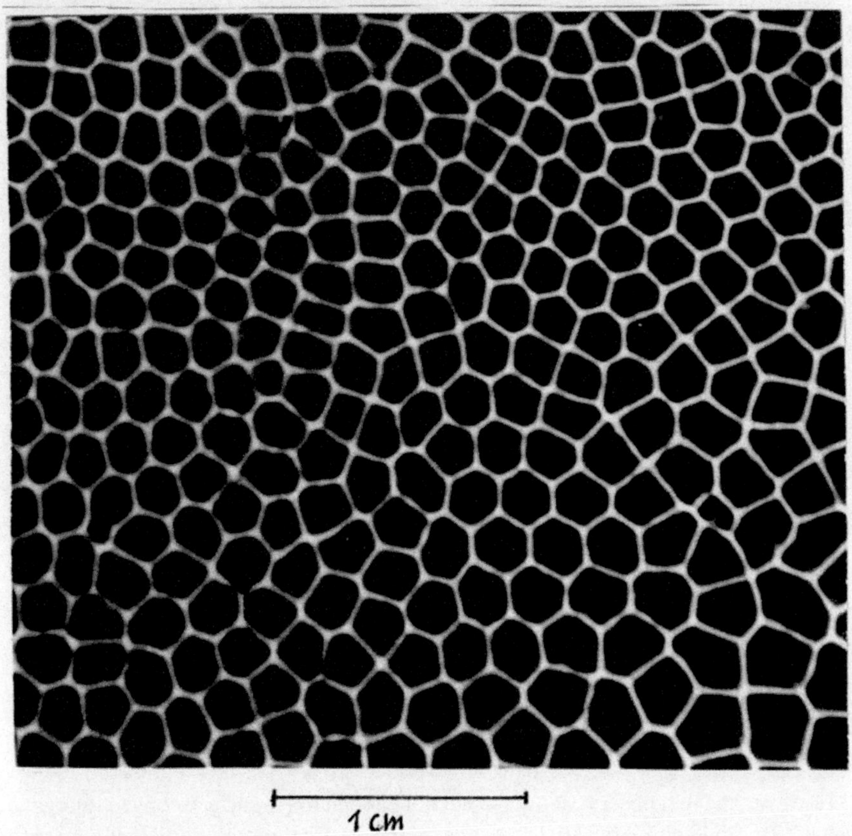

Fig. 10.2 Bénard cells, respectively roll cells of first order (RCI); mass-transfer from 5% butanol + cyclohexanol/water, 4 s after start

Bénard believed, that the convection due to the difference of the density was the driving force, but Rayleigh showed 1916, that the driving force of the density-difference was in this case not yet sufficient as driving force. Marangoni described 1865 on the other hand the observation, that the periodic occurring of the well-known rows of tears of wine results from an instable surface-tension distribution at the meniscus: That gives the name Marangoni-instability!

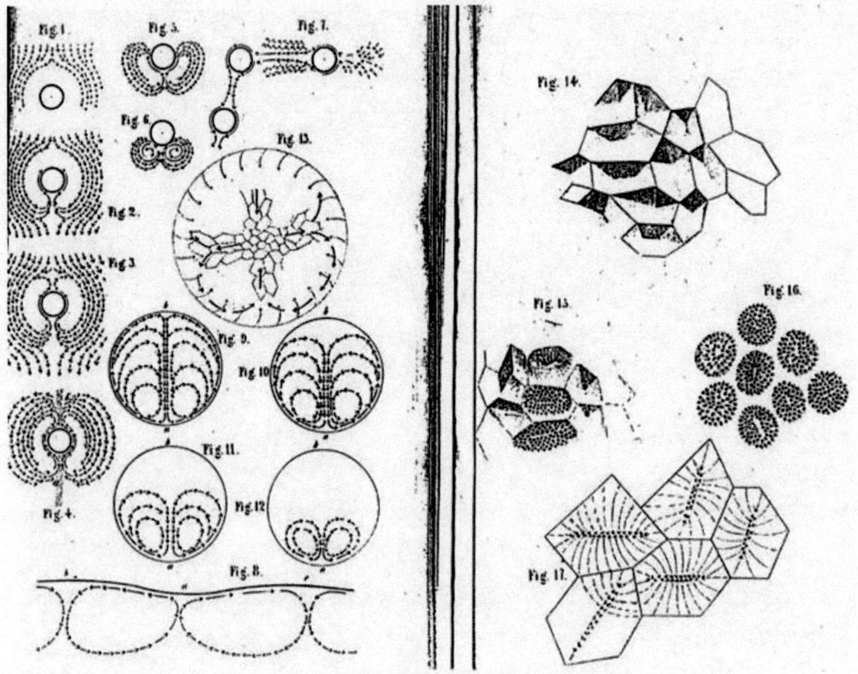

Fig. 10.3 First observations of surface convections given by Weber [3] above 1854

However, the first observations of—but not understood—surface convections gave Weber [3] 1854 with Fig. 10.3 "Microscopic observations of very low-governed movements (convections) which occur during the formation of the solution of resin into spirit". Spirit (ethanol) is the surface-active solvent and resin the matter of higher surface-tension! We see mainly circulating flows typical also for Marangoni instability [4–9].

Figure 10.4 shows the convection system of Fig. 10.2 in both phases in the side view in a capillary split of 320 μm with very small particles made 2 s after the start. (The other pictures are made with shadow-graph technique). The system is 1% $C_{16}H_{33}$ OSO_3Na+ iso-pentanol/water. The flow to the inter-phase occurs at the center of the roll cells and the flow back occurs at the border lines of the cells, see the arrows. The convection flows from spots at the interface with low surface tension to spots with higher surface tension.

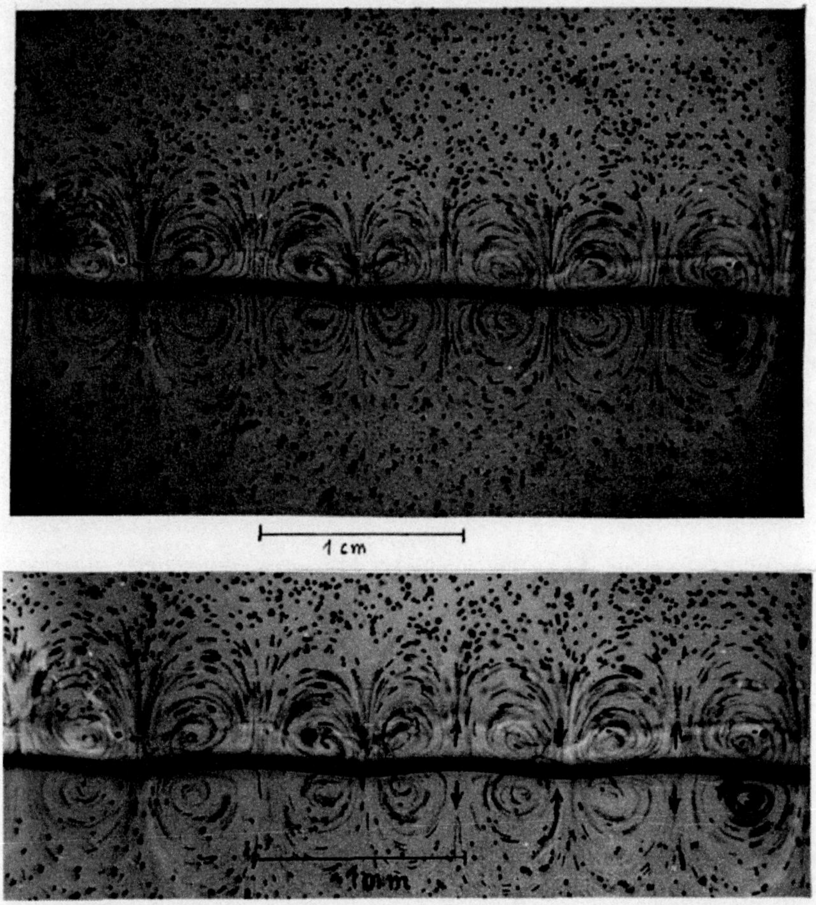

a)

b)

Fig. 10.4 Two convection systems of 1% $C_{16}H_{33}$ OSO_3Na + iso-pentanol/water with both phases in a capillary split of 0.32 mm 2 s after start of mass transfer

The—in the meantime in same cases outdated theory—of the so-called stationary regime contains not only stationary roll-cells of first order: they are hierarchically ordered of I. order (not sub-structured), second order (one times sub-structured) and third order (two-times sub-structured), see the sketch Fig. 10.5. It shows, that with increasing driving force the roll cells of first order transits into roll cells of second order and finally into roll cells of third order. (But there exist also chaotic and wave-like relaxation oscillations in such hierarchic system ordered and depending on the driving force and on the physical properties of the systems, as we show here.)

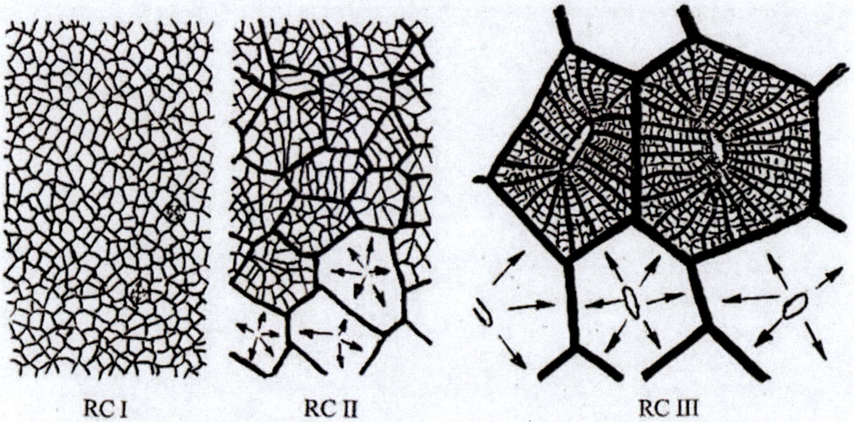

Fig. 10.5 Schematic representation of roll-cells of first order (RCI), second order (RCII) and third order (RCIII)

A typical RCII-system shows Fig. 10.6a with the system acetone/glycerin +12% water +1% CH_3COOH 4 min after start. In Fig. 10.6b we find also in a capillary split of 320 nm roll cells of second order in form of a central big roll cell surrounded by a chaotic oscillatory eddy-street, visualized by small particles. Figure 10.6c shows with a shadow-graph photo qualitatively the distribution of the concentration of the diffusing matter. Fig. 10.6d shows the streamlines and the calculated interfacial-tension distribution. Used systems are air/glycol saturated with chloroform 5 min after start or also 1% $C_{16}H_{33}OSO_3Na$/water at 7 min after start. (The central roll cell can show also relaxation oscillation.)

Furthermore, we show with Fig. 10.7 the center of an irregular RCIII with centrifugal flow from the center to the periphery. The system uses strong heat-transfer from a liquid sulphury layer with a temperature of 162 °C into nitrogen. Figure 10.8 shows at very strong driving force RCIII of the P-type, that means: with the flow from the periphery to the center of the RCIII! The system is 2% $C_{12}H_{25}OSO_3Na$+ iso-pentanol/water 25 s after the start.

With exhausting of driving force by the intense interfacial convection, we find at first regular relaxation oscillations: a periodic breakdown for instance of a single RCI suitable in a small facility like a capillary was observed. Breakdowns of RCI of a collective of RCI in a larger vessel appear not simultaneously but chaotically, and that happens for instance also with RCIII in large containers. Then the breakdowns of every unites of RCIII are independent from each-other developing a chaotic behavior. In a container for photometry with the measures height = 3 cm, widths = 2 cm × 1 cm, a single RCIII is possible with the mass-transfer system 1% $C_{15}H_{31}OSO_3Na$ + iso-pentanol/H_2O 45 min after start, shortly before the next breakdown, see Fig. 10.9a. Complete breakdowns of this single RCIII are observed at 12 min, 27 min and 39 min etc. after the start and the relaxation-times are 3–7 min. The striations in the photo are caused by the flows of the substructures.

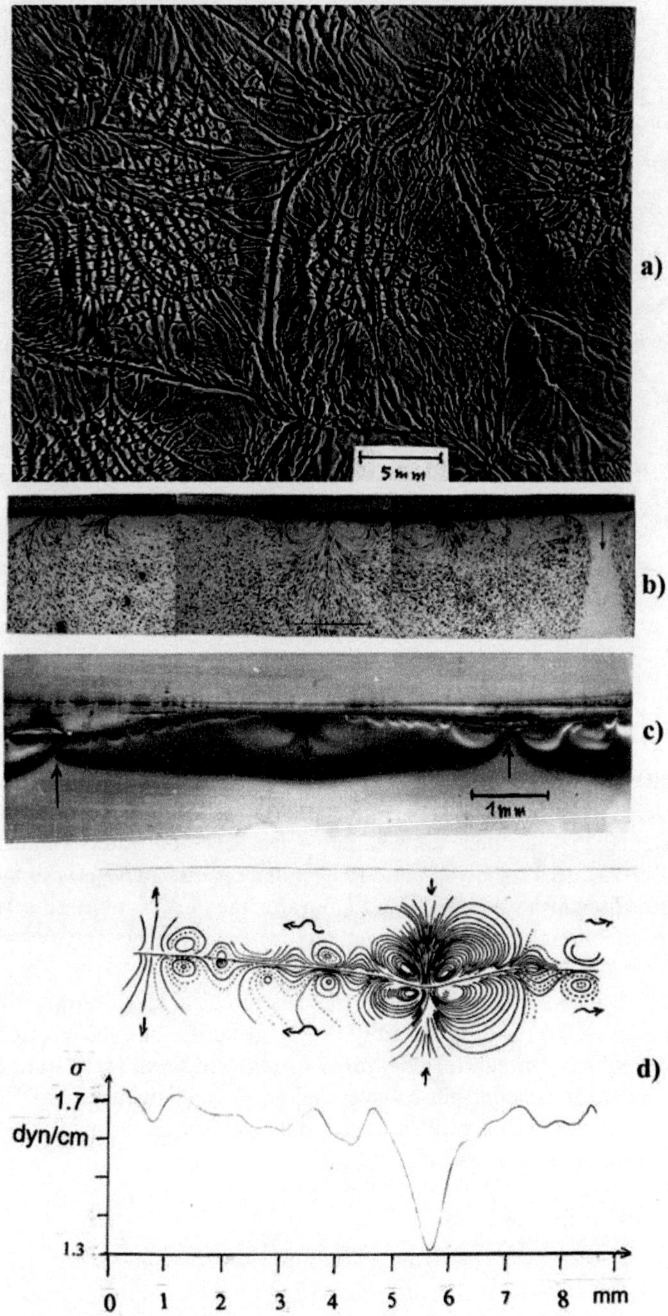

Fig. 10.6 **a** Typical RCII-system with the system acetone/glycerin +12% water +1% CH_3COOH 4 min after start; **b** capillary split of 320 nm roll cells of second order; **c** distribution of the concentration of the diffusing matter (shadow-graph photo); **d** streamlines and the calculated interfacial-tension distribution

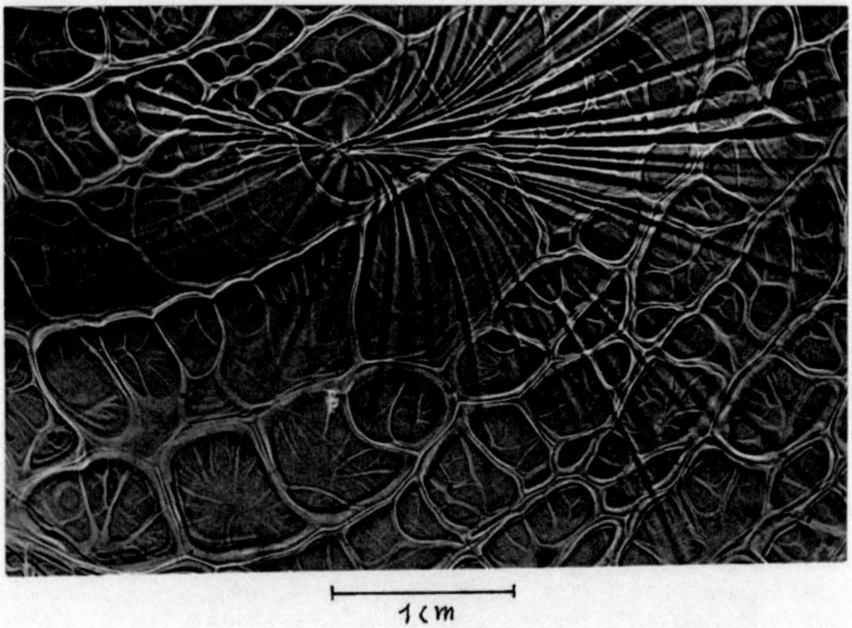

Fig. 10.7 Center of an irregular RCIII with centrifugal flow from the center to the periphery (strong heat-transfer from a liquid sulphury layer with a temperature of 162 °C into nitrogen)

Figure 10.9b explains this behavior with a scheme containing chaotic RO_{RCI}, chaotic RO_{RCII} and chaotic RO_{RCIII}. Figure 10.10a,b depicts snapshots of a chaotic RO_{RCI}, system 0,25% technical alkyl sulfate + iso-pentanol /water, 3 min after start. Figures 10.11 and 10.12 show chaotic RO_{RCII} and chaotic RO_{RCIII} with the system benzene / 20% dioxan + water 5 min and 7 min after the start. Note, that the two higher levels of the chaotic relaxation oscillation are one-times respectively two-times sub structured.

Figure 10.13 shows with a scheme the three hierarchical levels of wave-like relaxation oscillation: ROW I, ROW II and ROW III. a_1 until a_7 are ROW I from simple linear waves, which can be created with lowering of driving force from RCI (see a_3), which occur also as concentric waves and also as substructures of RCII (see a_5 and a_7). These waves react at broken wave-fronts with spinalization (see a_2) and can finally also form collective systems like a_6.

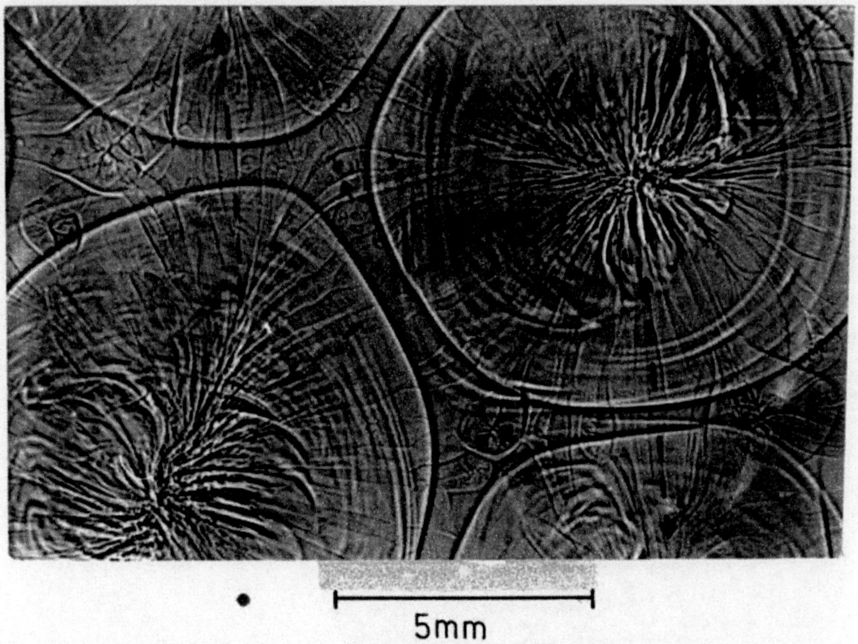

Fig. 10.8 System 2% $C_{12}H_{25}OSO_3Na$ + iso-pentanol/water 25 s after the start with a very strong driving force RCIII of the P-type flowing from the periphery to the center of the RCIII

Figure 10.14a,b shows simple linear waves created from RCI, system 5% butanol + cyclohexanol/water; Fig. 10.15a,b shows the trend to spinalization by disturbance and breaking the wave-fronts with the system 0,25% technical alkyl sulfate + iso-pentanol/water 19 s after start; Fig. 10.16 (small picture) shows collective behavior of ROWI (system butanol/5% $CuCl_2$ + water), (large picture) also collective behavior of ROWI (system air/CS_2 at 45 °C. In the bottom part Fig. 10.13 is shown, that ROW III starts "childlike" with ROW I at a_1, develops with one substructure then to ROW II at b_1, b_2, b_3 and finally at b_5 to ROW III. In Fig. 10.13, bottom part, we see in b_5 two substructures as well of RCI as of ROW II. With the scheme of b_4 we gave our expected streamline system of ROW II: The rotating convection system is not symmetric—because of a zone of relaxation of the driving force-, so that we have a travelling of the system.

Figure 10.17 shows at the right side ROW II, which travels to the left side under developing ROW III, which contains additionally also ROWII as substructure. The system is 20% CH_3OH (or acetone) + iso-pentanol/water.

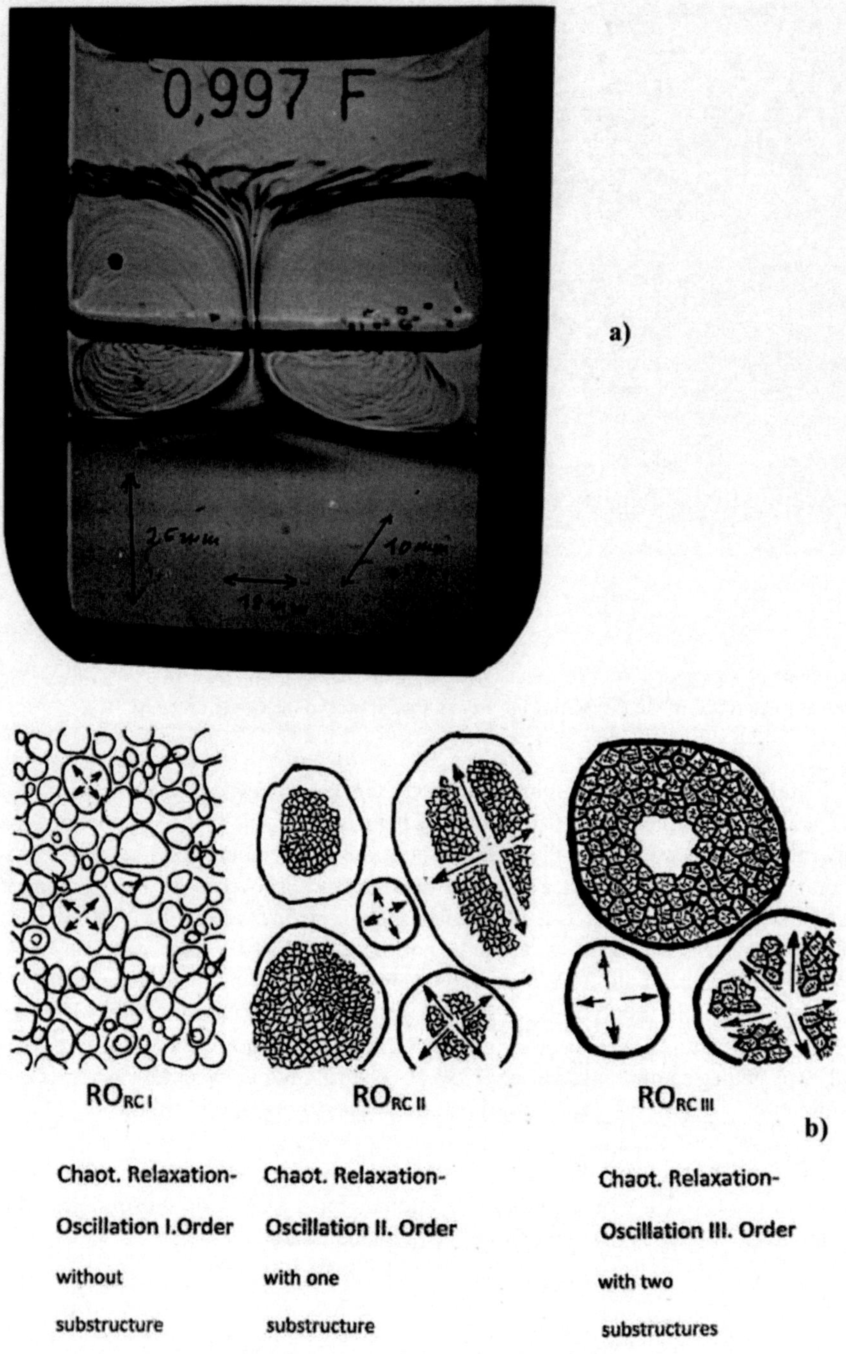

Fig. 10.9 **a** Single RCIII with the mass-transfer system 1% $C_{15}H_{31}OSO_3Na$ + iso-pentanol/H_2O 45 min after start, shortly before the next breakdown; **b** scheme containing chaotic RO_{RCI}, chaotic RO_{RCII} and chaotic RO_{RCIII}

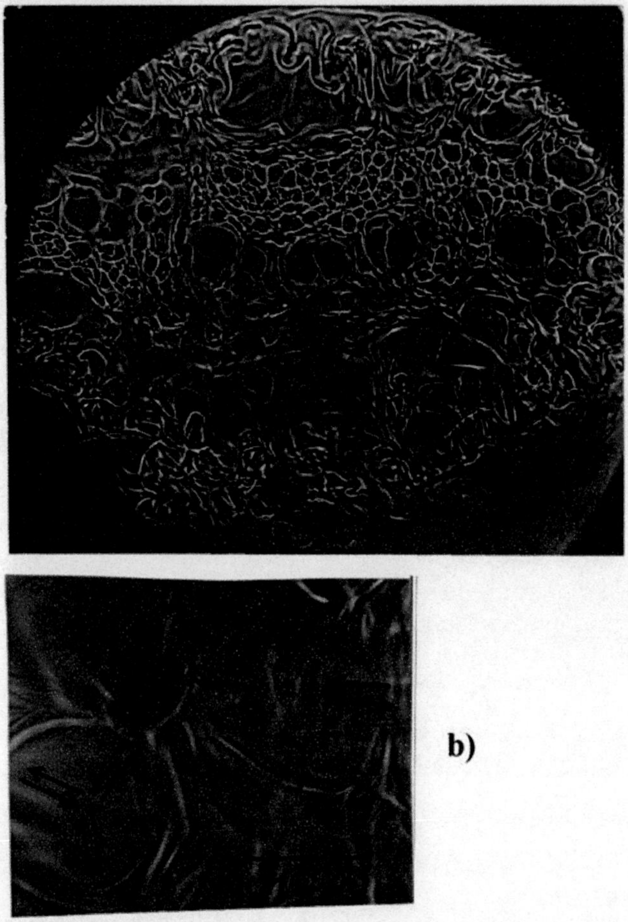

a)

b)

Fig. 10.10 Snapshots of a chaotic RO_{RCI}, system 0,25% technical alkyl sulfate + iso-pentanol/water, 3 min after start

Very fascinating is, that two (or even three) regimes can have in a hyper-cycle an alternating co-existence. For instance, this last regime in Fig. 10.17 (and some other systems too) can develop at first RCIII (but only over a childlike developing over RCI and RCII!) RCIII needs a relatively height critical driving force, however the effective convective transport of this regime leads after some min to the under-critical condition. Then the regime ROW III uses the opportunity to drive out respectively to displace RCIII (mostly by starting of ROW at the periphery) and develops (over the childlike forms ROW I and ROWII) finally ROW III! However, this regime is less intensive in using the driving force, which therefore can again increase and overcoming the critical value of RCIII. Then RCIII starts again—over RCI and RCII- mostly inside the ROW III-sculpture and displaces the ROW III-structure.

Fig. 10.11 Chaotic RO_{RCII} and chaotic RO_{RCIII} with the system benzene / 20% dioxan + water 5 min after the start

The analogue behavior happens in the system benzene/10–20% dioxan + water (and some others too), but this hyper-cycle alternates from RO_{RCIII} (from RO_{RCI} over RO_{RCII} to RO_{RCIII}) and then to ROW III (developed over ROW I and ROW II to ROW III) and that alternately.

We can arrange for demonstrating the hierarchical system of the "stationary" system of Marangoni instability in a scheme: Fig. 10.18, in which the actual driving force (a.d.f.) decreases stepwise—after our experience—in the regimes from the RC over RO to ROW and just as in the orders from the third order III over the second II to the first order I. I means not sub-structured, II means one-times sub-structured and III means two times sub-structured. The actual driving force decrease little on the way from RC over RO to ROW however stronger on the way of these regimes from the third order over the second order to the first order after our estimation.

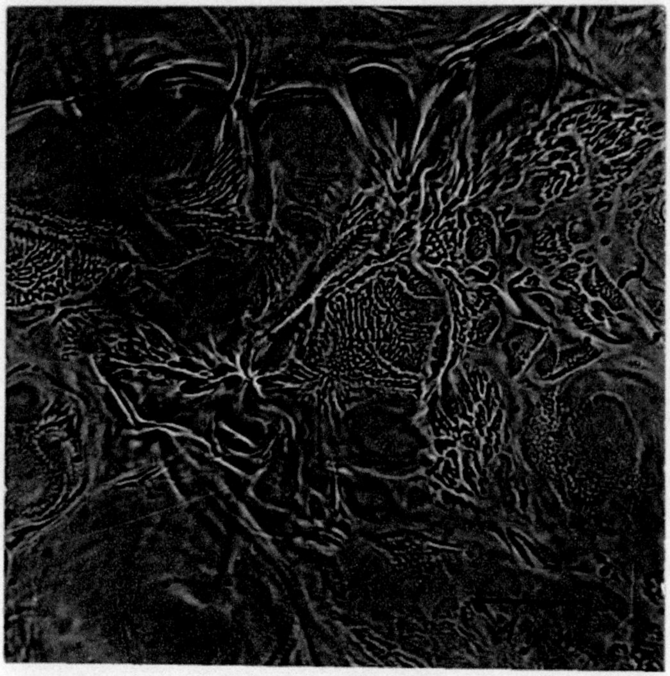

Fig. 10.12 Chaotic RO_{RCII} and chaotic RO_{RCIII} with the system benzene / 20% dioxan + water 7 min after the start

Figures 10.19 and 10.20 show the schemes of two hyper-cycles. In Fig. 10.20 the arrows as intersecting lines and as short lines show the paths of the regimes during the hyper-cycles.

From the theoretical point of view, we have here two extended Lotka-Volterra-systems with two predators (two the driving force consuming hydrodynamic regimes) and one prey (the driving force). Lotka and Volterra analyzed the phase-shifted alternating up and down of the number of coats of lynxes (the predator) and snow hares (the prey) found in the accounts of the Hudson's Bay.

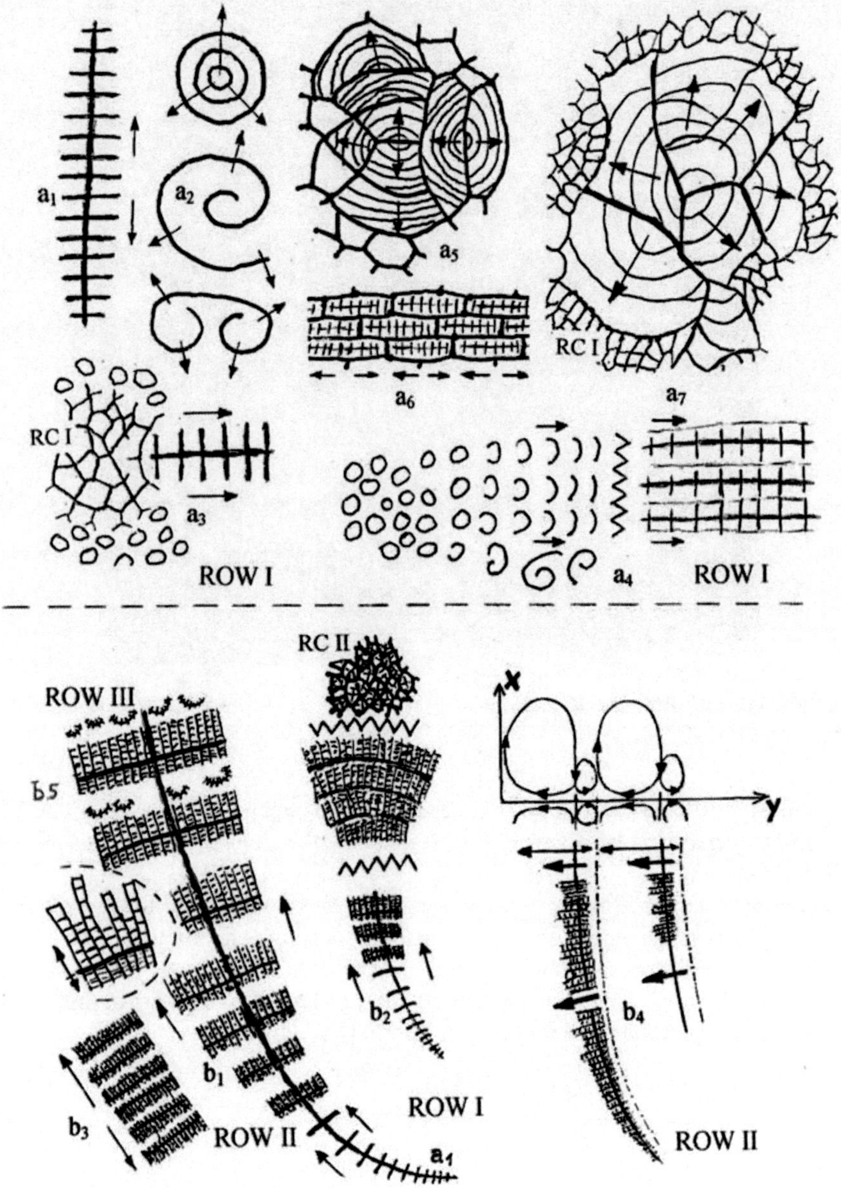

Fig. 10.13 Scheme with the three hierarchical levels of wave-like relaxation oscillation: ROW I, ROW II and ROW III

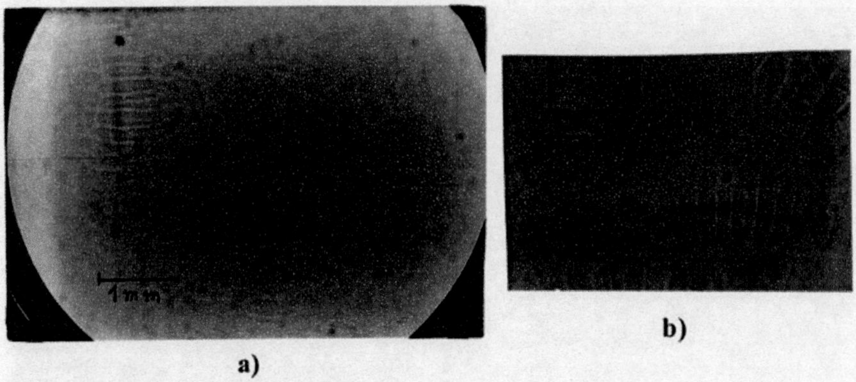

Fig. 10.14 Simple linear relaxation oscillation waves of first order created from RCI, system 5% butanol + cyclohexanol/water; **a** 6–7 min after start; **b** 7 min after start

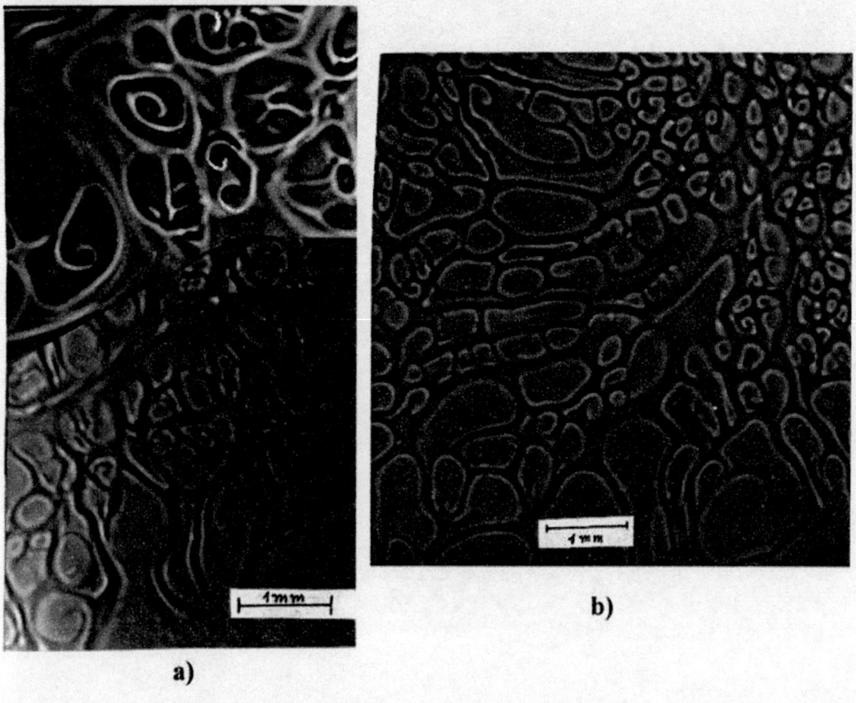

Fig. 10.15 System 0,25% technical alkyl sulfate + iso-pentanol/water showing the trend to spinalization by disturbance and breaking the wave-fronts (planar waves change into spirals); **a** spiral chaos following the breaks of fronts of waves; **b** transition from chaotic RO_{RCI} into ROWI

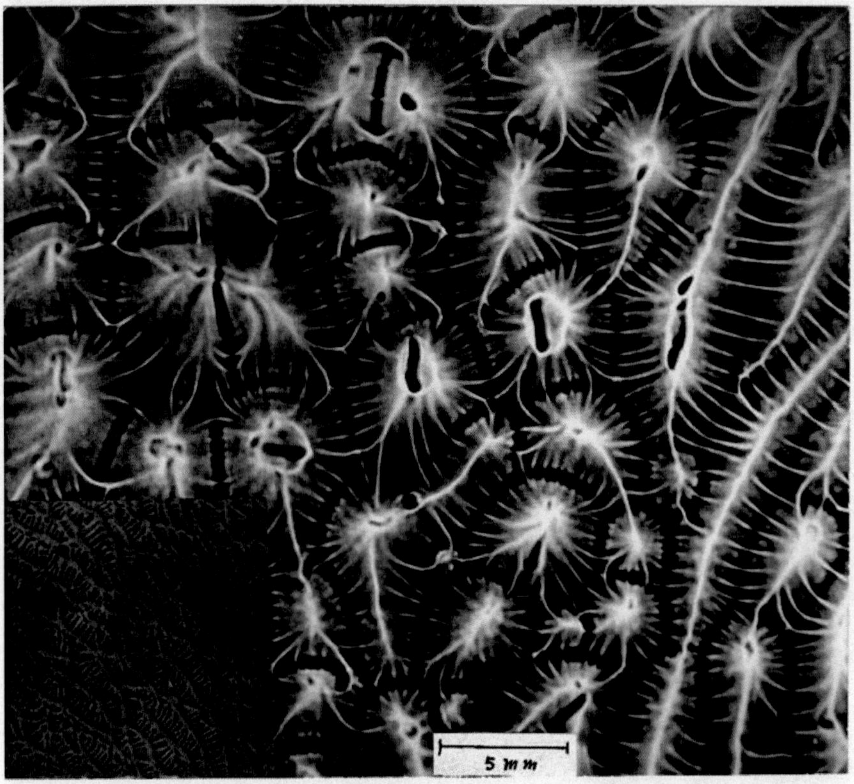

Fig. 10.16 Collective behavior of ROWI; large picture: system air/CS_2 at 45 °C; small picture bottom left: system butanol/5% $CuCl_2$ + water

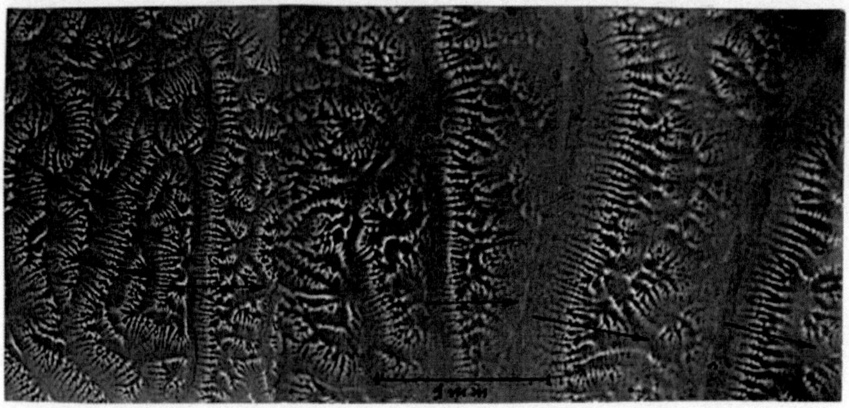

Fig. 10.17 ROWII at the right side travels to the left side developing ROWIII and ROWII as substructure (system 20% CH_3OH (or acetone) + iso-pentanol/water)

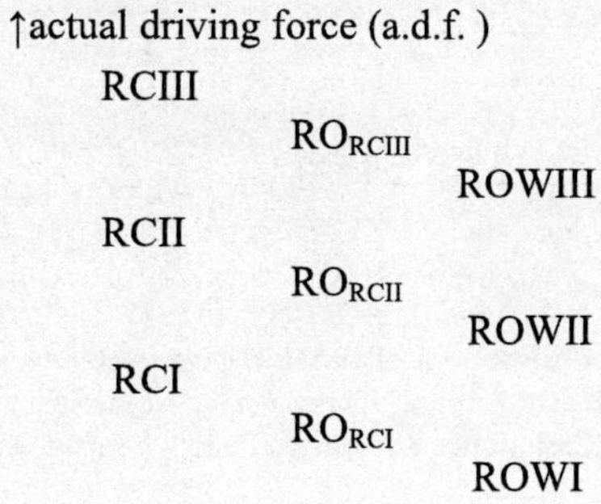

Fig. 10.18 Scheme of the regimes of the "stationary" regimes of Marangoni instability

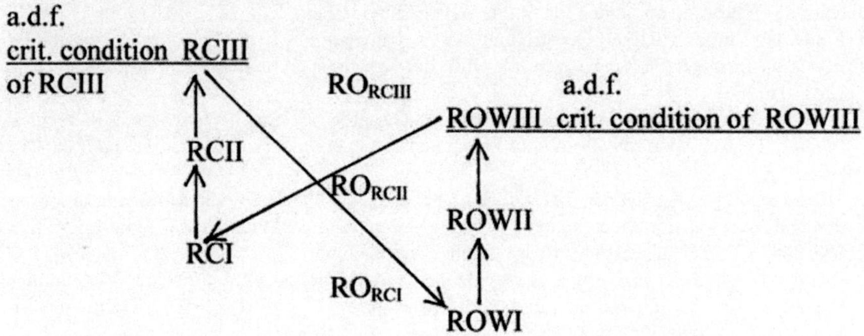

Fig. 10.19 Scheme of the hypercycle in the system like 20% CH_3OH + iso-pentanol/water. The alternating regimes are underlined

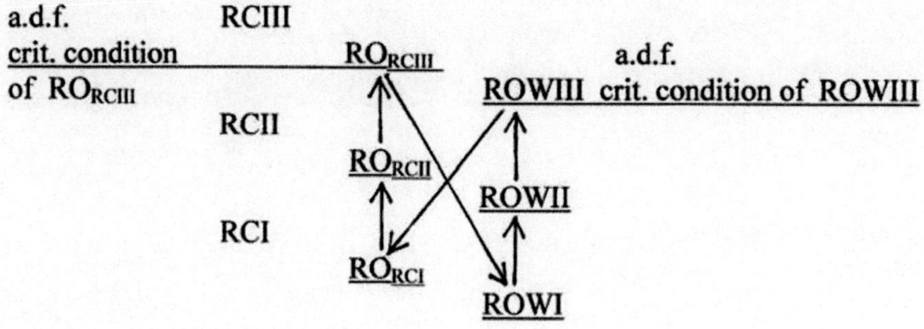

Fig. 10.20 Scheme of the hyper-cycle in the system like benzene / 10–20% dioxan + water. The alternating regimes are underlined

References

1. Bénard, H.: Les tourbillons cellulaires dans une nappe liquid. Rev. Gen. Sci. Pures Appl. Bull. Assoc. **11**, 1261–1271, 1309–1328 (1900). Methodes optiques d'observation et de enregistrement. J. Phys. Theor. Appl. (1901) 254–66
2. Marangoni, C.: (Pavia. tip. Fusi. Agosto 1865) Ueber die Ausbreitung einer Flüssigkeit auf der Oberfläche einer anderen. Ann. Phys. **219**, 337–354 (1871)
3. Weber, E.H.: Microscopic observations of very low-governed movements (convections), which occur during the formation of the solution of resin into spirit, Berichte über die Verhandlungen der königlich sächsichen Gesellschaft der Wissenschaften zu Leipzig, Mathematisch-physikalische Klasse (1854)
4. Sternling, C.V., Scriven, L.E.: Interfacial turbulence: hydrodynamic instability and Marangoni effect. AICHE J. **5**(4), 514–523 (1959)
5. Linde, H., Schwarz, E.: Untersuchungen zur Charakteristik der freien Grenzflächenkonvektion beim Stoffübergang an fluiden Phasengrenzen. Z. Phys. Chem. **224**, 331–352 (1963)
6. Linde, H., Pfaff, S., Zirkel, C.: Strömungsuntersuchungen zur hydrodynamischen Instabilität flüssig-gasförmiger Phasengrenzen mit Hilfe der Kapillarspaltmethode. Z. Phys. Chem. **225**, 72–100 (1964)
7. Linde, H.: Marangoni instabilities. In: Zierep, J., Oertel jr., H. (Eds.), Wissenschaft + Technik "Convective Transport and Instability Phenomena". G. Braun Verlag Karlsruhe, pp. 265–296 (1985)
8. Schwarzenberger, K., Köllner, T., Linde, H., Boeck, T., Odenbach, S., Eckert, K.: Pattern formation and mass transfer under stationary solutal Marangoni instability. Adv. Colloid Interf. Sci. **206**, 344–371 (2014), see 144 further references in this report
9. Linde, H., Kunkel, M.: Einige neue Beobachtungen beim oszillatorischen Regime der Marangoni-Instablität, Wärme- und Stoffübertragung **2**, 60–64 (1969) (Oscillation in liquid/liquid systems with mass-transfer and chemical reaction)

Chapter 11
Cooperation of Flow-Instabilities

Hartmut Linde

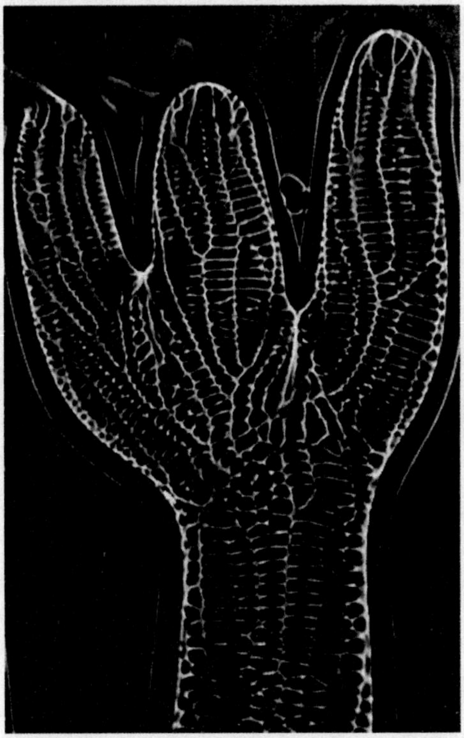

Fig. 11.1 Coupling of meniscus flow and Marangoni instability. The roll cell boundaries are displayed

Figures, captions and additional supplements inserted by E. C. Haß according to templates of H. Linde.

P. J. Plath et al., *Imagery Synergetics*, Understanding Complex Systems,
https://doi.org/10.1007/978-3-030-95607-3_11

An overview of articles concerning flow instabilities can be found exemplarily in [1–7] and the literature cited therein (see also Fig. 11.1).

11.1 Flow Instabilities with Streaks and Langmuir Circulation

The flow of gas or liquid along a flat solid surface—or of gas along a liquid surface—lead to streaks—produced by counter rotating longitudinal rolls—directed along the flow. These are observable by streaks of snow along a flat surface and as swimming lines of bubbles at the surface of lakes or of the sea. Figure 11.2a shows swimming artificial foams—and other sides also leaves and bubbles (see Fig. 11.2b)—at a lake ordered to streaks with distances of about 75–150 cm. These swimming particles are gathered at lines of convergent flow at the liquid surface, at which also the surface-active pollution is gathered and forms by damping of waves visible flat streaks by a better reflection of the sky. At higher velocity of the wind, the bubbles are unified to foam-streaks as shown in Fig. 11.3 with radial orientation due to the radial ordered wind of a helicopter.

This "Langmuir circulation" [8] consist of a pattern of counter rotating vortices orientated downwind and is a key process in the ocean or lake surface layer. After my own observation, there can be a hierarchical system, for instance observed at polluted lakes with foam-streaks of first order with streak-width of 0.5–1.5 m, of second order with streak-width of 50–75 m (with one sub-structure of first order) and of third order with streak-width of 200–300 m (with two substructures of second and first order). A sideways expanding of the streaks of lower order possibly forms the streaks of higher order. Figure 11.4 shows the row of surged sea grass at the beach by these streaks called also Langmuir circulation.

11.2 Flow Instabilities at Small Surfaces

At small by walls restricted surfaces of with surfactants polluted water a similar convection—but with hairpin-like convection—was observed at low flow velocity of air above the surface, see Fig. 11.5, which develops to turbulence with higher flow velocity, see Fig. 11.6. In a region of tailback of a smooth surface-flow of a lake, this kind of surface-turbulence forms an interesting collective behavior of such circulating flows at the surface. A radial laminar flow of water jet from below to a circular surface of the polluted water surface loses his radial flow at the surface and forms surface rotations like Fig. 11.7. Mostly we observe 4 vortices, but also 2, 3, and 5 vortices are observed, which remain their rotation direction at slow flow velocity. However, with increasing flow velocity, the rotation direction can be regularly and later irregularly changed and then transits into turbulent behavior.

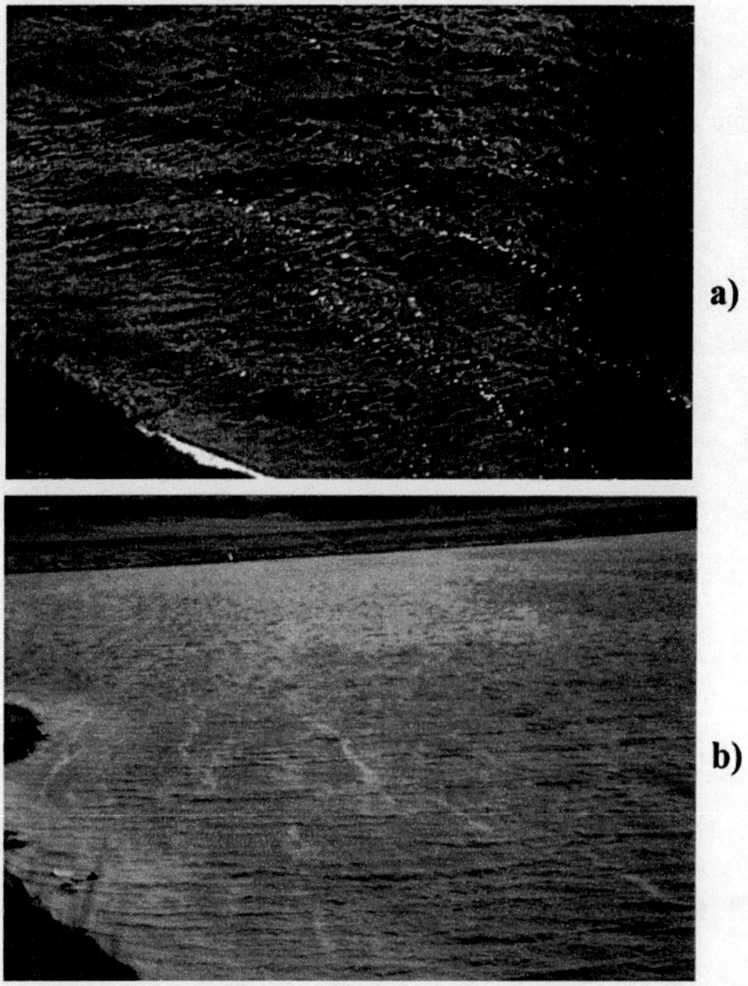

Fig. 11.2 Swimming lines of bubbles at the surface of lakes; **a** artificial foam-particles induced by wind-drift with a wavelength of about 75–150 cm; **b** flat streaks by a better reflection of the sky

11.3 Flow Instabilities at Coated Films and Solid Sheets

Coating of flat films or solid sheets like electronic wafers—for instance with a procedure like Fig. 11.8a—leads also to longitudinal rolls. v means here the velocity of over layering with the moving shaft. Narrow streaks at a thin liquid layer of these flow instabilities can be first increased by coupling with Marangoni-instability and then visualized by the shadow-graph technique like Fig. 11.8b and later developed to ladder-structures like Fig. 11.8c. Finally, even more complicated structures like Figs. 11.8d and 11.9 can result, if wide streaks are filled with small roll cells.

Fig. 11.3 Foam streaks by the radially ordered wind of a helicopter at the sea (downwash under helicopter); published in the Public Domain by the Defense Media Activity Enterprise Customer Services (DVIDS) of the US Department of Defense (DoD) [9]

Fig. 11.4 Row of surged sea grass at the beach possibly formed by a sideways expanding of higher-ordered streaks of lower order (Langmuir circulation [8])

Fig. 11.5 Hairpin-like convection at low flow velocity (flow direction left) of air above the surface of water polluted with Stearic acid $C_{17}H_{35}COOH$. (The photo was exposed for 5 s.)

Fig. 11.6 Same system as Fig. 11.5, but with higher flow velocity (flow direction left) developing to turbulence. (The photo was exposed for 5 s.)

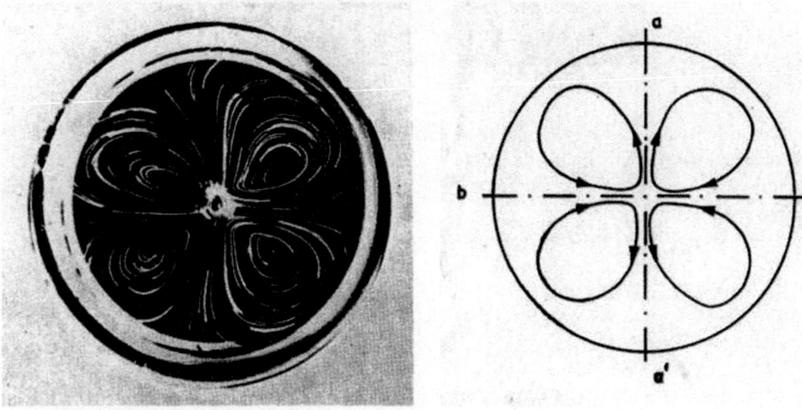

Fig. 11.7 Experimentally observed vortex rotation of 4 vortices and its schematic representation; two adjacent vortices are directed in opposite directions

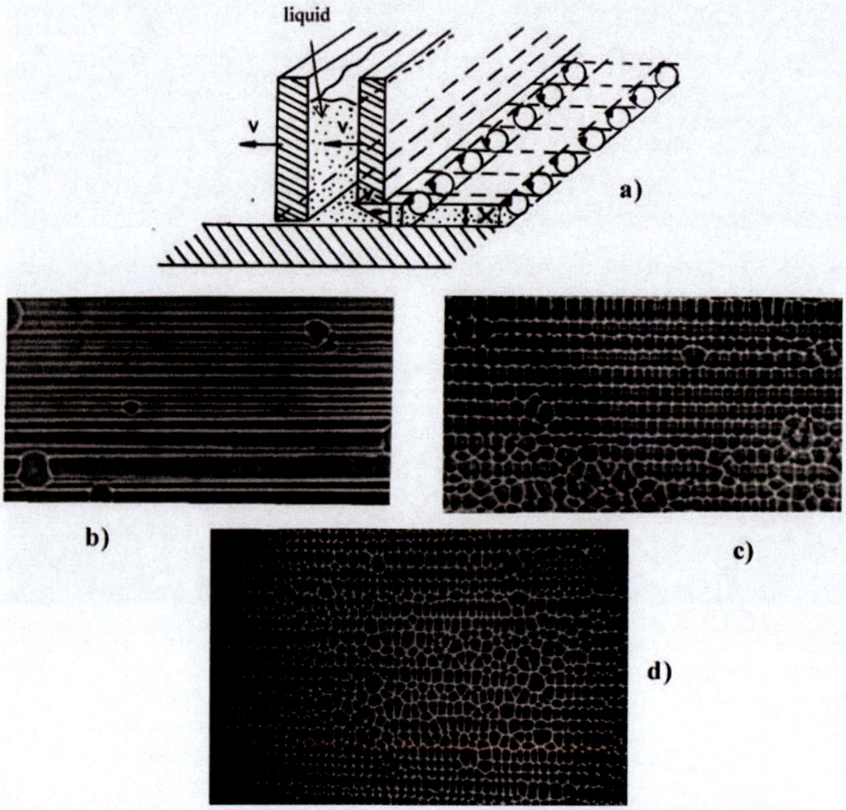

Fig. 11.8 Longitudinal rolls at coated films and solid sheets induced by flow instabilities; **a** schematic representation; **b** narrow streaks at a thin liquid layer of these flow instabilities visualized by the shadow-graph technique; **c** ladder-structures developed by coupling with Marangoni-instability; **d** complicated structure due to cooperation of flow instabilities with wide streaks filled with small roll cells

11.4 Meniscus Instability

Another easy to realize flow instability is the meniscus-instability, occurring if a meniscus is travelling by procedures like Fig. 11.10. (This instability is already used for the decoration of buildings!!) The travelling meniscus develops a wavy-structure and finally also a wavy substructure, resulting with the procedures of Fig. 11.10a,b finally by flowing into thickness structures like Figs. 11.11 and 11.12. The using of the procedure of Fig. 11.10c results into the structure of Fig. 11.13.

Now we open two round superimposed discs like Fig. 11.10a, which are filled in its gap with a liquid lacquer able to develop Marangoni-instability by the evaporation of the surface-active solvent. Then we observe first in Fig. 11.14a—15 s after start—the rough flow due to the meniscus-instability with the longitudinal streaks amplified by

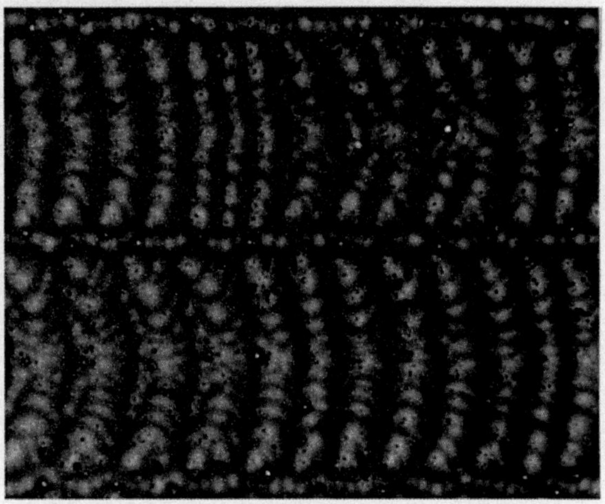

Fig. 11.9 Another complicated structure generated by cooperation of flow instabilities

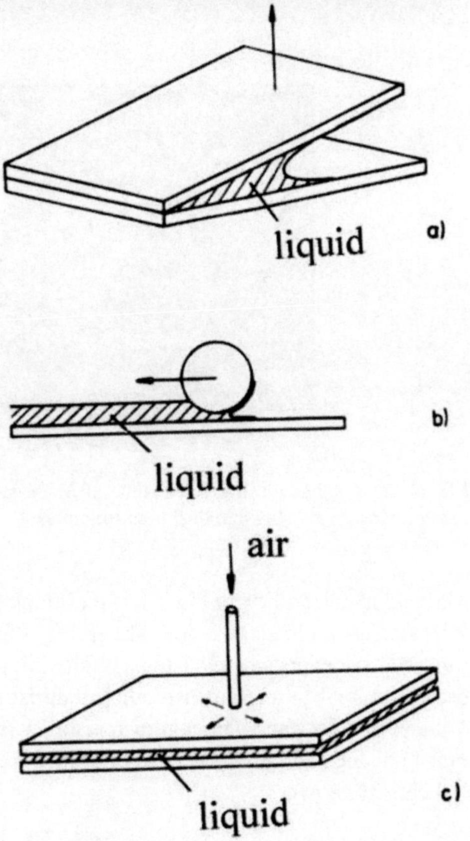

Fig. 11.10 Schematic representation of producing Meniscus-instability by a travelling meniscus; **a** and **b** flowing into thickness structures (see Figs. 11.11 and 11.12); **c** air displaces the liquid radially from the slit (see Fig. 11.13)

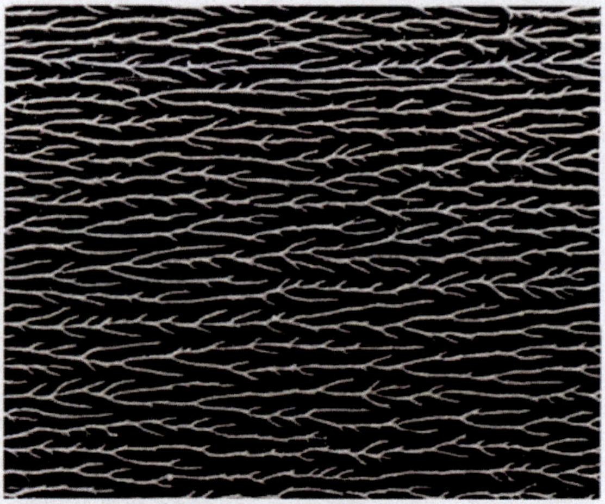

Fig. 11.11 Meniscus-instability due to multiple branching by a random process (travelling velocity: 8 cm s^{-1})

Fig. 11.12 Another illustrative example of Meniscus-instability with wavy structures and wavy substructures by flowing into thickness structures

Marangoni-instability. In Fig. 11.14b Marangoni-instability dominates further—20 s after start—and leads to ladder-structures, while 35 min after start in Fig. 11.14c the roll cells are everywhere visible. Figure 11.15 shows finally a kind of relaxation–oscillation by Marangoni-instability dominating the drying of lacquer. The coupling of meniscus flow and Marangoni-instability is illustrated in Figs. 11.1 and 11.16. In Fig. 11.1, the roll cell boundaries are displayed, and in Fig. 11.16, the center of the roll cells is shown.

Fig. 11.13 Meniscus-instability resulting from the procedure of Fig. 11.10c; air displaces the liquid radially from the slit

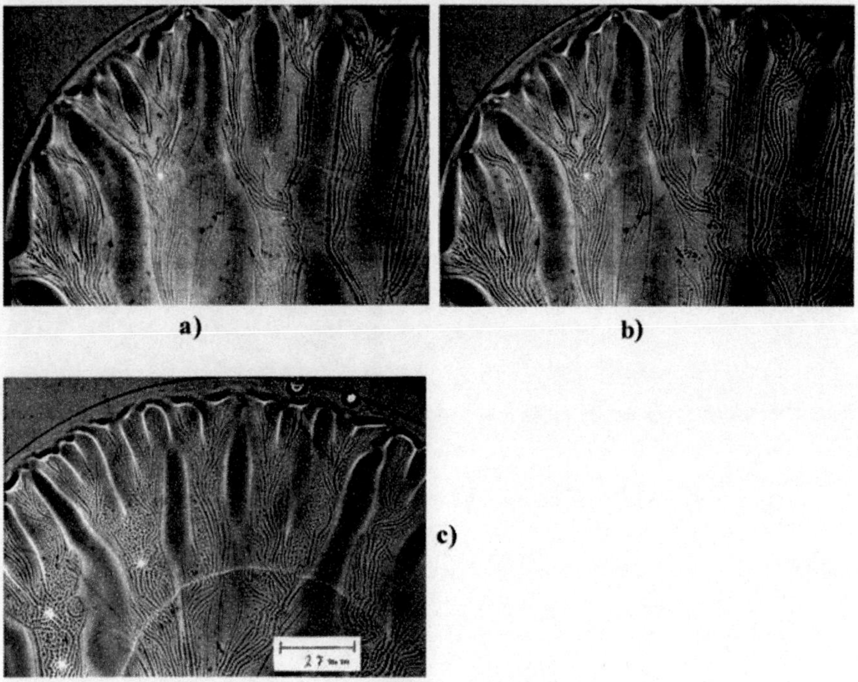

Fig. 11.14 Two round superimposed discs like Fig. 11.10a filled in its gap with a liquid lacquer developing Marangoni-instability by the evaporation of the surface-active solvent; **a** 15 s after start: rough flow due to the meniscus-instability with the longitudinal streaks amplified by Marangoni-instability; **b** 20 s after start: further dominating Marangoni-instability leading to ladder-structures; **c** 35 s after start: roll cells are visible everywhere

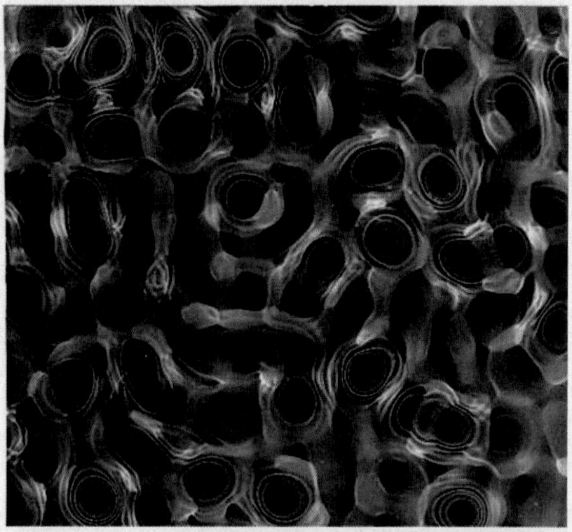

Fig. 11.15 Structure formation during stepwise drying of a lacquer: chaotic relaxation oscillations due to Marangoni-instabilities are observed

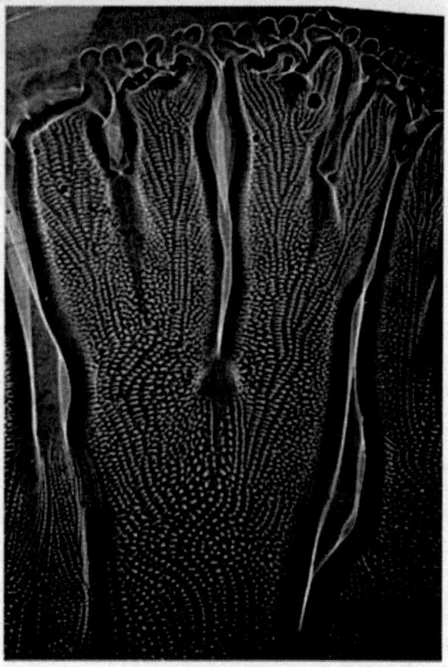

Fig. 11.16 Coupling of meniscus flow and Marangoni-instability. The center of the roll cells is displayed (compare also Fig. 11.1)

11.5 Further Examples of Flow Instabilities

Interesting and versatile structure types of flow instabilities are also ripple structures produced by the periodic wave wash overs on sandy beaches or dune structures in the desert sand generated by the wind. Figure 11.17a illustrates the sand waves of Mesquite Flat-Dunes in the Death Valley National Park in California and in Fig. 11.17b is shown the wonderful dunes of the Victoria Crater on the surface of the planet Mars. The flow intensity at Garni Crater on Mars, depicted in Fig. 11.18, is also remarkable. Possibly, this is a Langmuir-instability. In the past, before the time of the snow groomers, skiers could create ripple structures in the snow without wanting to, as the example in Fig. 11.19 from my own past shows.

a) b)

Fig. 11.17 Ripple structures due to flow instabilities; **a** sand waves of Mesquite Flat-Dunes in the Death Valley National Park in California generated by the wind. (Photo provided with kind permission by the U.S. National Park Service); **b** dunes of the Victoria Crater on the surface of the planet Mars (published in the Public Domain by National Aeronautics and Space Administration—NASA, USA [10])

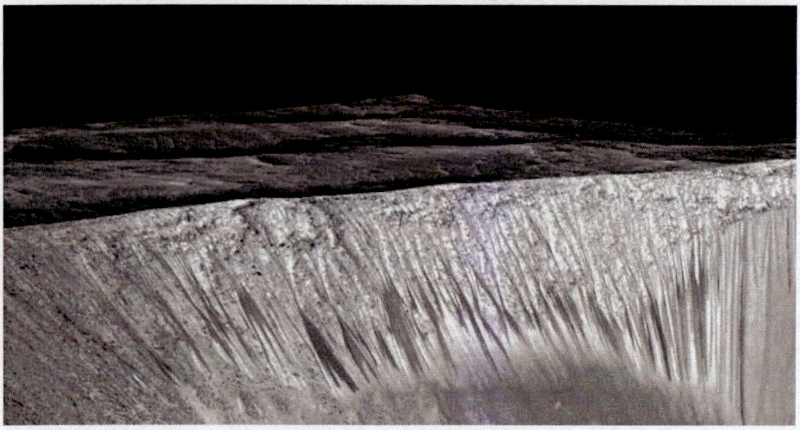

Fig. 11.18 Garni Crater on Mars with streaks of remarkable flow intensity, possibly caused by Langmuir-instability (photo (published in the Public Domain by National Aeronautics and Space Administration—NASA, USA [11])

Fig. 11.19 **a** Ripple structures in the snow caused by skiers without wanting to; **b** the author's wife creating Marangoni-instabilities

References

1. Gumerman, R., Homsy, G.: convective instabilities in cocurrent two phase flow: Part I, linear stability. AIChE J. 20(5), 981–988 (1974). Part II, Global Stability. AIChE J. 20(6), 1161–1167 (1974)
2. Linde, H., Friese, P.: Experimenteller Nachweis einer neuen hydrodynamischen Instabilität. Phys. Chem. **247**(5/6), 225–232 (1971)
3. Linde, H., Shulewa, N.: Eine neue hydrodynamische Instabilität an einer mit einem Tensidfilm bedeckten Wasseroberfläche bei Strömungen zur Oberflächenerneuerung, Monatsberichte der Deutschen Akademie der Wissenschaften zu Berlin, Band 12, Heft 11/12, 883–894 (1970)
4. Wassmut, F., Laidlaw, W.G., Coombe, D.A.: Interfacial instabilities: the linde instability. Chem. Eng. Sci. **45**(12), 3483–3490 (1990)
5. Linde, H.: Komplexe dissipative Strukturbildung durch Kopplung von Instabilitäten. Z. Chem. **27**(4), 154–155 (1967)

6. Linde, H.: Sequential action and simultaneous coupling of dissipative structures with complex structural interaction and memory effect. In: Velarde, M.G. (ed) Synergetics, order and chaos, pp. 301–306. World Scientific Pub., Singapore (1988)
7. Linde, H., Friese, P.: Experimenteller Nachweis einer neuen hydrodynamischen Instabilität. Z. Phys. Chem. **247**, 225–232 (1971)
8. Langmuir, I.: Surface motion of water induced by wind. Science **87**, 119–123 (1938)
9. Internet address of photo. https://upload.wikimedia.org/wikipedia/commons/thumb/c/c3/MISSIONS_DAY_DVIDS1081996.jpg/1024px-MISSIONS_DAY_DVIDS1081996.jpg. Accessed 28 June 2021
10. Internet address of photo. https://upload.wikimedia.org/wikipedia/commons/thumb/d/de/Victoria_crater_from_HiRise.jpg/477px-Victoria_crater_from_HiRise.jpg. Accessed 28 June 2021
11. Internet address of photo. https://upload.wikimedia.org/wikipedia/commons/thumb/0/0f/Garni_crater_Mars_HiRISE_Sep2015.jpg/800px-Garni_crater_Mars_HiRISE_Sep2015.jpg. Accessed 28 June 2021

Chapter 12
The Oscillatory Regime of Marangoni-Instability

Hartmut Linde

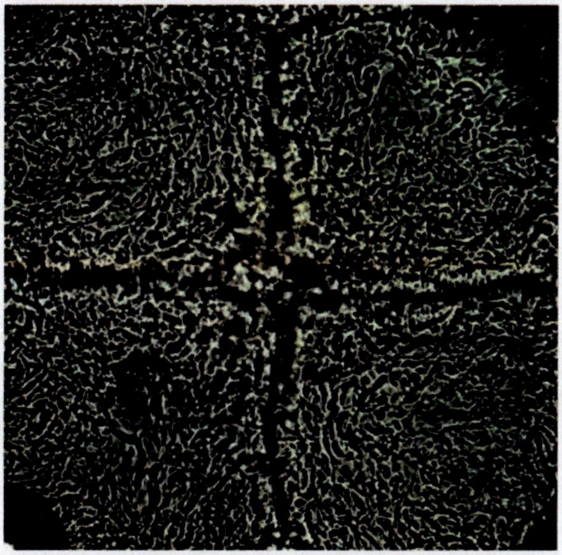

Fig. 12.1 Traces of head-on collisions in a liquid–liquid system with mass transfer and chemical reaction; dispersion-free oscillation of hexane +15 Vol. % propionic acid/water

An overview of articles concerning flow instabilities can be found exemplarily in [1–9] and the literature cited therein (see also Fig. 12.1).

Figures, captions and additional supplements inserted by E. C. Haß according to templates of H. Linde.

P. J. Plath et al., *Imagery Synergetics*, Understanding Complex Systems,
https://doi.org/10.1007/978-3-030-95607-3_12

12.1 Introduction

As we have reported on the recently discovered wave-like behavior of the relaxation oscillation ROWI, ROWII and ROWIII in the so-called stationary regime, there exists according to the theory of Sternling and Scriven a "direct" oscillatory regime, theoretically predicted by Sternling and Scriven and experimentally discovered by H. Linde, E. Schwarz, K. Loeschcke and completed by A. Wieschem and M.G. Velarde. This oscillation occurs, if a surface-active vapor (like pentane or even heat) becomes absorbed by a liquid of higher surface tension (like benzene) or, if a liquid of higher surface tension (like benzene) is desorbing from a liquid with lower surface tension (like octane).

There exist three kinds of waves: The first theory predicted an oscillatory regime showing (roughly spoken) waves with anomalous dispersion (decrease of wavelength with increasing velocity), the dispersion-free regime (constant wavelength with increasing velocity) found later by A. Wierschem in mass-transfer and even by H, Linde in K. Loeschcke's heat-transfer experiments. Furthermore, internal gravity waves occur in the density-stable liquid phase, which dominate the two waves above by coupling ("enslavement" after Haken) only at low driving force to normal dispersion (increase of wavelength with increasing velocity), see the scheme in Fig. 12.2.

12.2 Angle Crossing and Phase Shifts

At high driving force, both kind of waves—those with anomalous dispersion and dispersion-free ones—show surprisingly nonlinear behavior like "dissipative solitons", that means, with residence-time behavior with negative phase-shift at acute angle crossings or waves reflecting at the wall. At obtuse angle crossing or at under corresponding waves reflecting on the wall, we see a positive phase-shift with the Mach-Russel-third wave, see Fig. 12.3. In the first row of four pictures, the crossing of one indifferent angle and an acute angle (negative phase-shift) leads to a crossing of an also acute angle and an obtuse angle (with positive phase-shift) and with the Mach-Russel-third wave, which is also shown with four similar experimental examples.

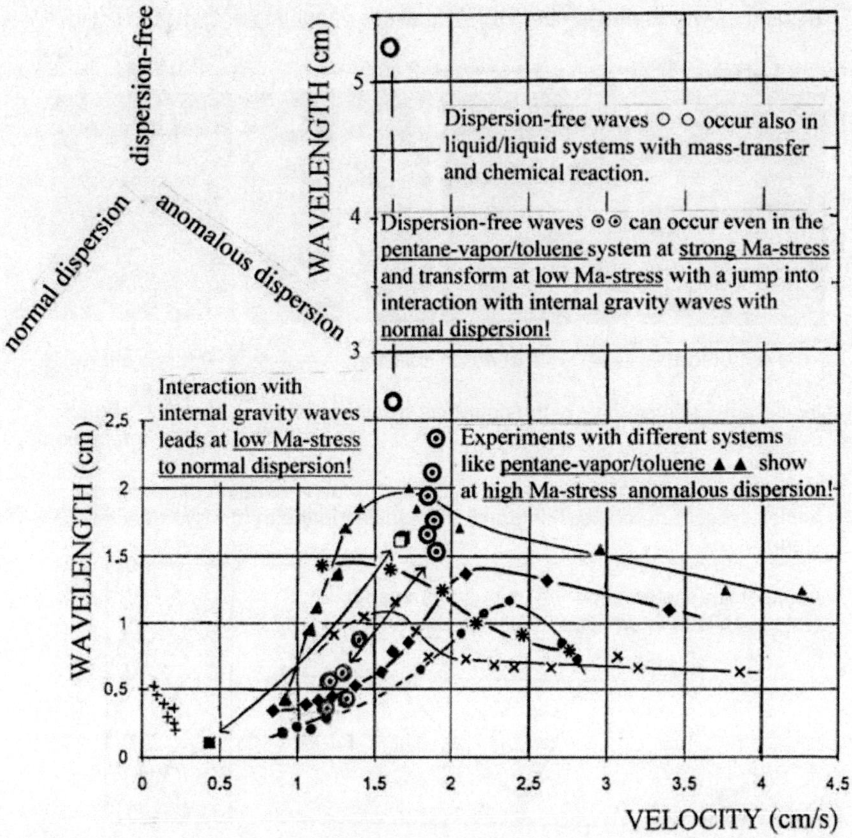

Fig. 12.2 Scheme of regions of different type of waves: anomalous dispersion, dispersion-free waves and internal gravity waves

Figure 12.4a–d again show the indifferent crossing with an angle near 90° (a), with acute angle (b), and for c, d with obtuse angle crossing and the corresponding reflection both with Mach-stem, known from ultrasound interactions. After these collision-behaviors, the waves continue their way with the pre-collision velocity if the intensity is not too strong.

Figure 12.5 shows a desorption experiment with $\frac{d\sigma}{dc} > 0$ with irregular reflections at the wall showing both kinds of crossings.

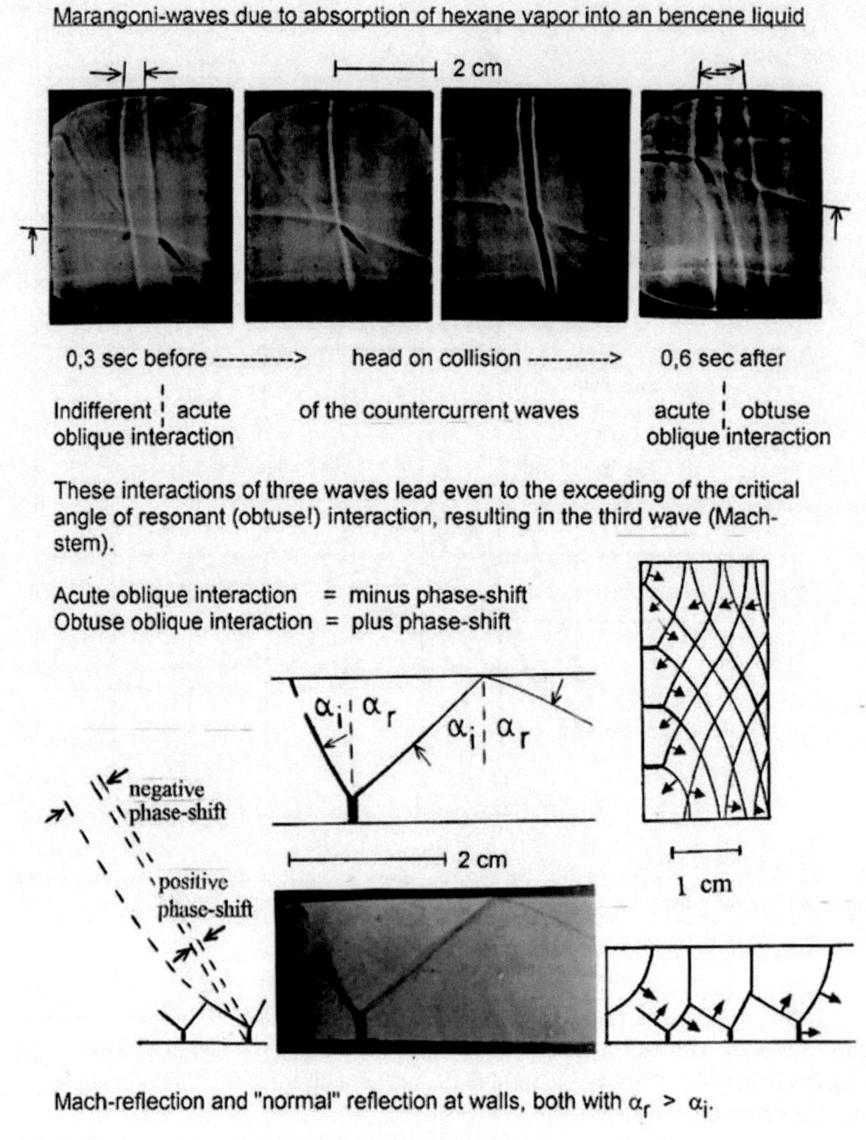

Fig. 12.3 Marangoni-waves due to absorption of hexane vapor into a benzene liquid and observed phase-shifts

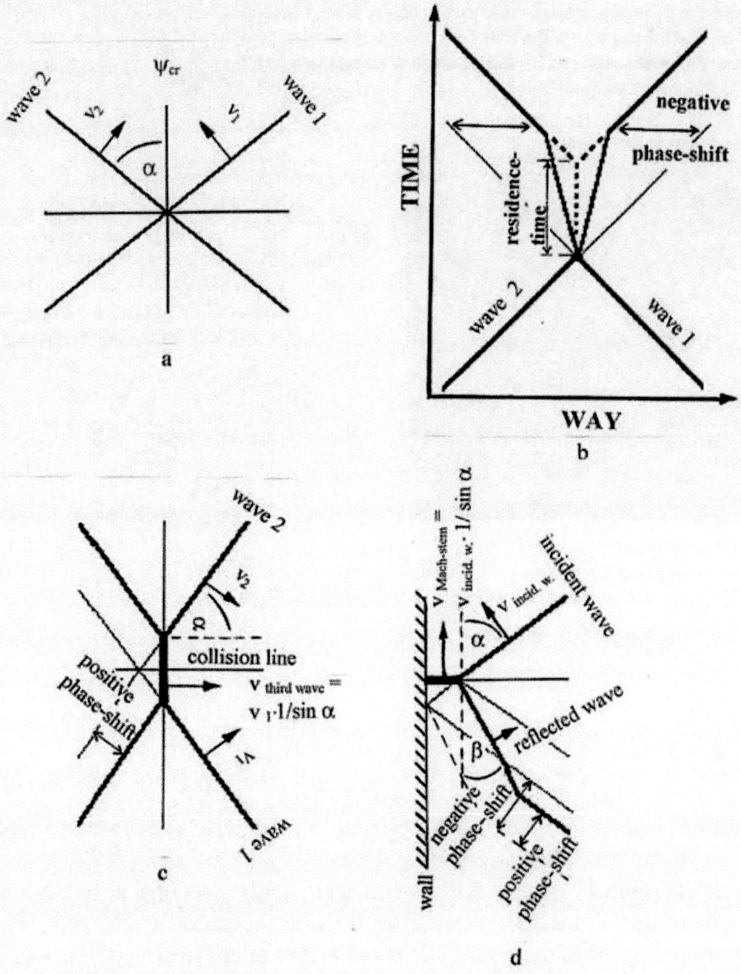

Fig. 12.4 Indifferent crossing with **a** an angle near 90°, **b** with acute angle, and **c** and **d** with obtuse angle crossing and the corresponding reflection

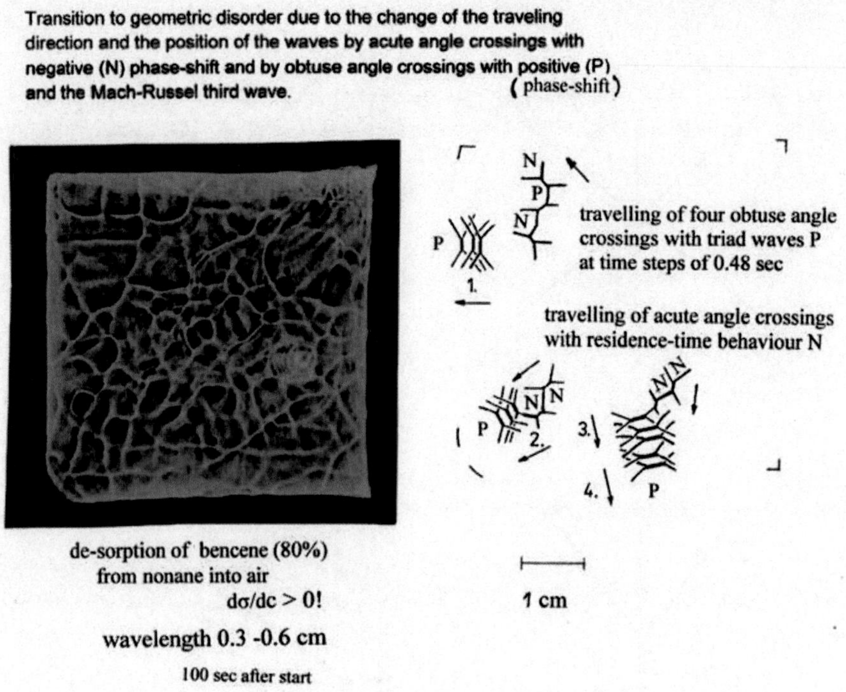

Fig. 12.5 Change of travelling direction and the position of waves observed at desorption of benzene from nonane into air

Figure 12.6 shows the same, but less chaotic and with network-forming behavior with negative and positive phase-shift-systems.

Figures 12.7 and 12.8 show waves with acute angle crossings at relative strong driving force: the waves fade away after crossing because of too much energy-dissipation. In Fig. 12.8, the waves that extinguishes after crossing produce damped internal gravity waves.

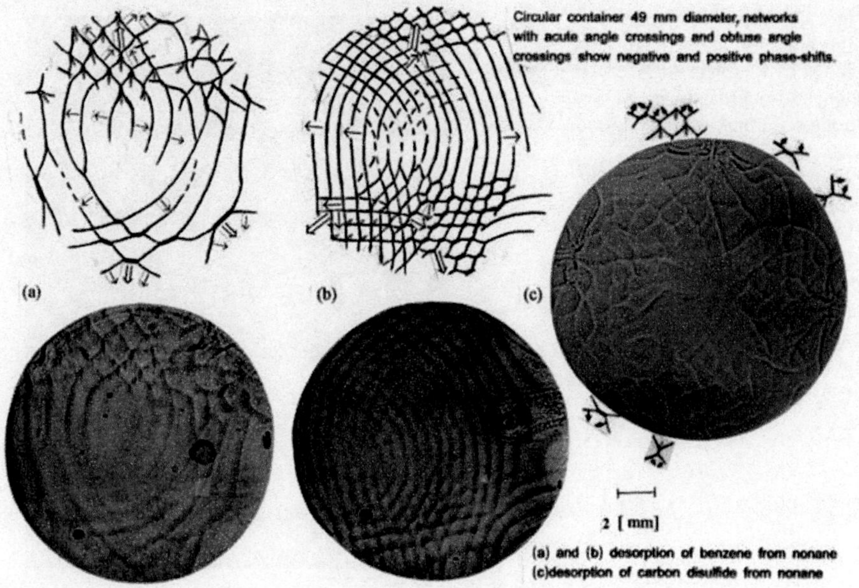

Fig. 12.6 Circular container with 49 mm diameter and corresponding networks with both angle crossings showing negative and positive phase-shifts; **a** and **b** desorption from nonane, and **c** desorption of carbon disulfide from nonane

12.3 Rotating and Counter-Rotating Waves

Figure 12.9 shows the start of the experiment of Fig. 12.10. The round surface of the circular facility was a little polluted by a fingerprint forming a thin film at the surface. Because the surface-active pentane vapor came not only from above, but also over the circular boundary, surface convection of the liquid from the boundary to the center compressed this film and has cleared a clean annular ring at the surface. This internal borderline, later called "Langmuir-ridge", arranged half-way to the center is originally circular, but here deformed as well by the surface-convection to the center with streaks as by the counter-rotating waves with counter-rotating light lines. Figure 12.9 already shows 5 and Fig. 12.10 even 6 waves rotating against each other in both directions (see the arrows) with periodic head-on collisions with residence-time behavior. The system is absorption of pentane-vapor into toluene. (The residence-time by these head-on-collisions was measurable.) The additional dark and light lines belong to triggered internal gravity waves which are soon damped.

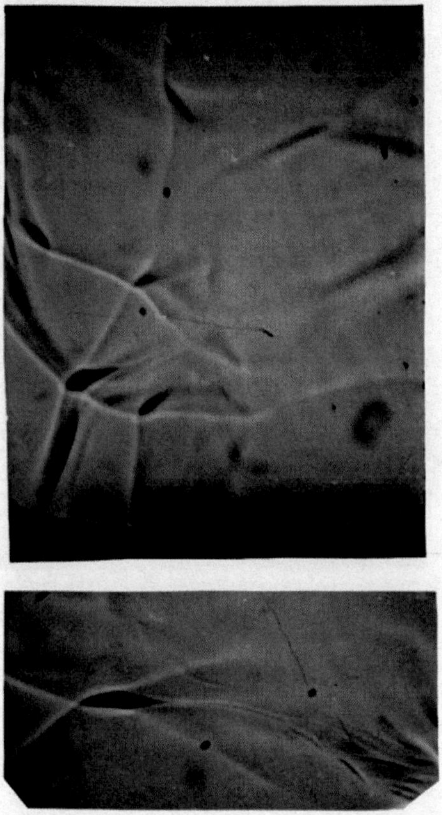

Fig. 12.7 Waves with acute angle crossings at relatively strong driving force: At relative strong Ma-stress: Strong disturbances after acute-angle crossings can lead to fading away of the after-collision waves, see (**c**). Also, in the case of obtuse angle crossings with the Mach-Russel third waves, the after-crossing waves can fade away due to loss of energy by the produced disturbances

Figure 12.11 shows different geometric forms of travelling waves in circular containers: Besides waves travelling in random direction as in Fig. 12.11a,b (and also in Figs. 12.7 and 12.8), we also found ring-formed waves starting from the borderline and disappearing in the center (Fig. 12.11c). There are also counter-rotating waves in one direction (Fig. 12.11d) and in two directions (Fig. 12.10). An extreme behavior was found at a wide annular channel in Fig. 12.11e: A short wave parallel to the walls works as an oscillator (see the arrows) and sends obliquely reflected counter-rotating waves around the channel with interesting interactions. The system is as in Fig. 12.10.

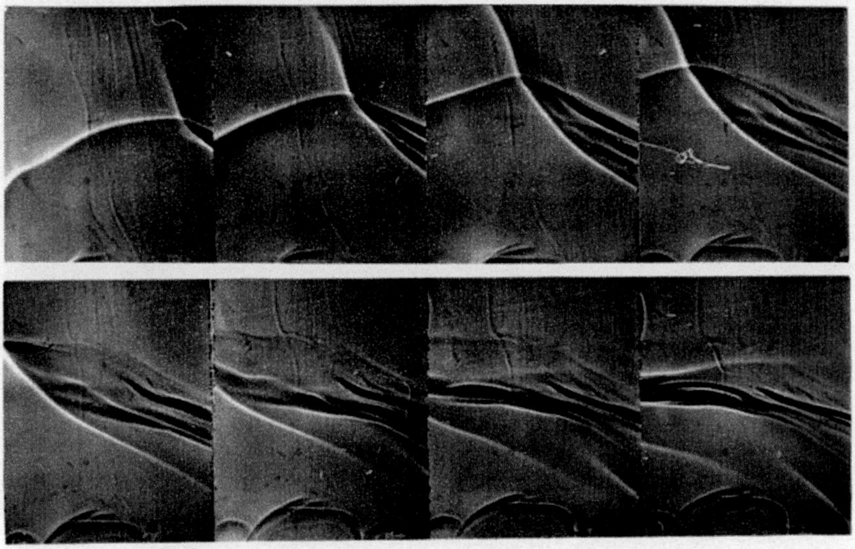

Fig. 12.8 Waves with acute angle crossings at relatively strong driving force: Absorption of diethyl-ether vapor into dichloroethane at relative strong Ma-stress, 8 s after start, shows on a scale of about 1:5 both the splitting of the after-collision waves and the triggering of damped internal gravity-waves

12.4 Structures With Completely Chaotic Behavior

Figure 12.12 shows in the acetone-vapor/octane system 5 s. after the start that there exist also oscillations (see the arrows in the scheme) with a substructure of RCI and RCII in the left column; 3 min later we observe a completely chaotic behavior, see the right column of pictures! That happens also at strong driving force with oscillations without substructure; see Fig. 12.13 with absorption of hexane-vapor into nonane. We see in motion pictures—to be read like a book—the rebirth of waves colliding with acute angle and their turbulent collapse by the same acute angle oblique collisions after they got too much energy over time. Such experiments (and mainly other experiments with unstable density distribution with chaotic behavior) gave the first—not correct—name"interfacial turbulence" to Marangoni-instability by Sternling and Scriven.

Fig. 12.9 Absorption of pentane-vapor into toluene at a circular facility which was a little polluted by a fingerprint forming a thin film at the surface. 5 waves rotating against each other in both directions with periodic head-on collisions with residence-time behavior can be seen

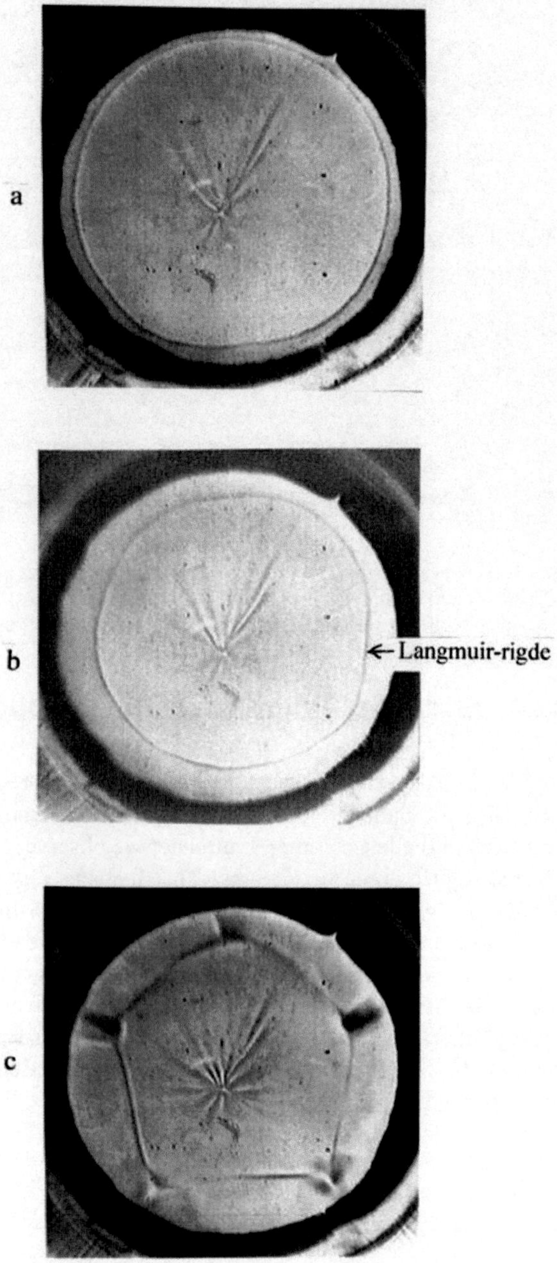

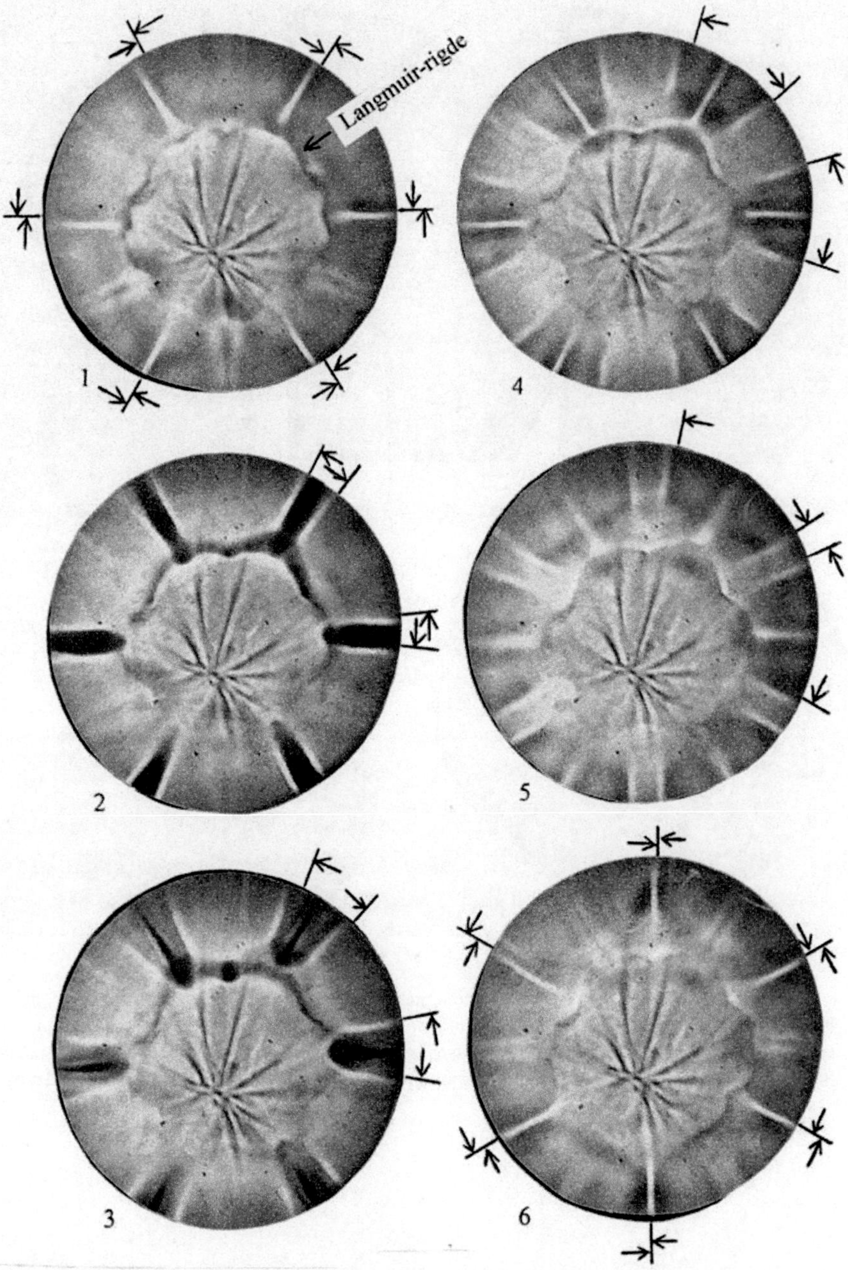

Fig. 12.10 From the wall to the center spreading of pentane (vapor) partially cleans the polluted surface of a circular container filled with the pentane-absorbing toluene. The cleaned ring-shaped channel shows six counter-rotating wave-trains. Traces after the head-on collisions indicate quickly damped internal gravity waves as disturbances

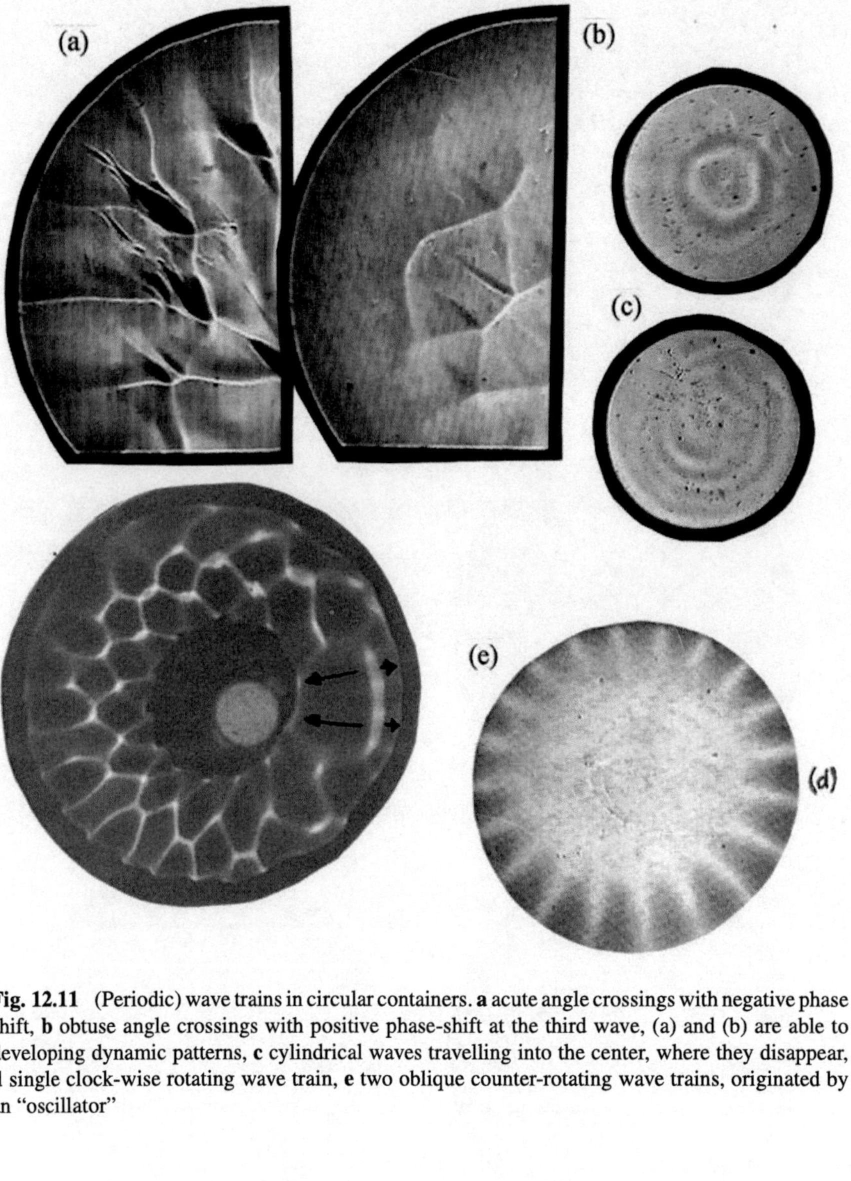

Fig. 12.11 (Periodic) wave trains in circular containers. **a** acute angle crossings with negative phase shift, **b** obtuse angle crossings with positive phase-shift at the third wave, (a) and (b) are able to developing dynamic patterns, **c** cylindrical waves travelling into the center, where they disappear, **d** single clock-wise rotating wave train, **e** two oblique counter-rotating wave trains, originated by an "oscillator"

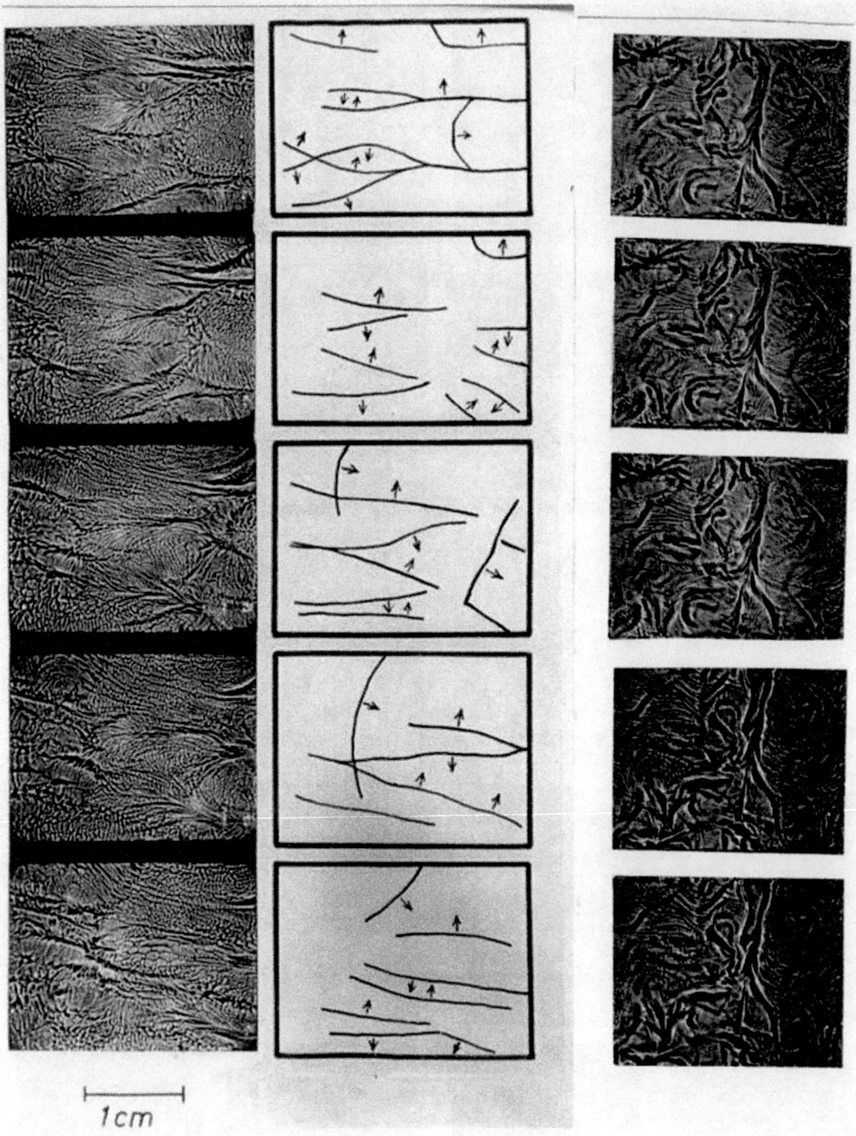

Fig. 12.12 System acetone-vapor/octane, 5 s after start, showing oscillations with a substructure of RCI and RCII roll cells (right column), schematically illustrated by the arrows in the middle column; 3 min later, a completely chaotic behavior is observed (right column)

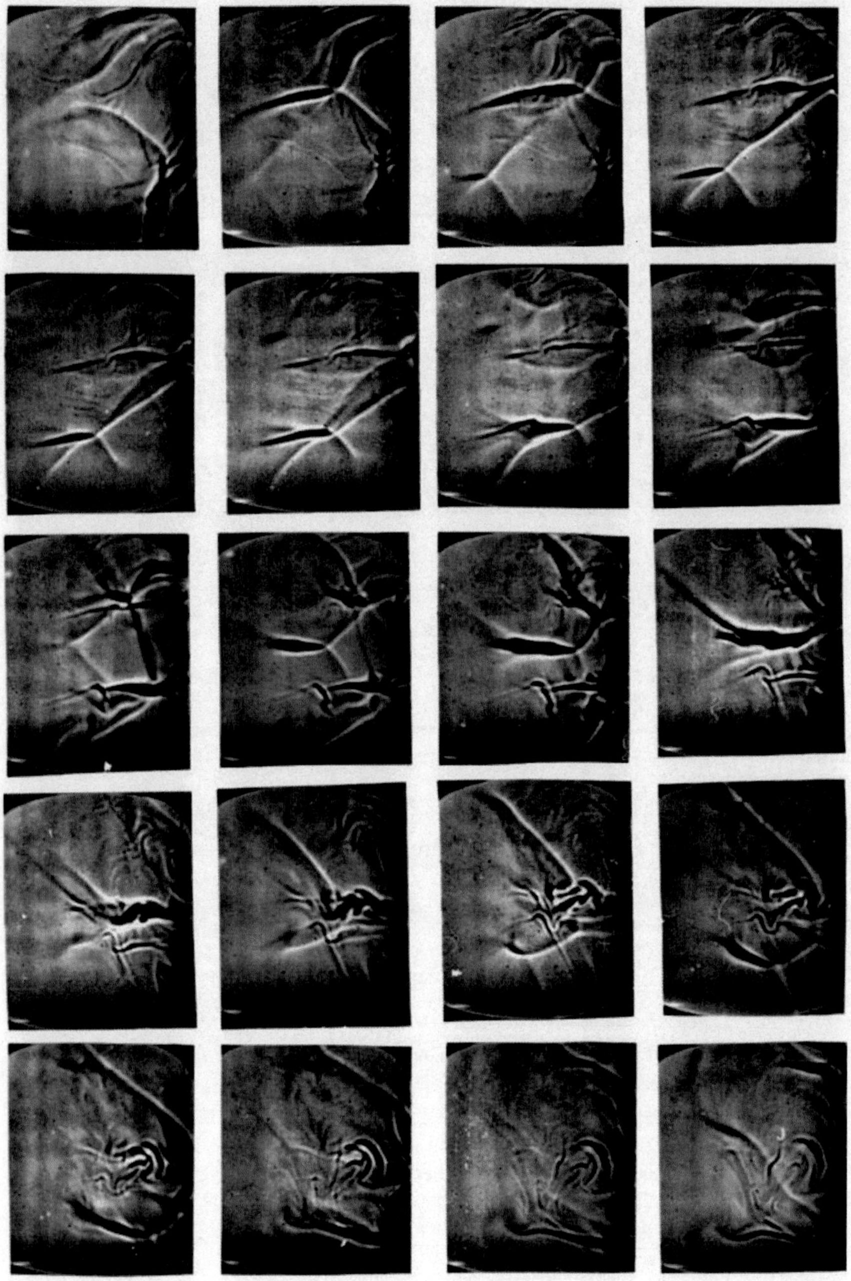

Fig. 12.13 Absorption of hexane vapor into liquid nonane at strong Ma-stress with 0.24 s/picture time-distance, at the scale 1:4.3, with collapses after collisions with negative phase-shift. New colliding waves are re-amplified after the collapses

References

1. Sternling, C.V., Scriven, L.E.: Interfacial turbulence: hydrodynamic instability and der Marangoni effect. AIChE J. **5**(4), 514–523 (1959)
2. Reichenbach, J., Linde, H.: Linear perturbation analysis of surface-tension-driven convection at a plane interface (marangoni instability). J. Colloid Interf. Sci. **84**(2), 433–443 (1981)
3. Linde, H., Loeschcke, K.: Rollzellen und Oszillation beim Wärmeübergang zwischen Gas und Flüssigkeit. Zellgrößen und Frequenzen im Vergleich mit der Theorie der Marangoni-Instabilität von Sternling und Scriven. Chem. Ing. Tech. **39**(2), 65–74 (1967)
4. Linde, H., Schwarz, E., Gröger, K.: Zum Auftreten des Oszillatorischen Regimes der Marangoniinstabilität beim Stoffübergang. Chem. Eng. Sci. **22**(6), 823–836 (1967)
5. Linde, H., Chu, X.-L., Velarde, M.G., Waldhelm, W.: Wall reflections of solitary waves in Marangoni-Bénard convection. Phys. Fluids A: Fluid Dyn. **5**, 3162 (1993)
6. Linde, H., Velarde, M.G., Wierschem, A., Waldhelm, W., Loeschcke, K., Rednikov, A.Y.: Interfacial waves motions due to Marangoni Instability I: Travelling periodic wave trains in square and annular containers. J. Colloid Interf. Sci. **188**(1), 16–26 (1997)
7. Wierschem, A., Linde, H., Velarde, M.G.: Internal waves excited by the Marangoni effect. Phys. Rev. E **62**(5), 6522–6530 (2000)
8. Linde, H., Velarde, M.G., Waldhelm, W., Loeschcke, K., Wierschem, A.: On the various wave motions observed at a liquid interface due to marangoni stresses and instability. Ind. Eng. Chem Res. **44**(5), 1396–1412 (2005)
9. Linde, H., Velarde, M.G., Wierschem, A.: Interfacial wave motions due to Marangoni Instability III. Solitary waves and (periodic) wave trains and their collisions and reflections leading to dynamic network (cellular) patterns in large containers. J. Colloid Interf. Sci. **236**(2), 214–224 (2001)

Part IV
Structure Formation in Social Systems

Chapter 13
Creativity—Comments to the Scientific Process

Knowledge Production Referring to Traditional Knowledge

Ernst-Christoph Haß and Peter J. Plath

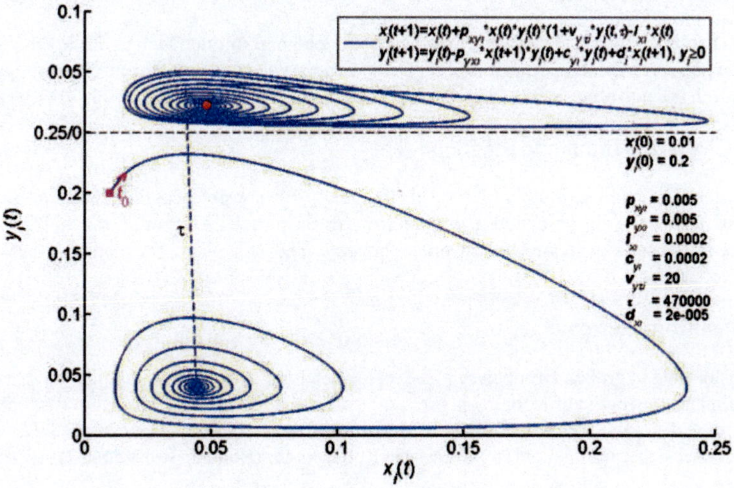

Fig. 13.1 Restricted creativity with a long-term discourse to previous knowledge—resulting toroidal boosts. Iterated function system based on an extended Lotka model where $x_i(t)$ means the actual knowledge and $y_i(t)$ are the problems which are aware of the i-th group at time t

P. J. Plath et al., *Imagery Synergetics*, Understanding Complex Systems,
https://doi.org/10.1007/978-3-030-95607-3_13

13.1 Introduction

13.1.1 Creativity in the Scientific Process

In the general understanding, the word *creativity* primarily means the property of an individual creating something that is new or original and useful [1–3]. This reference to the individual—especially the "genius"—leads to many problems in understanding the phenomenon of creativity. In particular, the attempt to understand creativity as a quality of an introverted, quiet and secluded working person, who is excellent and full of unusual ideas, seems to us to be very misleading [4].

But also the opposite thesis of brainstorming by Alex F. Osborn [5] does not work:

> People inspire each other to come up with new ideas and with the number of suggestions also their quality increases. You only have to put together a group and encourage them to express their ideas freely and without prohibiting thoughts.

Translated from German:

> Menschen inspirieren sich gegenseitig zu neuen Ideen und mit der Menge der Vorschläge steige auch deren Qualität. Man müsse bloß eine Gruppe zusammensetzen und sie ermutigen, ihre Ideen frei und ohne Denkverbote zu äußern. (Bund & Rohwetter, DIE ZEIT, 2019, p. 23)

Tomas Chamorro-Premuzic [6] found that

> Brainstorming in large groups was a 'waste of time'. ... Responsible was the group's intrinsic urge to mediocrity, the so-called regression towards the middle. As a result, the most imaginative minds soon expressed only average thoughts. They adapted to the mediocre level of their colleagues.

Translated from German:

> Brainstorming in großen Gruppen sei, Zeitverschwendung'. ... Verantwortlich sei der gruppenimmanente Drang zur Mittelmäßigkeit, die sogenannte Regression zur Mitte. Die führe dazu, dass die einfallsreichsten Köpfe bald nur noch durchschnittliche Gedanken äußerten. Sie passten sich dem mediokren Niveau der Kollegen an. (Bund & Rohwetter, DIE ZEIT, 2019, p. 23)

To be able to express and implement deviating and thus new ideas only works if committed confidants in the immediate vicinity of the idea provider take up this idea and work it out together [7, 8].

As a consequence, a protective and inspiring atmosphere is needed to develop new ideas and let them mature. In this sense, creativity is neither the property of a genius nor that of a brainstorming group, but a necessary nucleation phenomenon as with every phase transition [9]. In order for a new idea to spread into society, the society must also be excitable by the stimulation of this nucleus; i.e. the phase transition must be possible too, otherwise the nucleus of creativity vanishes.

Nuclei in this sense can be scientific institutes (e.g. the Center of Synergetics of H. Haken at the University of Stuttgart) or regular conferences (e.g. the Elmau conferences on Synergetics between 1972 and 1990 organized by H. Haken or the

Zeinisjoch seminars in Galtür/Austria between 1979 and 2008 organized by P. Plath); and the excitability of society can be achieved, for example, by appropriate research funding.

In their recent publication "Modellierungskonzepte der Synergetik und Theorie der Selbstorganisation", Ebeling and Scharnhorst [9] emphasize that the occurrence of innovations on the level of an overall system is always connected with destabilizing the current state and re-stabilizing the new one, i.e. the instability of the present system is a necessary pre-condition for the *New*. Then, an (external) impetus of a new invention into the pool of behavioral possibilities of the system may lead to a phase transition, and an innovation can prevail—or even not.

With respect to model various creativity approaches, we suggest to slightly modify the concept of Ebeling and Scharnhorst in a way that we consider a self-excitable system capable of an arbitrary number of impulses for new knowledge and new problems. In this context, we assume that knowledge and problems are arising again and again according to a Lotka-Volterra system, i.e. we consider a chain of (system-inherent) excitations which lead always to something new.

As already mentioned, such a system seems to be feasible only in a group of committed confidants, in which everyone can spontaneously bring in their ideas, without suppressing or deriding dissenting opinions in any way, e.g. in a relaxed tea or coffee round.

13.1.2 Modelling Creativity by a Lotka-Volterra Approach

This chapter is based on our presentation "Innovation und Interdisziplinarität", held in March, 2010, as part of the seminar "Interdisziplinarität und Institutionalisierung der Wissenschaft" at the Humboldt-Universität zu Berlin. In this lecture we presented an extended function system originating from the Lotka-Volterra [10–12] model in order to investigate the production of knowledge taking in account interdisciplinary cooperation. A remarkable result of this study was the occurrence of "bursts" in connection with innovations due to interdisciplinary cooperation (Fig. 13.1). The idea to use a generalized Lotka-Volterra system goes back to the work of Müller [13] in Greifswald who described the dynamics of the mutual dependence between problems and knowledge in the scientific process for the first time.

The suggestions from the discussion which followed our presentation lead us to the problem of "creativity". It seemed obvious to us that the approach taken by F. Müller, that the "*current problems*" and the "*current knowledge*" of a scientist or a group of scientists, or even of a whole scientific discipline, were the decisive variables of the dynamics of creation of new knowledge, that is knowledge capable of publication. Thus, they are very closely related to the problem of creativity.

To be clearly distinguished from current knowledge is the "*previously known knowledge*", which already exists in publications and is therefore no longer available for publishing.

In this way, we have created a rather operational approach to the idea of creativity, since the current, publishable knowledge is the new, just emerging knowledge, which has to be distinguished from the already known knowledge.

Analogously, the patent law is based on a very similar concept, i.e. a procedure is only patentable if the underlying idea is new, in other words it may not be considered as already known in lectures or publications including other patents.

In contrast to Amabile [14], however, we are not distinguishing between an operational definition and a conceptual one. For us, the process of generating new knowledge coincides with the newly identified product—the new knowledge. In other words, the process of creating new knowledge and new problems, as well as solving them, is itself a part of the problem and thus the object of our investigation.

For us it is crucial that the periodic process of the Lotka-Volterra system is limited by the "lifetime" of the researcher or research group, which can be achieved by an accordingly strong damping of the periodic function. In summary, it can be formulated:

In a first approximation, we consider **creativity as a periodic process** of creating new knowledge by solving problems.

For a group of scientists or a scientific institute headed by a scientific institute leader (usually a professor), creativity should be described by the classical equation system (13.1) of Lotka and Volterra, in which the problem function $y_i(t+1)$ is extended by the term $(+d_{xi}(t+1))$ corresponding to the damping caused by the current knowledge.

$$\begin{aligned} x_i(t+1) &= x_i(t) + p_{xyi}x_i(t)y_i(t) - l_{xi}x_i(t) \\ y_i(t+1) &= y_i(t) - p_{yxi}x_i(t+1)y_i(t) + c_{yi}y_i(t) + d_{xi}x_i(t+1); \\ & \qquad y_i(t) \geq 0, d_{xi} < 0\,. \end{aligned} \tag{13.1}$$

Here, $x_i(t)$ is the current knowledge and $y_i(t)$ are the current problems of i-th scientist or i-th group of scientists.

The concept of creativity also contains inherently the moment of surprise, i.e. the unexpected *New*. This goes beyond the conventional idea of the *New*, in which the *New* is indeed new, but somehow to be expected, since the problems from which it arises are certainly known.

But referring back to quite different problem areas, the *New* should no longer be associated with the omen of the already *known*. This can be accomplished by not only limiting to problems of the own discipline or of the own research group [15] as we have shown in the case of interdisciplinary and the innovations resulting from it [16], but by trying to tackle also problems of other disciplines or groups and treating them with the own methods. Then, one may obtain completely unexpected, eruptive events in the production of the *New*.

But one can also refer again to quite old problems which are almost forgotten. Then the resulting *New* is no longer the expected, foreseeable future, but the unexpected *New*, which no one seriously would have counted; and in this sense it is also emergent.

The technique of handling this recourse to previous problem areas is based on the use of the delay time τ. Instead of the current problem $x(t)$ at the actual moment—i.e., at time t—the problems $x(t - \tau)$ at time $(t - \tau)$ are now also taken into account, where τ should be sufficiently large.

13.2 Knowledge Reduces Problems

13.2.1 Natural Creativity

If one reduces the problem curve by a—negative—damping factor $d_{xi} = 1.4 * 10^{-6}$ (or even smaller), one obtains only a single maximum for $x_i(t)$ or $y_i(t)$, respectively, as the result of Eq. (13.1) (see Fig. 13.2). We will call this kind of creativity **natural** (or ordinary, normal) **creativity**. Then, the amount of the generated knowledge (integral) or the height of the maximum or the maximum slope of the knowledge function could serve as a quantitative measure of this type of creativity.

A crucial issue is how much the problem development of this leader or of his closest collaborators in their youth or during their studies has been—including their precocious, childhood or adolescent generation of problems. In this sense we speak here—without further differentiation, which would be quite possible—of natural creativity, since it is based on this natural creation of problems.

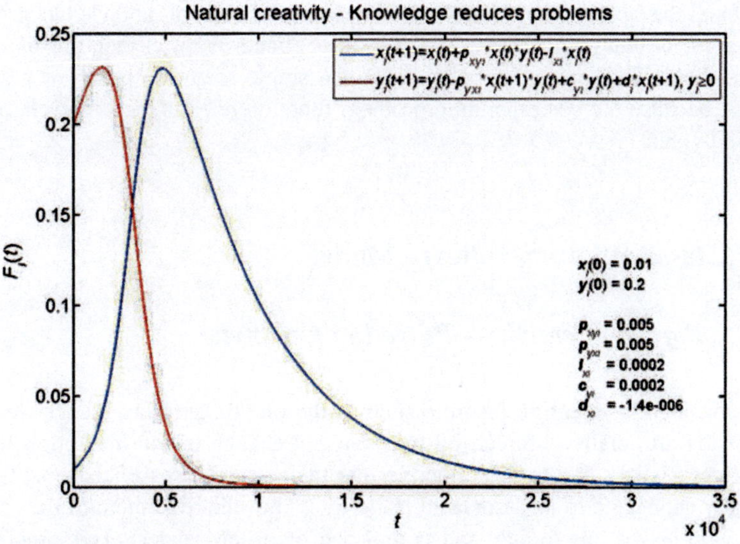

Fig. 13.2 Natural (ordinary, normal) creativity. Knowledge reduces the emergence of new problems. Time series of the problem and knowledge functions according to Eq. (13.1). The damping by the actual knowledge is so large in this case that only one problem/knowledge cycle is traversed

In the publications by Dobrov [17] and by Müller [13] an empirical investigation can be found of important scientists from the USSR and the USA regarding their "productivity", which comes fairly close to our conception of "natural creativity".

13.2.2 Autonomous Creativity

If the damping—i.e. the negative influence onto the problems due to the just created knowledge—decreases, several maxima (damped oscillations) can arise as well—even in the case of only one scientist/scientific leader. These multiple maxima of knowledge production correspond to the same number of maxima of problem generation. They cannot only be ascribed to the creativity or problem generation in the youth of the scientist, but correspond to further creative shifts due to the scientific process in which the scientist is involved. For this type of creativity we use the term **autonomous creativity**. An appropriate experimental evidence of such a behavior can be found in an article by Plath and Haß [16], where the "productivity" of selected scientists from the University of Bremen—which was just founded at that time—was studied.

An excellent example therefore is the group *aka* (angewandte Katalye = applied catalysis) which was part of the Physical Chemistry Department of the University of Bremen.

The principle course of knowledge production via autonomous creativity is exemplarily represented in Fig. 13.3 for a damping factor $d_{xi} = 5 * 10^{-7}$.

At first, the actual knowledge stimulates the process of knowledge production before its damping character comes into effect. We consider the damping term inevitable for knowledge generation of a single scientific group or a single scientist, because the corresponding problem function reflects the life cycle of the group/individual.

13.3 Classical Lotka-Volterra Model

13.3.1 Forced Creativity—Pulsating Creativity

If one considers a scientific institute such as the one of the Max-Planck-Society, creativity is artificially enforced by the "forced" change of the institution leader (**forced creativity**), due to the principle that the research area of that institute is personally connected to its associated leader (i.e. until the retirement of the former leader, who leaves due to age, and is replaced by a new leader). This practice is intended to ensure a constant—even though pulsating—creativity of the respective institute. It seems therefore obvious to describe this approach approximately by the classical Lotka-Volterra system.

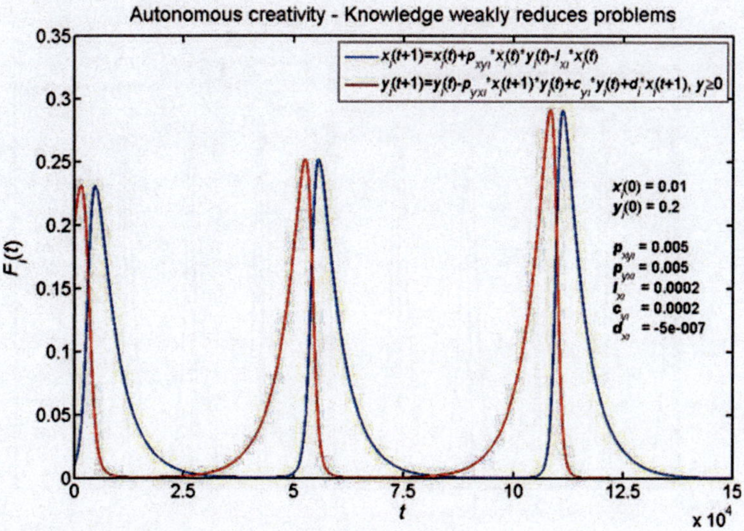

Fig. 13.3 Autonomous creativity. Knowledge reduces the emergence of new problems. Time series of the problem and knowledge functions according to Eq. (13.1). The damping by the actual knowledge is small in this case, such that several cycles of the problem and knowledge functions are passed through before the time series breaks down. The curves of maxima of the time series increase exponentially in the beginning before finally the damping prevails

In the ideal case of no damping ($d_{xi} = 0$), our equation system takes the form of the classical Loka-Volterra model and leads to a periodic solution for arbitrarily long times (Fig. 13.4). Let us denote in the following the distance between two knowledge maxima as a period of problem/knowledge interaction (about 30,000 time steps in the example shown in Fig. 13.4).

It is quite clear that there exists no damping in the principle of the Max-Planck-Society, since upon a change of the institution leader no accumulation of knowledge by means of knowledge transfer to the new leader occurs. The accumulated knowledge of the institute and the formerly existing problems are almost completely negated if the leadership of the institute is altered. At the best, old contracts continue until they end.

13.3.2 *Large, Free Systems—Fully Developed Creativity*

It looks totally different if we consider a society with a sufficiently large, diverse, and highly developed science landscape (e.g. the old Federal Public of Germany), in which parallel research—at different institutions—is possible. This would also lead to a periodic solution of the Lotka-Volterra system (no damping, see Fig. 13.4) and would thus correspond to a highly creative knowledge production.

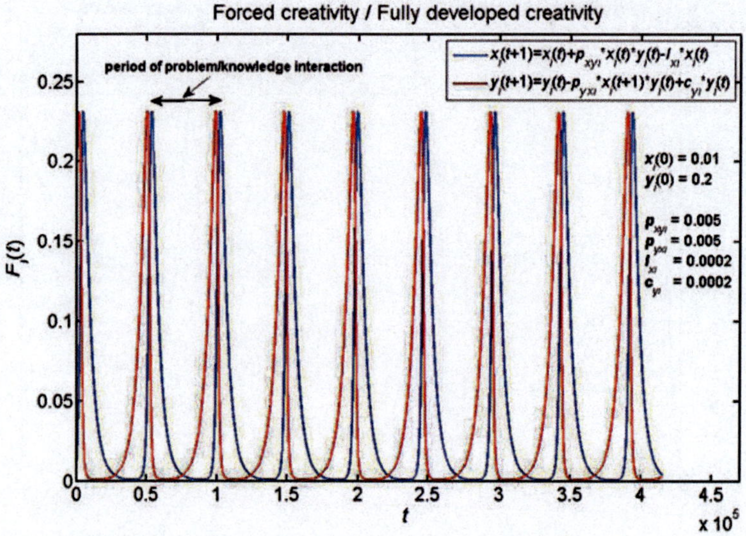

Fig. 13.4 Forced creativity/fully developed creativity. Classical periodic solution of the Lotka-Volterra model. As an example, the second period of problem/knowledge interaction is marked

13.4 Knowledge Enhances Problems

13.4.1 Restricted Creativity

A different situation exists, if—based on the argument of "comparability"—a canon of knowledge is in introduced with general obligation, as—for example—in the case of the unification of the education system by the Bologna and Pisa reforms in the new FRG or as practiced in the former GDR.

Then, a certain accumulation of knowledge and problems may happen, which leads to a positive damping of the system by the damping term $(+d_{xi}(t+1); d_{xi} > 0)$ (see Eq. (13.2)).

$$\begin{aligned} x_i(t+1) &= x_i(t) + p_{xyi}x_i(t)y_i(t) - l_{xi}x_i(t) \\ y_i(t+1) &= y_i(t) - p_{yxi}x_i(t+1)y_i(t) + c_{yi}y_i(t) + d_{xi}x_i(t+1) \\ & y_i(t) \geq 0, d_{xi} > 0\,. \end{aligned} \tag{13.2}$$

In such a case, the system oscillates to a fix point (first quadrant in the *x*/*y* space) which is different from the zero point. This takes place the faster the stronger the damping factor d_{xi} is, i.e. the more the system is forced to constrain itself (see Figs. 13.5 and 13.6).

It would not produce anything new with respect to creativity or it could be even counterproductive, if one would modify the above-mentioned MPG principle, so

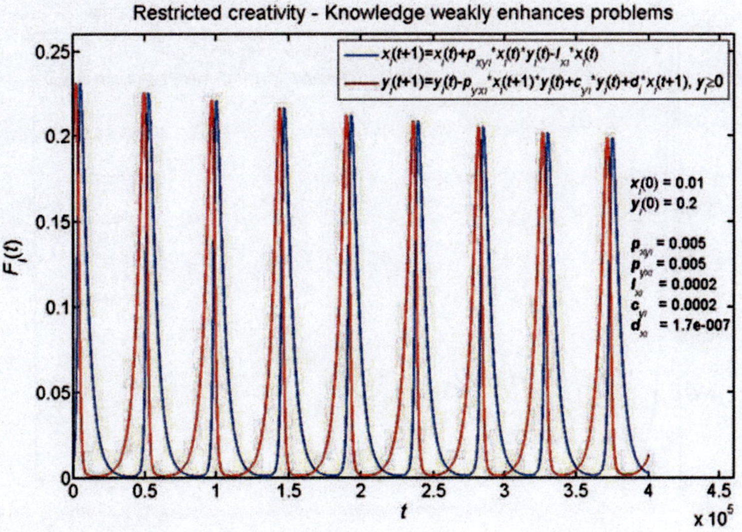

Fig. 13.5 Restricted creativity—weak damping: $d = 1, 7{*}10^{-7}$. Knowledge reinforces the emergence of new problems. Time series of the problem and knowledge functions according to Eq. (13.2). The strengthening by the actual knowledge is quite small in this case, such that the maxima of the problem and knowledge curves diminish only slowly

that the previous research is continued during an age-related change of the institution leader (e.g. by electing a suitable new successor) and thus the accumulated knowledge and the—thereby resulting—established problems are passed on the new leader. The damping cannot be avoided by this approach (see Fig. 13.6).

This is also true if one extends the function $x_i(t+1)$ of the knowledge production by an additional term $p_{xyi} x_i(t) y_i(t) \boldsymbol{v_{yi}\, y_i(t)}$ which is quadratic with respect to the problems (see Fig. 13.7).

$$x_i(t+1) = x_i(t) + p_{xyi} x_i(t) y_i(t)\big(1 + v_{yi} y_i(t)\big) - l_{xi} x_i(t)$$
$$. \, y_i(t+1) = y_i(t) - p_{yxi} x_i(t+1) y_i(t) + c_{yi} y_i(t) + d_{xi} x_i(t+1); \qquad (13.3)$$
$$y_i(t) \geq 0, d_{xi} > 0.$$

Such an additional term could be justified by the assumption that the knowledge production (generation of new knowledge) does not only result by the simple product of knowledge and problems $x(t)y(t)$, but is also affected to a certain degree by higher powers of $y(t)$, e.g. in form of the product $x(t)y^2(t)$. This would mean that an intensified pressure of problems is created which contributes to a higher solution capability, i.e. creativity.

However, it results—as expected—that thereby the activity of knowledge production in the first period is increased (and also the frequency of the decaying creativity is slightly larger), but—on the other hand—the damping is strengthened and thus

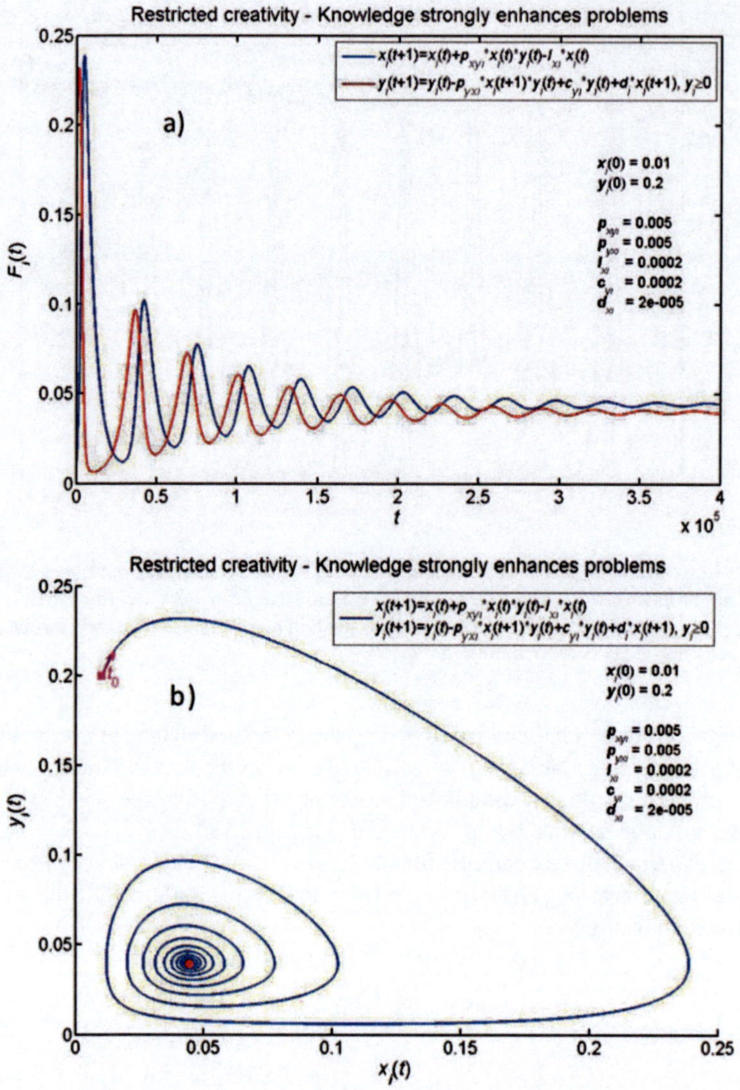

Fig. 13.6 Restricted creativity—strong damping: $d = 2{*}10^{-5}$. Knowledge reinforces the emergence of new problems. **a** Time series of the problem and knowledge functions according to Eq. (13.2). The enhancement by the actual knowledge is fairly large in this case, such that the maxima of the problem and knowledge curves diminish fast, **b** Phase diagram of the time series above. The trajectory clearly shows the oscillation to a fix point (red). It is also possible to exponentially approximate the decay curves of the maxima; for $t \to \infty$ these approximation functions approach values greater than zero of knowledge and problems, the value for the knowledge being greater than that for the problems

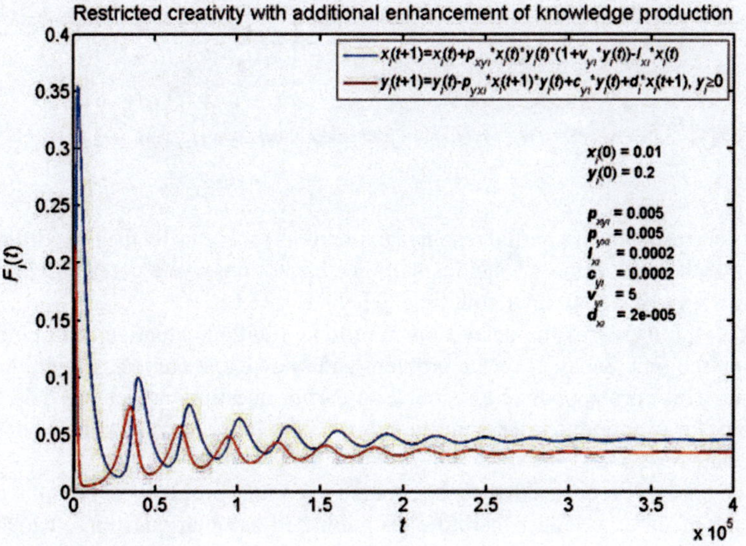

Fig. 13.7 Restricted creativity with enhancement of knowledge production by an additional term which is quadratic with respect to the problems. Knowledge reinforces the emergence of new problems. Time series of the problem and knowledge functions according to Eq. (13.3). The enhancement by the actual knowledge is quite large in this case, such that the maxima of the problem and knowledge curves diminish very fast

the fix points of the knowledge and problem functions are reached in less time t (compare Figs. 13.6 and 13.7).

13.4.2 Restricted Creativity with Recourse to Previous Knowledge

13.4.2.1 General

Let us now consider the question, how the system behavior is changed if the quadratic term

$$x(t)y^2(t)$$

is replaced by the expression

$$x(t)y(t)y(t-\tau)$$

according to Eq. (13.4):

$$x_i(t+1) = x_i(t) + p_{xyi}x_i(t)y_i(t)\left(1 + v_{y\tau i}\left\{\begin{matrix} 0, t \le \tau \\ y_i(t-\tau), t > \tau \end{matrix}\right\}\right)$$
$$\cdot \qquad -l_{xi}x_i(t)y_i(t+1) \qquad (13.4)$$
$$y_i(t+1) = y_i(t) - p_{yxi}x_i(t+1)y_i(t) + c_{yi}y_i(t) + d_{xi}x_i(t+1);$$
$$y_i(t) \ge 0, d_{xi} > 0.$$

This corresponds to a partial recourse to previous problems by the time difference (t − τ). If the delay time τ is quite small, one cannot observe a noteworthy change as compared to a situation without time delay, i.e. τ = 0.

But if one increases the delay time τ sufficiently, then situations can be found in which the time behavior of the problem and knowledge curves are significantly changed. This corresponds to a recourse to earlier questions which are now taken up again, for example to document previous results in an overview article or in an appropriate textbook.

Of course, this requires that such recourse to former problems is possible at all. Problems of this type must therefore be published as open questions, which—in contrast to very old publications from the 19th and in the early twentieth century—is rather unlikely in the modern scientific publishing process.

Another possibility to recourse to former problems is to talk with scientists who are retired from the actual academic activities or to read their books or memoirs. In this way, senior scientists could play an important role—not so much as consultants, but rather as preserver of problems—a fascinating perspective. This could certainly be an interesting aspect of gerontology—virtually a cultural gerontology.

13.4.2.2 Short-Term Recourse to Previous Knowledge

If a delay time τ is chosen such that the recourse occurs between the two maxima of the knowledge and problem curves, one obtains a (transient) oscillation behavior to a limit cycle. This is shown exemplarily in Fig. 13.8 for the first occurrence of this situation after one period of problem/knowledge interaction (after about 39,000 time steps using the parameters chosen in the previous examples). We propose to describe this kind of knowledge production as (restricted) **creativity with short-term recourse to previous knowledge**.

As an exception, an oscillation behavior to a limit cycle is also obtained if the recourse takes place shortly after the first maximum of the knowledge curve when the problem curve already falls to a minimum (see Fig. 13.9 as an example for τ = 8000 time steps). In contrast to the previous case, the second maximum of the knowledge curve is significantly smaller and the convergence to the limit cycle occurs faster.

Such a recourse shortly after the first knowledge maximum may correspond to the fact that other researchers or research groups immediately take up a new idea and expand it with additional results.

On the other hand, if the number of iteration steps is selected in a way that the recourse takes place close to the minima of the problem and knowledge curves, the

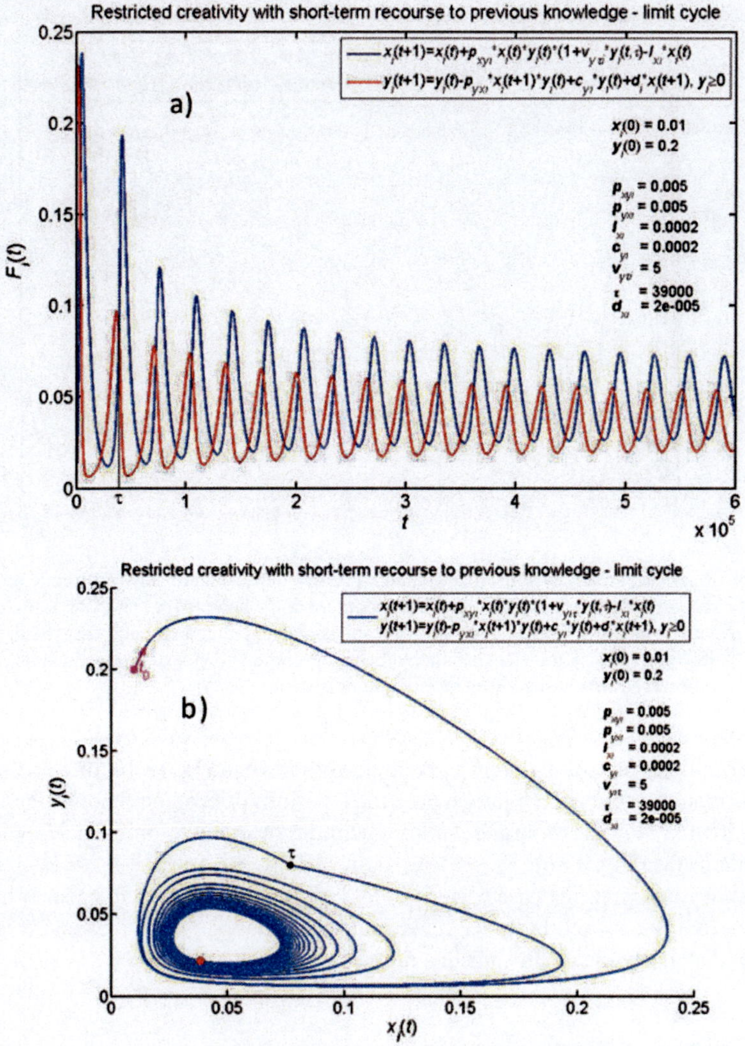

Fig. 13.8 Restricted creativity with enhancement of knowledge production by a quadratic problem term referring to problems which go back to about one period of problem/knowledge interaction. **a** Time series of the problem and knowledge functions according to Eq. (13.4) with $\tau = 9000$ time steps. A transient oscillation behavior of the knowledge production into a limit cycle can be observed. **b** Phase diagram of the time series above. The trajectory clearly shows the oscillation to a limit cycle (where the red point is located). Note that the trajectory exhibits a kink at delay time t when the recourse takes place

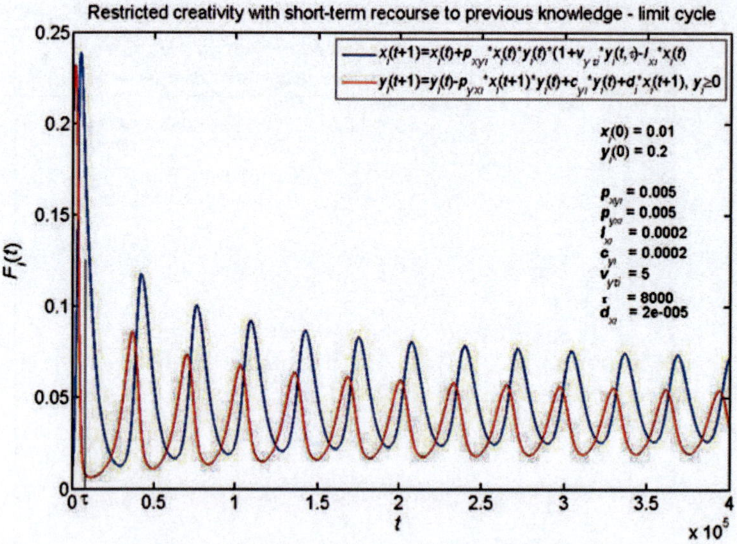

Fig. 13.9 Restricted creativity with enhancement of knowledge production by a quadratic problem term if the recourse occurs shortly after the first maximum of the knowledge curve, i.e. just after the invention of a new idea. Time series of the problem and knowledge functions according to Eq. (13.4) with $\tau = 8000$ time steps. A transient oscillation behavior of the knowledge production into a limit cycle can be observed with fast convergence to a limit cycle

system oscillates fast to a fix point as exemplarily shown in Fig. 13.10 for $\tau = 23{,}000$. This oscillation occurs much faster (only three periods of problem/knowledge interaction) than in the case of strong damping without recourse to previous knowledge or problems (see Fig. 13.6).

One may interpret this case as a (fast) loss of memory. This may occur if a new idea is strongly opposed by other scientists or by science policy reasons. It is not unusual that a new idea then ends in a drawer.

13.4.2.3 Long-Term Recourse to Previous Knowledge

A new situation results if τ is an n-fold ($n > 1$) of iteration steps between two maxima (or minima) of the knowledge or problem curve, respectively. Figure 13.11 shows exemplarily the corresponding time series for $n = 3$ and $\tau = 104.000$. After an initial decrease of the maximum values of both curves, a renewed amplification of the maxima is obtained after the time τ—in particular of the maximum value of the knowledge curve—followed by a rapid transient oscillation to a limit cycle.

This could be interpreted in such a way that, at the end of the creation process of a scientist, once again a creativity push takes place whereby the problem situation has changed only slightly.

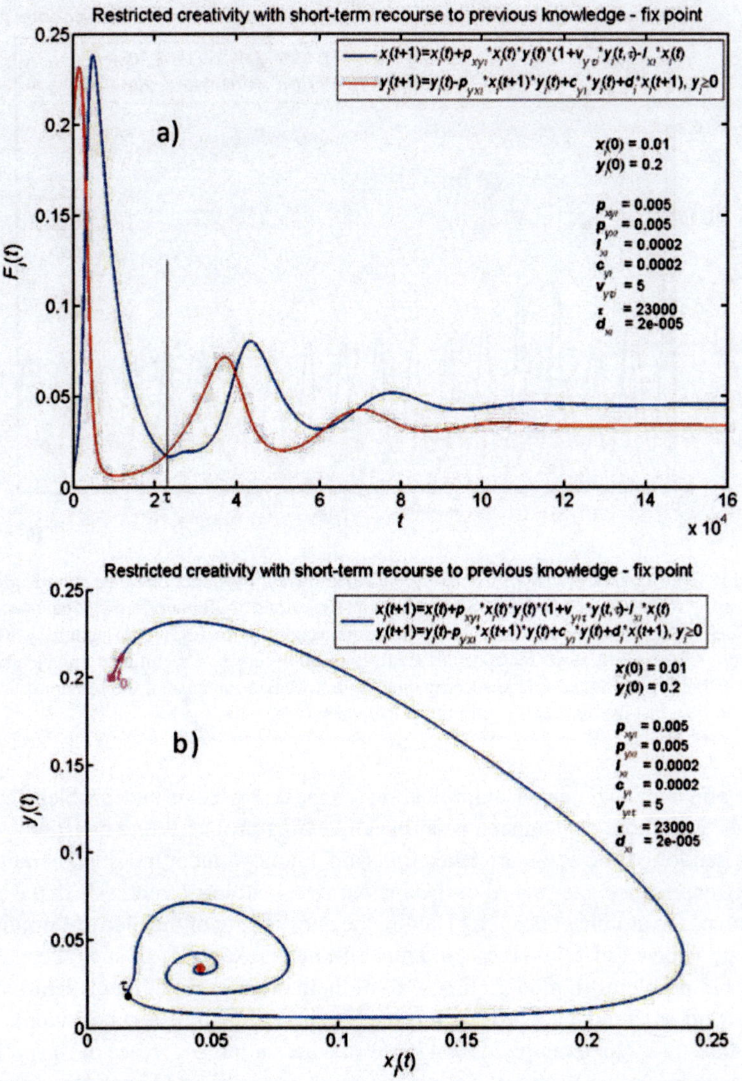

Fig. 13.10 Restricted creativity with enhancement of knowledge production by a quadratic problem term if the recourse occurs close to the first minima of the problem and knowledge curves. **a** Time series of the problem and knowledge functions according to Eq. (13.4) with $\tau = 23{,}000$ time steps. A very fast decay behavior of the knowledge production into a fix point can be observed. **b** Phase diagram of the time series above. The trajectory shows that the fix point (red) is reached after only three periods of problem/knowledge interaction

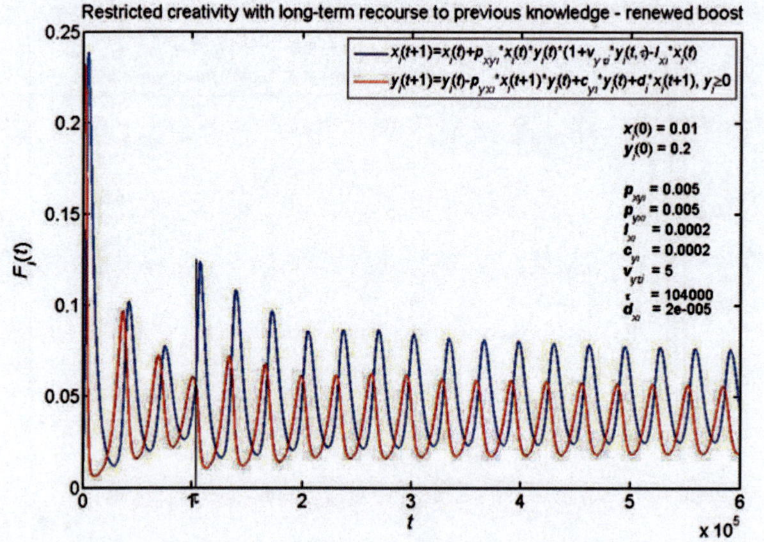

Fig. 13.11 Restricted creativity with enhancement of knowledge production by a quadratic problem term referring to problems which go back to about three periods of problem/knowledge interaction. Time series of the problem and knowledge functions according to Eq. (13.4) with $\tau = 104.000$ time steps. After an initial decrease of the maxima of both curves, a significant enhancement of the knowledge function takes place accompanied with a weaker increase of the problem function. Subsequently, a fast oscillation into a limit cycle occurs

The effect of a renewed amplification of the knowledge and problem curves becomes even more pronounced if far back reaching problems and questions—that were raised long time ago—are taken up again, but now under modified conditions.

The Renaissance, the revival of the ancient Greek atomic hypothesis in the reformulation of the atomic concept by Dalton, the questioning of the idea of "simultaneously" by Albert Einstein—these all are well-known examples of such "recourses" to very old problems in modern times, all of them combined with high creativity.

In our iterative model approach, this leads to a toroidal transient behavior at very large values of τ (for example: $\tau = 470.000$), where after a decrease of the problem and knowledge curves to almost fix point behavior, a new boost of knowledge production takes place according to a second, small circular frequency on a long-term scale (Fig. 13.12). In the (toroidal) transient range of this long-term oscillation, the maxima of the problem and knowledge curves oscillate similarly to a superimposed beat rhythm with a larger frequency due to the periods of problem/knowledge interaction. This may be interpreted as a toroidal creativity process.

The long-term behavior of the toroidal transient processes becomes particularly clear when the time series are extended over a considerably longer time interval, e.g. ten periods of long-term oscillations (see Fig. 13.13).

This long-term behavior which results from our model approach is indeed mathematically of interest, but it has probably no meaning for the modeling of creativity

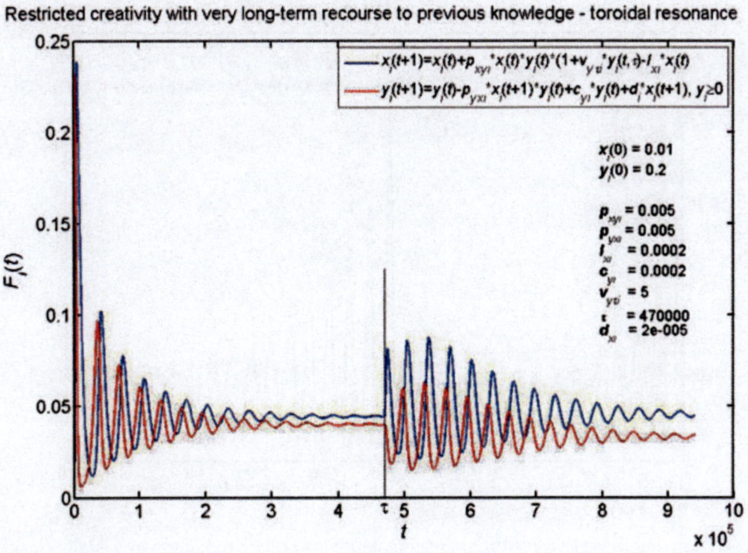

Fig. 13.12 Restricted creativity with enhancement of knowledge production by a quadratic problem term referring to far back reaching problems. Time series of the problem and knowledge functions according to Eq. (13.4) with $\tau = 470.000$ time steps over two periods of long-term oscillation. A toroidal transient behavior of the creativity can be observed, where—similarly to a superimposed beat rhythm—the problem and knowledge curves oscillate with a larger frequency within the small frequency of the long-term oscillation

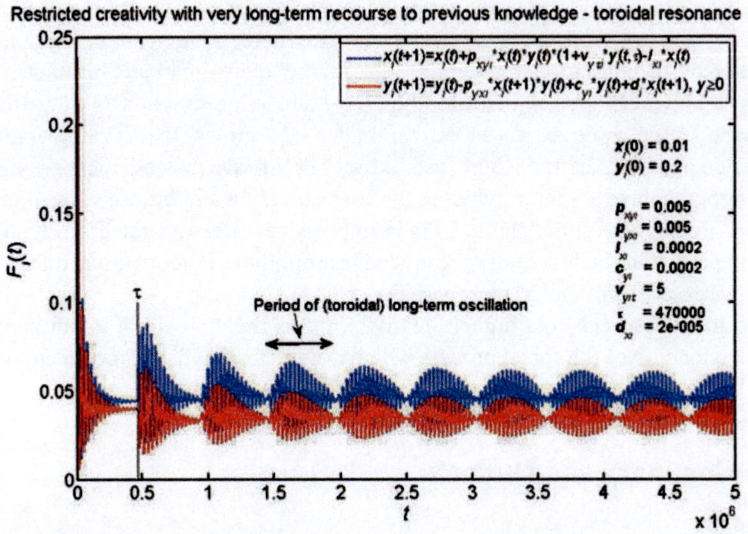

Fig. 13.13 Restricted creativity with enhancement of knowledge production by a quadratic problem term referring to far back reaching problems. Time series of the problem and knowledge functions according to Eq. (13.4) with $\tau = 470.000$ time steps over a very long time (about ten periods of long-term oscillation). As an example, the fourth period of (toroidal) log-term oscillation is marked

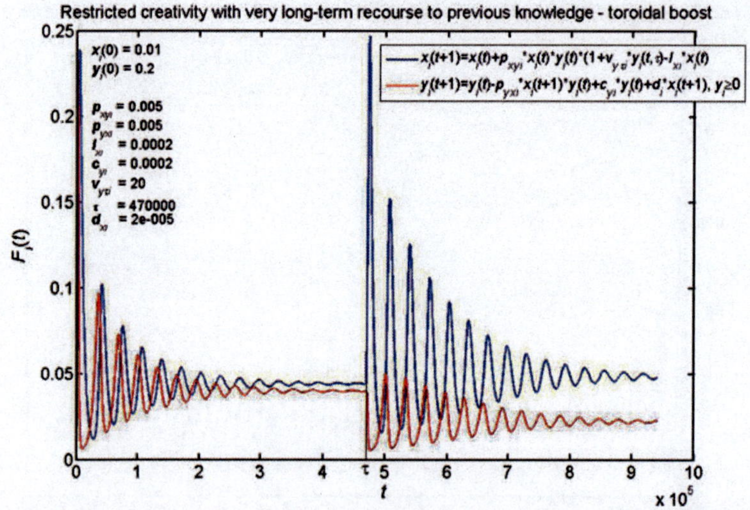

Fig. 13.14 Restricted creativity with enhancement of knowledge production by a quadratic problem term referring to far back reaching problems. Time series of the problem and knowledge functions according to Eq. (13.4) with $\tau = 470.000$ time steps over two periods of long-term oscillation. As can be seen in comparison with Fig. 13.12, an increase of parameter $v_{y\tau i}$ from 5 to 20 leads to an enhancement of the first amplitude by a factor three after recourse at time τ

in the real, present science process, since the conditions in our social system are nowadays no longer at least approximately constant over such long time periods.

Rather, the toroidal transient processes are here of essential meaning. In this context, the variation of weight $v_{y\tau i}$ in the quasi-quadratic problem term $p_{xyi} v_{y\tau i} x(t) y(t) y(t - \tau)$ which modifies the knowledge function is of particular influence. To the same extent as the parameter $v_{y\tau i}$ grows, the amplitudes of the knowledge curve of **the transient (oscillatory) creativity process** increase significantly after recourse at time τ, whereas the corresponding amplitudes of the problem function are slightly diminishing. This is impressively demonstrated in Fig. 13.14 where in particular the first of these knowledge amplitudes is about three times larger if $v_{y\tau i}$ increases from 5 to 20 (compare Fig. 13.12).

This all corresponds to a higher creativity and is thus of central importance for the understanding of the social process which is described with our model approach.

13.5 Summary and Outlook

As a consequence of our previous discussion, we propose the following classification of the concept of creativity:

- Natural creativity: negative damping by knowledge
  → Fix point (0|0) in the origin

- Social creativity: positive damping by knowledge
  → Fix point $(x_\infty | y_\infty) with x_\infty, y_\infty > 0$
- Cultural creativity: positive damping with recourse to previous knowledge after time τ
  → Fix point, limit cycle, toroidal oscillations

With a suitable choice of time delay $\tau \gg 0$, it is possible to transfer a system which is running to a fix point into a new state in which it can – oscillatory self-exciting – take the form of a limit cycle or a toroidal oscillation.

Finally, let us conclude with some general remarks about the concept of creativity.

Usually, creativity is conceived as a property of an individual person, whose respective thinking and action is targeted at producing originary, mostly unexpected, changes or solutions in the scientific process. Csikszentmihalyi [18] goes beyond this in so far as he considers domains which have to be changed, but which themselves—or the therefore responsible authorities—have to agree too. In this sense, he introduces a social component—the socially responsible environment—into the concept of creativity.

Steiner [19, 20] goes even one step further by granting creativity also to a group or to a network—more general to a cooperative system. This is similar to our approach which we are pursuing in our paper, although we distinguish between natural, social and cultural creativity, respectively. For us, Steiner's approach is particularly interesting in that it also attributes creativity to a network. The inclusion of networks opens up the possibility—analogous to our perception—to describe global processes in the field of sciences or arts in terms of creativity.

As an example for such processes, we consider highly networked and globally distributed groups of scientists who, on the one hand, are in close, almost direct exchange with each other—e.g. associated by public funding programs—and who, on the other hand, communicate and cooperate with groups far-away in other countries or cultural regions, but with a certain, noticeable temporal delay. First thoughts in this direction, based upon cellular automata, we have presented 2006 at a winter seminar in Galtür/Austria. These results were also published online in a short form [21].

Due to the abstract nature of our approach, however, it is also possible to treat any highly cross-linked network, where, apart from the direct neighborhood, there exist also areas, which are "temporally far-away" and which thus communicate only with a considerable delay. Such systems are of central importance in brain research, where appropriate networks have been discovered by Deco et al. [22] only a few years ago.

References

1. Runco, M., Jaeger, G.: The standard definition of creativity. Creat. Res. J. **24**,1.1, 92–96 (2012)
2. Mumford, M.: Where have we been, where are we going? Taking stock in creativity research. Creat. Res. J. **15**, 107–120 (2003)

3. Kreativität. (16. 04. 2020). Information retrieved on May 11th 2020, from https://de.wikipedia.org/w/index.php?title=Kreativit%C3%A4t&oldid=198954807.
4. Bund, K., Rohwetter, M.: Die Macht der Stillen. DIE ZEIT **49**, 23–24 (2019)
5. Osborn, A.: Your Creative Power: How to Use Imagination. Charles Scribner's Sons, New York (1948)
6. Chamorro-Premuzic, T.: Personality and individual differences (Vol. 4 of BPS Textbooks in Psychology). Wiley and 3rd edition ISBN: 978-1-118-77303-1 December 2016 (2007)
7. Swart, E., Plath, P.: Simulation of individual behaviour. In: Encarnação, J., Peitgen, H.-O., Sakas, G., Englert, G. (eds) Fractal Geometry and Computer Graphics, pp. 144–161. Springer-Verlag, Berlin, Heidelberg, New York, London, Paris, Tokyo, Hongkong, Barcelona, Budapest (1992)
8. Plath, P., Haß, E.-C., Swart, E.: Kooperatives Verhalten von Individuen bei destruktivem Wachstum in einer globalen Welt. Leibniz Online-Archiv, Jahrgang 2020, Nr. 39, 07 (2020), Plath.pdf. Information retrieved on May 11th 2020, from https://leibnizsozietaet.de/wp-content/uploads/2020/02/Portfolio-LO-39-2020.pdf
9. Ebeling, W., Scharnhorst, A.: Modellierungskonzepte der Synergetik und der Theorie der Selbstorganisation. In: Braun, N., Saam, N. (eds) Handbuch Modellbildung und Simulation in den Sozialwissenschaften. Springer, Wiesbaden VS (2015)
10. Lotka, A.: Contribution to the theory of periodic reactions. J. Phys. Chem. **14**(3), 271–274 (1910)
11. Lotka, A.: Zur Theorie der periodischen Reaktionen. Z. Phys. Chem. **72**, 508–511 (1910)
12. Volterra, V.: Variations and fluctuations of the number of individuals in animal species living together. ICES J. Mar. Sci. **3**(1), 3–51 (1928)
13. Müller, F.: Fortschritt der Wissenschaft - mathematisch modelliert. Wissenschaft und Fortschritt **22**(4), 162–165 (1972)
14. Amabile, T.M.: Creativity and Innovation in Organizations. Harvard Business School Background Note, 396–239 (1996)
15. Plath, P.J.: A Study on Knowledge/Problem Dynamics in Science. In: Ebeling, W., Peschel, M., Weidlich, W. (eds) Mathematical Research, Vol. 64, , pp. 60–76. MOSES. Akademie-Verlag, Berlin (1991)
16. Plath, P., Haß, E. Interdisziplinarität oder vernetzte Wissenschaften. In: Plath, P., Haß, E. (eds) Vernetzte Wissenschaften - Crosslinks in Natural and Social Sciences, pp. 153–179. Logos Verlag, Berlin (2008)
17. Dobrov, G.: Wissenschaftswissenschaft, p. 104. Akademie-Verlag, Berlin (1969)
18. Csikszentmihalyi, M.: Creativity; Flow and the Psychology of Discovery and Invention. Harper Collins Publishers, New York (1996)
19. Steiner, G.: The concept of open creativity; Collaborative creative problem solving for innovation generation—A systems approach. J. Bus. Manag. 5–33 (2009)
20. Steiner, G.: Das Planetenmodell der kollaborativen Kreativität - Systemisch-kreatives Problemlösen für komplexe Herausforderungen. Gabler Verlag | Springer Fachmedien, Wiesbaden (2011)
21. Haß, E., Plath, P.: Ein Modell der globalen Ausbreitung von Wissen. Leibniz-Online, Jahrgang 3/2007. Information retrieved on May 11 2020, from https://leibnizsozietaet.de/wp-content/uploads/2012/11/05-Hass_Plath_Ausbreitung.pdf
22. Deco, G., Jirsǎ, V., McIntosh, A., Sporns, O., Kötter, R.: Key role of coupling, delay, and noise in resting brain fluctuations. Proc. Natl. Acad. Sci. **106**(25), 10302–10307 (2009)

Chapter 14
Mother Hulda and the Blue Sky Catastrophe

Literary Coding of Deep, Ancient Knowledge of Social Dynamics

Peter J. Plath and Ernst-Christoph Haß

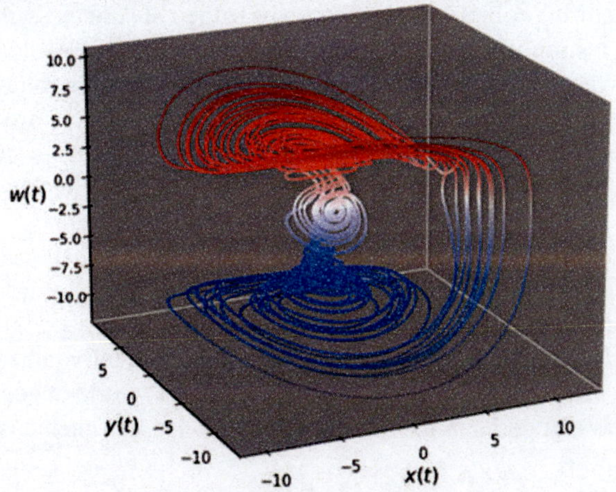

Fig. 14.1 Visual representation of the complex dynamics of the fairy tale “Mother Hulda”. Attractive chaotic trajectory in the three-dimensional sub-space of the four-dimensional space of the socio-economic dynamics with an instable Blue-Sky-Catastrophe in between the two anti-cyclic partial attractors in the upper and lower half spaces

14.1 Some Introductory Remarks

This chapter is largely a direct translation of the corresponding section on “Frau Holle” and the associated “mathematical appendix” in the book “Vom Märchen zur Mär” by Peter J. Plath [1], which was published in 2012 by Logos Verlag Berlin. The translation was done after consultation with Dr. Volkhard Buchholtz, the managing

P. J. Plath et al., *Imagery Synergetics*, Understanding Complex Systems,
https://doi.org/10.1007/978-3-030-95607-3_14

director of Logos Verlag Berlin GmbH. Since it was first published, many discussions and public readings have meant that a number of more in-depth changes, particularly with regard to the interpretation of mathematical modeling, have become necessary. Many of the graphics that were still carried out on Atari computers at that time have also been recreated. On the other hand, a number of mathematical details and etymological comments have been omitted here.

When the Brothers Grimm collected fairy tales at the beginning of the nineteenth century, they turned to people who were known to be able to tell old stories. Of course, they told them the fairy tales in their own way, for there was no fixed canon how to tell them. Moreover, these ancient stories varied from person to person and with the area where these narrators lived. The Brothers Grimm then also told fairy tales in their very own way. The old stories now became their fairy tales. However, they were able to concentrate on the essential features of the stories of the old storytellers, because they not only collected old fairy tales but also began to collect old legends [2] (1816 and 1818) around the same time, and started to work on the"German Mythology" [3] (1835–1878) and the "German Grammar" [4] (1819) [5].

Instead of capturing complex social processes in their abstractness, they are generally mapped in a simplistic way to high-ranking people or gods and their actions. In this way they are understandable and comprehensible. If one follows the events described in the myths in the corresponding realm of gods, then, conversely, conclusions can be drawn about corresponding events in the respective societies. Robert von Ranke Graves demonstrates this using the example of the Greek worlds of gods [6].

With all personification of social events, it is precisely this that makes it possible to disregard the concreteness of the historically acting people in order to extract the underlying dynamics (Fig. 14.1).

We follow this approach here in our interpretation of the fairy tale by Frau Holle. To do this, we have to make Mother Hulda's (Frau Holle's) world of gods visible, i.e., understand her as a goddess in the first place. The concept of meme is an excellent tool for this purpose.

When designing a socio-economic model of ancient societies, we use the order parameter approach of synergetics [7]. In addition to the resources and tools, the knowledge inherent in the use of the tools and the abstract knowledge developed in the priestesses' colleges and their academies play the decisive, regulating role.

14.2 The Great Goddess Mother Hulda—Frau Holle

From our point of view, it was possible for them to combine the *memes* contained in the various stories in a certain, logical and poetic form in their fairy tales.

We understand by a meme an encrypted, highly concentrated information that is independent of the carrier or transmitter of the information. Memes can be deciphered. A meme is a generating system, a logic or dynamic of a social process.

In connection with Frau Holle's story, for example, the sentence: "*If you make my bed good and that you shake it up diligently so that the feathers fly, then it will snow on the world: I am Frau Holle*" is a meme in the sense understood here.

Among all the fairy tales recorded by the Brothers Grimm, "Frau Holle" or "Mother Hulda" [8] occupies a special position: This is the story about the great goddess—and she frankly reveals herself as the "old woman with the big teeth" when Marie met her first:

Endlich kam es (Marie) zu einem kleinen Haus, daraus guckte eine alte Frau, weil sie aber so große Zähne hatte, wurde ihr Angst, und es wollte fortlaufen. Die alte Frau aber rief ihr nach:

Fürchte Dich nicht, liebes Kind, bleib bei mir, wenn du alle Arbeit im Hause ordentlich tun willst, so soll es dir gut gehen. Du mußt nur recht darauf achtgeben, daß du mein Bett gut machst und es fleißig aufschüttelst, daß die Federn fliegen, dann schneit es auf der Welt: **Ich bin die Frau Holle.**

At last, it (Marie) came to a small house, an old woman was peeping out of it, but because she had such big teeth it was afraid and wanted to run away. But the old woman called after it:

Don't be afraid, dear child, stay with me, if you want to do all the work in the house properly, you should be fine. You just have to make sure that you make my bed good and that you shake it up diligently so that the feathers fly, then it will snow on the world: **I am Frau Holle**.

However, who still thinks today that Frau Holle—the "old woman with the big teeth"—was a goddess? What kind of goddess was she anyway? No day of the week is named after her, such as Thursday bears the name of the god Donar (Thor) (English Thursday). However, there is no fairy tale in which Donar plays such a prominent role as the goddess Frau Holle, but there are a number of legends about Donar and he also appears powerfully in the Edda. Frau Holle must have been a very old goddess even at that time, since she was no longer taken into account with a day name in the new world of the gods of the Aesir, to which Donar belonged.

What was known to the Brothers Grimm, who collected this fairy tale and wrote it down, that Frau Holle was a goddess? But of course, Frau Holle was known to them as a mighty goddess, as "Mother Earth" and goddess of Fire or as the mother of the god Thor, as Jakob Grimm describes her functions and appearances in every detail in the three-volume edition (1835) of his "German Mythology" [9].

Frau Holle: "Aus dem, was uns die Tradition noch bewahrt hat, ergeben sich folgende Züge:

Frau Holle wird als ein himmlisches, die Erde umspannendes Wesen vorgestellt; wenn es schneit, macht sie ihr Bett, dessen Federn fliegen. Sie erregt den Schnee wie Donar den Regen; die Griechen legen ihrem Zeus die Hervorbringung des Schnees und Regens bei, … Holda erscheint darum als hehre Göttin. Die Vergleichung der Schneeflocken und Federn ist uralt, die Scythen erklärten die nördliche Weltgegend, weil sie mit Federn angefüllt sei, für unnahbar. Holda muß sich also durch die Lüfte bewegen können, wie Frau Herke.

Sie liebt den Aufenthalt in *See* und *Brunnen*; zur Mittagsstunde sieht man sie, als schöne weiße Frau, in der Flut *baden* und *verschwinden*, dieser Zug stimmt zu Nerthus. Sterbliche gelangen durch den Brunnen in ihre Wohnung, vgl. die Benennung *wazzerholde*." [10]

Mother Hulda: "The following features emerge from what tradition has preserved for us:

Frau Holle (Mother Hulda) is presented as a heavenly being that encompasses the earth; when it snows, she makes her bed, the feathers of which fly. She excites the *snow* as Donar excites the rain; the Greeks attribute the production of snow and rain to their Zeus ..., Holda therefore appears as a noble goddess. The comparison of snowflakes and feathers is ancient, the Scythians declared the northern part of the world to be inaccessible because it was filled with feathers. Therefore, Holda must be able to move through the air like Frau Herke.

She loves staying in the *lake* and the *well*; At noon you can see her, as a beautiful white woman, *bathing* in the flood and *disappearing*, this trait agrees with Nerthus. Mortals get into her home through the well, see the name *wazzerholde*."

The Brothers Grimm retell the fairy tale of Frau Holle or Mother Hulda (Hluodana), respectively, by combining the many different and scattered aspects of this goddess, some of which can be found as memes, as good narrators to form a self-contained representation.

14.3 The Spinning Meme

Wilhelm Grimm also takes this knowledge into account in the later editions of the fairy tale collection in the 7th edition of "Last Hand" (1857) by inserting correspondingly clear additions—e.g., at the beginning of the fairy tale, with an almost overly clear reference to Frau Holle as a goddess:

Das arme Mädchen mußte täglich bei einem Brunnen sitzen und dort so viel spinnen, daß ihm das Blut aus den Fingern sprang. Nun trug es sich zu, daß die Spule einmal ganz blutig geworden war. Da bückte sich das Mädchen und wollte die Spule im Brunnen abwaschen. Sie sprang ihm aber aus der Hand und fiel hinab.

The poor girl had to sit by a well every day and spin there so much that the blood leapt from her fingers. Now it happened that the spool had once become quite bloody. Then the girl bent down and wanted to wash the spool in the well. But it jumped out of her hand and fell down.

The girl sits with a spindle or coil at the well, the entrance to the world of Mother Hulda (Frau Holle), and thereby performs a function that was very important for society at that time—she spins. Spinning is also an activity that is under the special care and protection of the goddess Frau Holle. This addition by Wilhelm Grimm only reinforces what is already laid out in the old story.

Holla, die spinnende Frau: "*Holla* wird wiederum als *spinnende* Frau dargestellt, der Flachsbau ist ihr angelegen. Fleißigen Dirnen schenkt sie *Spindeln* und spinnt ihnen nachts die Spule voll; faulen Spinnerinnen *zündet* sie den Rocken an und *besudelt* ihn. Dem Mädchen, dessen Spindel in ihren Brunnen fiel, lohnte sie durch Begabung. Wenn sie Weihnachten im Land einzieht, werden alle Spinnrocken reichlich angelegt und für sie stehen gelassen; Fastnachts aber, wenn sie heimkehrt, muß alles abgesponnen sein, die Rocken stehen dann vor ihr versteckt; trifft sie alles an, wie es sich gehört, so spricht sie Segen aus, im Gegentheil ihren Fluch; ... [11]"

> **Holla, the spinning woman:** "Holla is shown as a spinning woman, growing flax is important to her. She gives spindles to hard-working young woman and spins them the spools full at night; lazy spinners she lights the skirt and sullies it. She rewarded the girl whose spindle fell into her well with her talent. When she moves into the country at Christmas time, all distaffs are laid out in abundance and left for her; But at Shrovetide, when she returns home, everything must be spun off, the distaffs are then hidden from her; if she finds everything as it should be, she pronounces a blessing, on the contrary her curse; ..."

The girl falls or jumps into the well—the entrance to the second world, the world of Mother Hulda (Frau Holle). This statement alone is completely clear.

As a boy who liked to swim and dive in the lake in the Jungfernheide in Berlin, it always seemed completely unbelievable to me (P. Plath) that if you jump into the water of a well, you wake up on a meadow and can walk on it. I just didn't get it.

Let us therefore take jumping into the well as a metaphor for the transition from the "local" first world to the "otherworldly" second world. It is only important that at the time when Frau Holle reigned unchallenged as the goddess, one had an idea of the second world that was neither a hell nor a heaven in the Christian sense.

However, what kind of world was this, the world of Frau Holle? This is exactly what is described in the fairy tale, which is almost an important tale even today:

> In the meadow the girl went away and came to an oven that was full of bread, but the bread called: "Oh, pull me out, otherwise I'll burn, I've long since baked!" Then she came diligently and took everything out.

Bread is baked here; but let's consider what was meant by that!

Bread (in German: Brot), as we know it today, which only appeared in the Iron Age, is "bread" baked with sourdough, in contrast to "unleavened bread", the "loaf" (in German: Laib), which was baked as early as the Neolithic Age. But already in Old High German, the word *bread* is also used for the loaf. Today, we only know the word *loaf* in terms such as "a loaf of bread" (ein Laib Brot) or "loaf of cheese", as a specially shaped bread or cheese. The loaf, which is baked only from water and flour, is better known to us today as flatbread.

In the English-speaking world, the terms *Lady* (Lady: Herrin, Frau, originally aengl. Hlaef-dige: "Bread Kneader") and *Lord* (Lord: Herr, originally hlaford, hlafward, bread protector) go back to the great cultural and religious significance that the baking of the loaf also took place in our culture up to the Bronze Age and probably also into the Iron Age. Terms for flat dishes such as plinsen, blini and bread (Russian, Хлеб, chleb) show that the unleavened bread, the flatbread or the loaf was of central importance in Indo-European languages.

We can and must therefore assume that the story with the oven in the realm of Frau Holle, the goddess, is actually about the sacred act of baking unleavened bread.

The term *oven* (vessel for cooking or storing embers) does not refer to the stove (charcoal hearth), but originally to the device or the stone plate/the stone shape on or in which the loaf was baked (Fig. 14.2).

Baking the loaf was, of course, a woman's business, but much more important—in the realm of Frau Holle—it was a priestly activity; and the product—the loaf—was used for ritual purposes in the service of Frau Holle as the goddess of fertility and rebirth.

Fig. 14.2 **a** "Temple, from the Cucuteni culture in the Moldovan Sabatinivka, which was dedicated to the snake goddess; from the period 4800–4600 B.C.E. The building took up 70 m^2 and contained (1) a stone slab floor, (2) a bread oven, (3) a pedestal/altar, (4) a clay chair, (5) statuettes (6) several vessels grouped around the oven. Next to the altar was a seat of normal size, the back of which is decorated with horns, and which was presumably used by the priestess who directed the cult activities. **b** Examples of the figurines found in the temple of Sabatinivka. Sixteen snake-headed figures were found on the altar at the front of the room, sitting on chairs with horn backs" [12].

Goddess of fertility: "The pregnant goddess who unfolds and dies with the vegetation is a metaphor for death and the renewal of plant life. She was venerated as a bread donor in the courtyard or in the house." [13](Marija Gimbutas, "The civilization of the goddess", publisher: Zweausendeins, Frankfurt / Main (1996) p. 342)

14.4 The Apple Tree Meme

But in the fairy tale the girl went on and came to an apple tree.

"Then she went on and came to a tree that was full of apples that called out to him:

<Oh, shake me, shake me! We apples are all ripe together!>

Then she shook the tree, so that the apples fell as if they were raining, until no one was up, and went away."

Fig. 14.3 **a** Oldest crab apple tree (malus sylvestris) in Germany from the municipality Stubbendorf (Mecklenburg-Vorpommern), age 400–500 years, Trunk circumference 4.42 m (*Photo* D. Antoni, 2006) [14]; **b** Crab apple from the garden of P. Plath (Lychen, Brandenburg), (*Photo* P. Plath, 2021); the fingers demonstrate the smallness of the apple

Does this apple tree episode go back to a similar old story as the bread story? The apple can be found in Germany in a settlement near Heilbronn at the time of the band ceramic culture (4000 B.C.E.). Furthermore, in archaeological investigations of the pile dwellings in Switzerland and on Lake Constance—from the time of the Cortaillod culture (3800–3600 B.C.E: location on Lake Neuchâtel, Switzerland)—apples were found. Among them were not only wild crab apples, but also apples that are much larger than crab apples (diameter of the crab apple approx. 3 cm), which proves that this is not one of the wild crab apple varieties (Fig. 14.3).

In addition, some of these fruits were found to contain substances that are only found in cultivated apples.

Cultivated apples: "The fruits found in Mondsee were found to contain anthocyanins, a substance that occurs exclusively in cultivated apple varieties" [15]. (Marija Gimbutas, "The civilization of the goddess", publisher: Zweitausendeins, Frankfurt / Main (1996) p. 196)

Whether the cultivated apple varieties from the local culture come from the many variants of the crab apple (malus sylvestris) or were imported from the Kazakh-Persian region as apples that were already cultivated under the influence of the Kurgan I culture, is irrelevant for the time being. It is crucial for understanding the fairy tale that the apple was already at the end of the 5th millennium B.C.E. was cultivated as a cultivated plant in the realm of Frau Holle, i.e., during the transition of the people living in the foothills of the Alps from the Mesolithic culture to the Neolithic horticultural culture [16]. (Marija Gimbutas, "The civilization of the goddess", publisher: Zweiausendeins, Frankfurt/Main (1996) p. 195).

The cultivation of the European wild apple (crab apple) or the cultivation of already cultivated apple varieties (e.g., from the former Kazakh wild apple malus sylvestris) always required the protection of the young plants against browsing by wild animals by means of hedges (hedge: thorn bushes, fencing, enclosure), fences (fence: enclosure, enclosure, garden) or palisades. Protected apple groves—as sacred groves (grove: thorn bushes, enclosed space) of Frau Holle—were probably suitable forms for cultivating this important fruit (as well as other fruits such as plums and pears).

It is ultimately the Garden of Eden (garden: cattle corral; Indo-European: wattle, fence, enclosed), the paradise (paradise: loan word from Persian—fencing), which is described in the fairy tale by Frau Holle. However, the people (Adam and Eva), or the women here are not driven out of paradise and replaced by guardians (kerubime) or gardeners; but rather the "white goddess" Frau Holle invites girls to stay with her in the garden and serve as priestesses, i.e., to cultivate, raise and harvest the apple. The cultivation of a plant is truly a priestly function.

But what made the apple so important for people 6000 years ago? It could—well stored or dried (dried apples)—also be kept in winter (from the harvest in autumn until spring) and thus provided essential substances (e.g., vitamins) for healthy survival.

14.5 The Meme of Divine Snowmaking: It Snows When Frau Holle Shakes the Feather Beds

The girl finally comes to Frau Holle's house and is frightened by the old woman. As already mentioned, however, the old woman immediately identifies herself as Frau Holle and invites the girl to take care of her household.

> Do not be afraid, dear child, stay with me. If you want to do all the work in the house properly, you should be fine: You just have to be careful that you make my bed good and that you shake it up diligently so that the feathers fly, then it will snow in the world: I am Frau Holle.

Above all, she should shake up the beds well so that it will snow in the world! Here, Frau Holle appears in her function as a snowmaker—as the protector of the new life that is just emerging—the germinating seeds in the earth are protected by the snow from drying out and being destroyed by the frost—so that a new year can arise from the dead year.

The loaf has to be baked, the apples have to be harvested and dried and the snow has to fall so that people can relive the cycle of life-death-life over and over again.

14.6 The Well Meme or the Second World

So far, the jump into the well has been completely ignored. The girl jumps into the well and wakes up on a meadow in another world. It is a second world, the existence of which, at least for the people of the time, in which the fairy tale originated, must have seemed completely natural.

In his book about the beginnings of intelligence, J. Kuczinski [17] states—quoting V. G. Childe [18]—, that the cave art originating from the Paleolithic Age, as it was found in the caves of Aurignac (Haute Garonne, France), only makes sense if one assumes that it served a magical purpose. "For the way of thinking of pre-scientific minds, such a creation (of the artist, the authors) had to have a counterpart in the outside world that one could both taste and see. Just as certain as the artist drew a bison in the dark cave, so sure would be a living bison out in the steppes, which his comrades could kill and eat."

The basic idea of this highly abstract conceptual construction is that one assigns a twofold reality to the thought or the creative activity that expresses it: the creation of the artistic product is itself a reality and at the same time creates a second reality in one other world. When the worlds are exchanged by going into the cave, the second reality of artistic activity arises in the world from which one came, namely the real world known to us. For these people the world consisted of this duality of the two conditioned worlds that were interwoven through creative activity.

This second world was—and still is today—necessary from a physical point of view, as O.E. Rössler explains in his book on endophysics [19]. It is practically our known world, but a second time, and in constant change with the first world indissolubly connected to it.

The exciting thing now is how one—according to the ideas at the time of the creation of the fairy tale—could get into this second world, which both the "golden girl" (Goldmarie) and the "dirty girl" (Pechmarie) of the fairy tale succeeded in doing.

The solution is the "well mystery", the remnant of which we still encounter today in baptism (baptize: actually "deepen", i.e., immerse and submerge), but also in a completely degenerate form during "waterboarding".

Access to the "second world" was nothing more than the acceptance into the secret covenant of Frau Holle's priestesses, because this was seen as the realization of the second world through which it can be experienced and grasped.

This, of course, can only happen through a secret mystery in which the female test person has to pass a difficult test that can change and re-shape her consciousness. Submersion or immersion in the well—and a proper resuscitation of the almost drowned people would certainly be an effective part of this secret mystery, about the secret aspects of which we know nothing. However, Robert Grave refers to the secret initialization ritual of the Orphic priests in Crete, which the mathematician Pythagoras underwent, according to the biographer Porphyrius, in order to get their secret knowledge. For this he had to be included in the group of these priests.

Secret Mysteries - Pythagoras "According to his biographer Porphyrius he (the Pelasgian Pythagoras from Samos; the authors) went to Crete, the seat of the purest Orphic doctrine, for initiation by the Idaean Dactyls. They ritually purified him with the thunderbolt, that is to say they made him pretence of killing him with either a meteoric stone or a neolithic axe popularly mistaken for a thunderbolt; which he lay face-downwards on the sea shore all night covered with black lamb's wool; then spent 'three times and nights nine hallowed days in the Idaean Cave'; finally emerged from this initiation. Presumably he then drank the customary Orphic cup of goat's milk and honey at dawn (the drink of Cretan Zeus who had been born in that very cave) and was garlanded with white flowers. Porphyrius does not record exactly when all this took place that Pythagoras saw the thrown annual decorated with flowers for Zeus; which suggests that the twenty-eight days that intervened between this thunderbolt death and his revival with milk and honey were the twenty-eight-day month R, (R = Ruis; Irish: elderberry the 13th tree of the Irish tree calendar, the last, 13th month of the tree year; the authors), the death-month ruled by the elder or myrtle; and that Pythagoras was reborn at the winter solstice festival as an incarnation of Zeus – a sort of Orphic Pope or Aga Khan – and went through the usual mimetric transformation: bull, hawk, woman, lion, fish, serpent, etc" [20].

The meadow on which the girls wake up after the "baptism" is the garden or the real apple grove of Frau Holle, the sacred grove that her priestesses tend and cherish (cherish: make a fence around, around-fence). In this academy of priestesses, the priestesses, like the novices, lack nothing, provided that they follow the rules of the academy, the rules of a very own "second world".

The stay in this academy of Mother Hulda (Frau Holle) is limited in time for the novices and ends with an evaluation of their work by Frau Holle. This work includes baking the sacred bread, tending to the apple orchard and shaking the beds so that it snows.

Mother Hulda (Frau Holle) or her priestesses lead the girls to the gate of paradise—this is not the well through which they entered paradise—and release them back into the familiar "first world" with a corresponding reward.

Golden Girl (Goldmarie):

"She then took it (Marie) by the hand and led it to a large gate. It was opened, and as the girl was standing under it, a tremendous golden rain fell, and all the gold clung to her, so that she was covered all over with it."

Dirty Girl (Pechmarie):

"Frau Holle also led her to the gate, but when she stood under it, instead of the gold, a large kettle full of pitch was poured out. <This is to reward your services>, said Frau Holle and closed the gate."

Golden Girl (Goldmarie) receives the promise to continue to count on the full support of the priestesses of Frau Holle, the Dirty Girl (Pechmarie), however, goes with their contempt or even hostility.

But we have serious doubts that the two girls could have left the second world quite differently, but in a very simple way. That sounds a bit simplistic and harmonizing. After all, after their stay in Frau Holle's realm, they are the bearers of secrets, and they are not simply allowed to run without any further examination! Even today, passing a thorough examination is a necessary prerequisite for leaving higher educational institutions.

It is a wonderful portrayal of religious life and the importance of Paradise 6000 years ago, but is that the whole story? Wouldn't that, despite all the truth that lies in this fairy tale, be a little too simple, too specific to this time, to survive 6000 years in almost pure form—that is, as a fairy tale—until today?

There are good and not so good people—some the goddess rewards, others she despises. That would be just a moral story that would soon be forgotten or strongly adapted to the respective circumstances. The former story would no longer be recognizable today.

Goldmarie and Pechmarie go through the same initiation rite, experience the same mystery, with both then their social position in the first world is completely reversed due to these experiences.

They come from the same house:

"A widow had two daughters, one of whom was beautiful and hardworking, the other ugly and lazy."

Therefore, they had the same upbringing. The only small difference that made the big difference in later life was that one was hardworking and unloved by the mother, the other lazy and loved by the mother. And happiness should now be with the hardworking but unloved girl? Is that what the fairy said?

This can not be it!

14.7 Dynamics of the Tale

But let's look at the same situation in a somewhat more abstract way.

Two young, closely related people walk the path that leads them to the second world of Frau Holle's academy, behave very differently there and then return to the first world, whereby their future paths in life develop completely differently than could predicted before the event of the Fountain Mystery.

Small differences in the upbringing of children can have big consequences for their future life!

This is a social observation that people were able to make in the Neolithic, and they found a very abstract logical term for it—chaos—and a very common, understandable, linguistic description of this very term—the tale, i.e., the meaningful story of the triple goddess Frau Holle, who is the basis of the fairy tale of Frau Holle.

One of the central findings of the chaos theory is the "sensitivity to the initial conditions". Sometimes this finding is referred to as the "butterfly effect" in an advertising but misleading way. It means that there are situations in a chaotic system where the paths (trajectories) of the system are so close to each other, that minimal disturbances of the system lead to initially hardly distinguishable situations—the system is thrown off the track on an immediately adjacent track. However, in their further development this leads to completely unpredictable, completely different paths or results.

Such an "initial situation" is created in the Fountain Mystery and the first subsequent tests—on the oven and on the apple tree—with the result developing from it, the Golden Girl (Goldmarie) and Dirty Girl (Pechmarie).

It is most remarkable that this highly abstract mathematical concept of "Chaos" was preserved as early as in the Neolithic, in a story told over six millennia. One might therefore be inclined to regard this as a newfangled over-interpretation.

We want to consider, however, that at the same time with the elliptical solar observatory in Meisternthal (see also the tale of Sleeping Beauty) a highly abstract mathematics for calculating calendars had been developed—why not also the concept of chaos in a form which corresponds to our today's conception of chaos to a large extent.

In addition, through Pliny and Hesiod, we know the creation stories from the Neolithic, non-Indo-European, pre-Greek society, in which "chaos" was a central logical term:

Pelasgian creation myth: "In the beginning there was Eurynome, the goddess of all things. She rose naked from the **chaos**. But she found nothing solid to put her feet on. Therefore, she separated the sea from the sky and danced lonely on its waves. She danced southwards…". (Pliny, natural history; quoted from R. von Ranke-Graves [21])

Philosophical creation myth: "Some say that there was darkness first and that **chaos** arose from darkness. From the pairing of darkness and **chaos** arose the night, the day, Erebos and the air. ... ". (Hesiod, Theogony; quoted from R. von Ranke-Graves [22])

Back then, people were not more stupid than we are. They were excellent at observing their environment and describing it mathematically, and that included not only the movement of the stars, but also that of their own society. This was probably the most complicated system available to them at that time, which they could observe and describe, trained on the mathematical descriptions of their astronomical observations.

In northern Germany there is the "Geest" (Geest: "high-lying dry land" as opposed to the wet marshland)—which is, etymologically, the German as well as English word for "chaos". These are the areas of land where after the last ice age strongly loamy soils remained, which, after getting wet, when dried resulted in a cracked earth crust (in spring and summer) that evaporated in the sun. To designate such a region as chaos—*geest* is nothing more than a further way of making the difficult, abstract term tangible: chaos—just as the tale of Frau Holle is an illustration of this term.

This term was so central and significant that it could only be explained in connection with the great "white goddess", Frau Holle, in her mysteries and even survived albeit not understood.

We know we will be immediately contradicted here, as the term "chaos" is nowadays very strictly defined. It could not be assumed that at the time of the Neolithic goddess Frau Holle one already understood what we mean by it today—and anyway, what does our current concept of chaos have to do with religion?

But let us remember that in the 1970s the choice of the word "chaos" for certain physical phenomena was not made without recourse to pre-Greek and Hebrew mythology. The mathematical term "chaos" [23] was first introduced by Li and York [24] in 1975.

Modern concept of chaos: "'Chaos' is the canonic translation into Greek of the Hebrew term 'tohu-wa-bohu' found in the first chapter of the Bible. 'wa' means 'and', and 'bohu' certainly means the same thing as 'bohu'. However, since the word appearce only once in the bible, and there is no continuous oral tradition, one can only guess, what 'tohu' means. A possible English analogue is ‚topsy(-and-)turvy.' (Tohowabohu: in German "Wirrwarr,

Durcheinander"; Duden Etymologie; "wüst, öd" und "leer" im gesellschaftlichen Sinn; in English: Tohuwabohu: "confusion, disorder"; Duden Etymology; "Desolate, dreary" and "empty" in the social sense; the. authors). A mathematical redefinition thus seems admissible. This has been done by Yorke, who proposed the word 'chaos' as a label for a kind of dynamical behaviour characterised by the triad: infinite number of periodic trajectories; uncountable number of non-periodic trajectories; hyperbolicity (instability) of all (or the overwhelming majority of; as is proposed here) trajectories in the regime." (Cited from O.E. Rössler: "Chaos"23)

Let us also consider that our sixteenth-century word "gas" also represented a deliberate use of the Greek mythological term "chaos"—as empty space. Later, under the influence of Maxwell and Boltzmann, it was understood as the "confusing, irregular movement of gas molecules" in analogy to the reception of the concept of chaos in the Olympic creation myth in the sixteenth century as a "disordered, confusing movement". And, it should be noted here, it is very reasonable to understand gas as a system of colliding, more or less spherical, rigid bodies, the molecules, which all move chaotically in the sense of today's mathematical-physical understanding. However, that was not known neither in the 16th nor in the nineteenth century!

In addition, something else should be noted. The reception of the concept of chaos in the Olympic creation myth of the already at that time much older concept of chaos, corresponding to mathematical logic, was a contemporary reception in its time, reflecting the patriarchal misunderstanding of the matriarchal concept of chaos. This misinterpretation, which was modern at that time, can also be found in the Germanic language area. Chaos was understood as an incomprehensible tangle and assigned to the goddess Frau Werre, but later also to Frau Holle in the fairy tale, as the old woman with the tousled, tangled hair.

Older chaos term: "Some say that there was darkness first and that chaos arose from darkness. From the pairing of darkness and chaos arose the night, the day, Erebos and the air." (Hesiod; Theogenie 211-232, quoted from: R. von Ranke-Graves [25])

But what could one have understood by chaos in Mrs. Holle's time? The tired yawning with a gaping throat? Or did it mean the evaporation of the morning mist, from the wet clay soils of northern Germany that cracked when drying—the Geest, or the gas emissions from the Omphalos in the Greek Temple of Apollo in Delphi, the former temple of the "White Goddess" Gaja? [26]

All of these were only images of strong symbolic meaning for the much more abstract concept of chaos!

Frau Holle was the triune goddess, the not yet childbearing woman or the virgin Hulda, the woman Holle of childbearing potential and the old woman Hel in one person, who carries out these transformations anew within each year, just as a woman in her biological development goes through these stages within their life. These metamorphoses of Frau Holle, or her trinity, represent the concept of becoming, the transition from being to non-being [27], whereby each transition is one of her emanations in itself, represented by its own goddess or fairy, e.g., the Mechte known in Northern Germany.

But that includes the idea of rebirth—the transition from death to life, which is symbolically represented in connection with Frau Holle in a variety of ways (see e.g., frog prince, frog goddess, the birth process, or the mystery of birth). Both, the concept of becoming, as well as that of death or nothingness, as they are described in the preface of my book "Vom Märchen zur Mär" (From Fairy Tale to Tale), were often regarded as meaningless concepts in the development of philosophy. Thus, Kant knows the concept of time [28] but not that of becoming. Both terms are, however, inseparably linked with the matriarchal, ancient European philosophy through the creation myths and thus also with the logical unfolding of the old concept of chaos, which clearly relates to the idea of development.

Let us ask ourselves once again what was meant by chaos at the time of the great goddess Frau Holle. It is obvious to understand by it, for example, the miracles or the changes in the path of life caused by the oracle or the secret mysteries, which were provided with the metaphor/the promise of rebirth, the emergence of life from death. To make this tangible, we have to try to translate the dynamics of the fairy tale into the language of mathematics.

If one want to grasp the dynamics of Frau Holle's fairy tale in the language of today's mathematics, you have to note that the two girls, golden girl (Goldmarie) and dirty girl (Pechmarie) enter the second world through the same well mystery. However, they leave the second world through the judging gate that leads out of the garden, and return to the first world.

We must also bear in mind that it was essential for Frau Holle's second world that it was—albeit isolated—a real world in which other values applied, in which life was the other way around, since the simple works are sanctified in it. These are demands on the dynamic description of this complex social system, which are met quite well by the assumption of a symbolic, four-dimensional, socio-economic world. The system of equations with the four variables $\boldsymbol{x}$, $\boldsymbol{y}$, $\boldsymbol{z}$ and $\boldsymbol{w}$ (14.1), has the required chaotic dynamics [29].

$$\begin{aligned}\dot{x} &= -(y+z)w + g \\ \dot{y} &= (x+ay)w \\ \dot{z} &= -cz^2 + xz + b \\ \dot{w} &= dxy + zy - ew + f\end{aligned} \tag{14.1}$$

The three-dimensional subspaces $\{x, y, z\}$ and $\{x, y, w\}$ of the four-dimensional phase space $\{x, y, z, w\}$ have the desired properties. In the 3D sub-space $\{x, y, w\}$ with the variables x, y, and$w$ (see illustration in Fig. 14.4), the space can be divided into two half-spaces $\{x, y, w \geq 0\}$and$\{x, y, w < 0\}$, which can be entered or exited via different paths and which are symbolically represented in the fairy tale by the well mystery and the gate rating.

In the context of this picture, the life path of a single person would only constitute a relatively small part of the infinitely long path of the chaotic trajectory. But many people, who all, each for himself, only go through such a short section at completely different times, would all taken together cover a long way on this trajectory. For example, some would pass through the well mystery on their way, others as priests

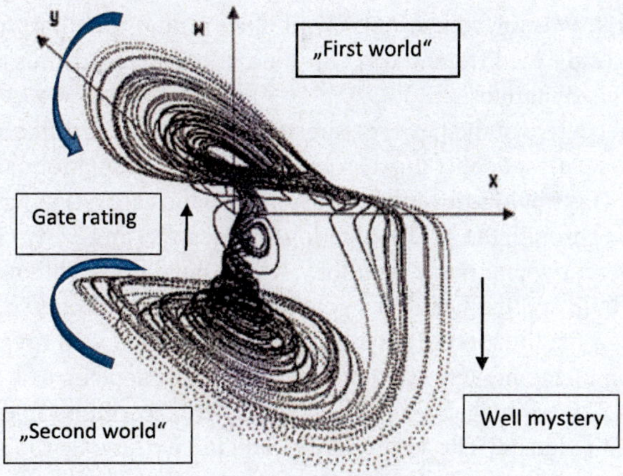

Fig. 14.4 Chaotic trajectory of the system of equations in the $\{x, y, w\}$-subspace with the two half-spaces above and below the x/y-plane. If you look upwards in the direction of the w-axis, the trajectories in the two half-spaces are traversed in opposite directions, with ↑ indicating the direction of the gate rating and ↓ indicating the well mystery

would practically only move in this second world for almost their entire life, others again never enter the sacred area (see e.g. Fig. 14.5).

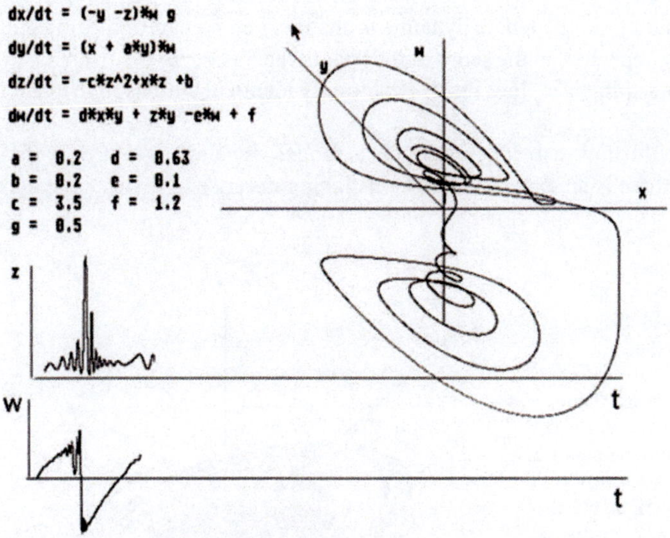

Fig. 14.5 A "girls's trajectory" in the three-dimensional subspace, which represents a possible section from the infinite traction of the system of equations

The two Girls, who of course only symbolize a dual girl (pars pro toto), pass through both worlds in different ways in their ideal life and thus symbolize the complexity of this dynamics.

From the perspective of the $\{x, y, z\}$-subspace it almost seems like a miracle that a trajectory or a life path suddenly turns around after many circulations and almost runs back into itself. This obvious reversal corresponds to the gate rating when leaving the "second world". Through the goal evaluation, Golden girl and Dirty Girl go through completely different social paths than they would have done without a stay in the realm of Frau Holle and the wondrous goal evaluation. But of course, the trajectory is also reversed in its direction of rotation while passing the well mystery.

The inversion of the movement almost in itself only appears as a miracle in the "non-religious $\{x, y, z\}$-space" (Fig. 14.6), since here, according to experience, the processes are not reversed: The water does not suddenly flow up the mountain. The sea does not divide, in order to be able to cross it with dry feet, the rock does not open of its own accord to get to the cave of Mother Hulda (Frau Holle), and the shot arrow does not suddenly turn around to hit the archer.

The divine miracle is not understood differently in Christianity either: it is God who intervenes in human history and, as if by a miracle, reverses his future path in life. The path of suffering, godly life and the miracle take over the function of the well mystery, the stay in the paradise of Frau Holle and the gate evaluation.

The almost reversal of the trajectory or its "reflection" should also be noticeable in the time series of the individual variables (Fig. 14.7).

You go back your way, so to speak, but in a new way—by revaluing previous values. In addition, the whole dynamic is chaotic, i.e., no part of the trajectory can be copied or is repeated; in the sense of the "panta rhei" (Greek: πάντα ῥεῖ, "everything flows") everything is in flow, i.e., is structurally identical with itself in that it is always different.

One could now ask whether such a model dynamic is not very far-fetched. However, there is an experimentally verifiable reference to reality in dynamics with

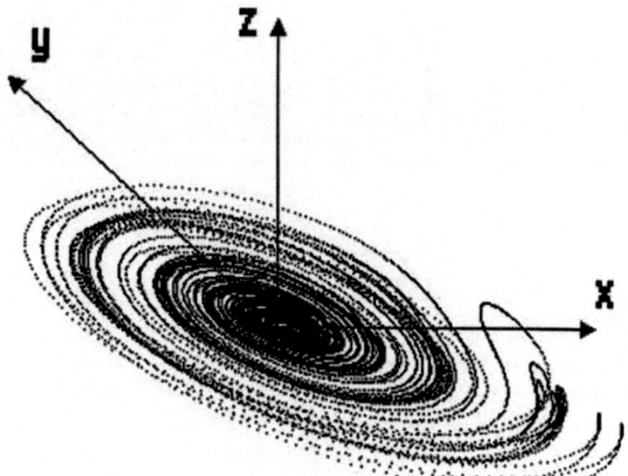

Fig. 14.6 Chaotic trajectory—"miracle trajectory"—of the system of equations in the non-religious $\{x, y, z\}$-subspace. The trajectory almost returns to itself at the turning points of the "Fountain Mystery", visible on the right, but also at the turning points that lie inside this "disc" and correspond to the "gate assessment"

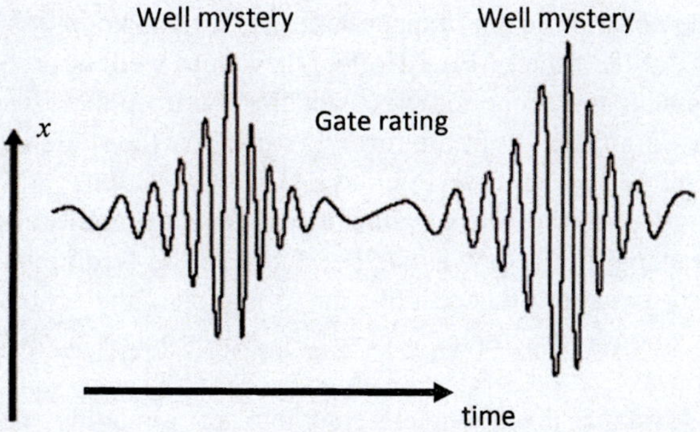

Fig. 14.7 Chaotic time series (excerpt) of the x-variable of the system of equations. The time series is almost "reflected" both on the well and on the gate

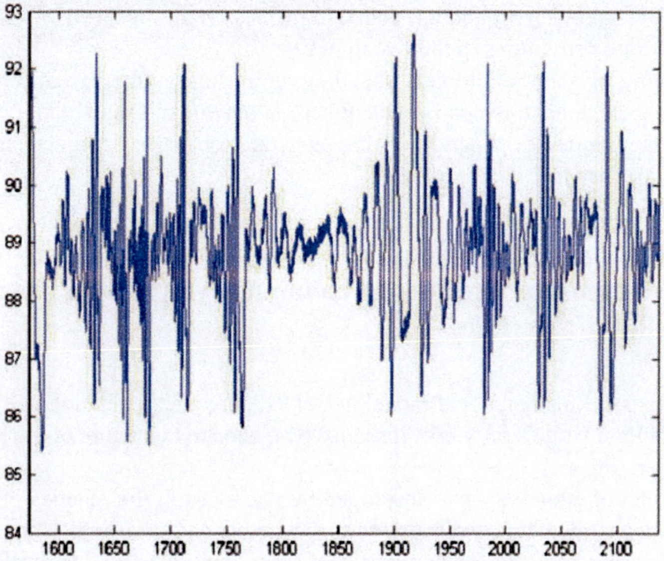

Fig. 14.8 Electrochemical dissolution of copper in alcoholic phosphoric acid (85%). Time series of the electrochemical potential, measured with an Ag/AgCl reference electrode. The chaotic time series shows an unusual "mirror symmetry" (Experiment: by U. Sydow/P. Plath, March 2000)

the term trajectory, which can be represented as a "time series". The reversal of the trajectory in the phase space "almost in itself" corresponds to a very characteristic "mirror structure" of its time series.

At least in electrochemistry, there is a system in which we have been able to detect a "mirror structure" of its time series (see Fig. 14.8) and which could therefore be described with a symbolic dynamics—expressed by corresponding kinetic equations—as we have given them here.

Even if we do not want to claim that all of this was known in this way to people in the Neolithic "at the time of Frau Holle", they were well aware of the complex social, psychological and economic processes that made up their lives. They could also express this linguistically in the form of stories. A fairy tale like that of Frau Holle could therefore also be understood as *poetic mathematics*.

It should be emphasized at this point that neither the molecules in chemical systems nor the people of today, as well as those in the Neolithic, need to know anything about the equations that describe the formation of their chemical structures or their social-economic structures, respectively. Nevertheless, such structures arise and can be observed.

In the various systems, the individuals enter into corresponding relationships with each other. The molecules react chemically, people plant, bake, weave, produce, trade, etc. Structures result from self-organization, without any awareness of these processes in the respective individuals.

However, unlike molecules, people can perceive and recognize the structures in principle. When they report about it coded in memes or tales respectively, then they are aware of these structures in their societies.

The capture of such a complex, social structure, that is the extraordinary, the astonishing achievement, which is evident here in the tale of Frau Holle.

This is the unique key function of this fairy tale for the basic understanding of other fairy tales as well.

14.8 Reflections on the Socio-Economic Dynamics in Frau Holle's Fairy Tale

As beautiful as the mathematical model presented here phenomenologically grasps the events in Frau Holle's fairy tale, it seems to be alien to the nature of the fairy tale and the underlying tale.

This system of equations was developed by one of us in the nineties of the last century in connection with the discussion as to whether and, if so, which new properties attractors could have in four-dimensional space that go beyond the well-known chaos of three-dimensional space. As physical chemists, we were convinced that chemistry possessed the necessary complexity to realize such structures. Even then, we thought social structures or social myths were just as suitable to provide examples of such a complex dynamic.

In this context, we called the structures that can be generated with this system of equations: "Wirrwarr" [30] (English: tangle or confusion) with reference to the goddess "Werre", who, like Frau Holle (Mother Hulda), confuses the not yet spun flax or the hair of the "lazy" spinners at Christmas. Unfortunately, this term "Wirrwarr" for the differential equation system is slightly misleading in this context, because this system has a trajectory. However, the Wirrwarr should no longer have a trajectory or it should not be able to be embedded in one. But this is another story.

At the beginning of the nineties, especially in physics, the question was discussed whether and how social systems could be described with mathematical-physical models [31]. Above all, the question of the appearance of new properties (emergence problem) played a decisive role [32]. All of these problems are closely related to the development of synergetics [33, 34], and the processes of self-organization and self-structuring [35] it deals with, which from the beginning were also applied to economic and other social issues.

In addition, it is of great interest for our question that in the recently published book by Wolfgang Wildgen, "Myth and Religion—Semiotics of the Transcendent" [36], the dynamic semiotics of religion plays a central role in the description of the emergence and self-organization of religious forms.

14.9 Cyclic Dynamic of Resources and Tools for Their Use

Now, the system of equations discussed here, consisting of four coupled, nonlinear differential equations, is very abstract and one would like to know what is hidden behind the variables x, y, z and w. What might these variables represent in a social context?

An attempt should be made here to describe at least qualitatively, which socio-economic content could be assigned to these abstract variables.

Let us consider a very simple social structure in which people process the resources x available to them and thus use them to produce the products of daily life, i.e., food, clothing and shelter, but also the necessary tools y. These wear out during work and always have to be repaired or replaced with new ones.

Then the expression $\frac{dx}{dt}$ describes the temporal change of the resources due to the working process as the processing of the resources. The greater the intensity of the work, the more tools are used up and destroyed. It is therefore true that the change in resources is proportional to the number of tools used: $\frac{dx}{dt} \propto y$ or:

$$\frac{dx}{dt} = -y \tag{14.2}$$

where the minus sign refers to the consumption of tools by work. The proportionality constant is set to one here for the sake of simplicity and is not mentioned any further.

One could imagine this process in such a way that, for example, during the winter the tillage tools, hunting weapons and fishing nets were repaired and recreated and then in the summer slowly used up again in the garden and for hunting and fishing. The less intact hunting equipment is still available, the fewer animals can be killed with it. However, this model would also apply to nomadic farming societies who only practiced simple agriculture or horticulture for a short time and then moved on.

Therefore, the number of tools also changes over time $\frac{dy}{dt}$ in the course of the work process. They wear out, break and tear, and always have to be repaired or replaced.

To simplify matters, the more accessible resources x are available, the more tools y can be developed (Eq. 14.3).

$$\frac{dy}{dt} = x \tag{14.3}$$

It is not the edible dishes, the ingestible food that determines this process, but the resources and the production of the necessary tools. They form the decisive variables for the dynamics. Only what can be turned into tools in the most general sense are resources! Examples of such resources are, e.g., animals whose bones or antlers could be used to make hoes and arrows. The flint mines [37] in north-western Europe also play an important role as a resource for flint, from which a myriad of different tools could be produced.

A C14 dating of Charcoal in a shaft showed an age of 3,150 years before the calendar. A display board in the research tunnel mentions C14 dates between 3750 and 3940 B.C. The pits can thus be clearly placed in the Neolithic. The Neolithic in Europe covers the period between about 5500 and 2000 BC. Metal tools were not yet known at that time. The mining industry in Rijkholt is estimated to have lasted around 500 years. Longer interruptions are not to be accepted. The mining period was likely sometime between around 3950 and 2650 B.C. It may also be possible to assume several periods. The scientifically investigated area covers an area of 2436.6 square meters with 75 shafts. The flint obtained was transported upstairs and the unusable pieces were sorted out. Usable material was turned into weapons and tools. For example, smaller blades and larger axes were produced. Only rough forms were made on the spot. Fine machining and possible grinding of the tools took place elsewhere. The tools and weapons from the Rijkholter flint were coveted trade items and have been proven within a radius of 400 km.

The production process outlined here in a greatly simplified manner is thus described by the coupled system of Eqs. (14.4):

$$\begin{aligned} \frac{dx}{dt} &= -y \\ \frac{dy}{dt} &= x \end{aligned} \tag{14.4}$$

The system of Eqs. (14.4) also describes the movement of a rigid, undamped pendulum at location x at time t with velocity $\frac{dx}{dt} = v$ and acceleration $\frac{dv}{dt}$. Its solution is a circle in the two-dimensional $\frac{x}{y}$- or x/v-phase space (see Fig. 14.9). The speed v of the pendulum thus corresponds to the quantity y of tools in the socio-economic process.

A simple transformation of the two equations leads us immediately to the well-known pendulum Eq. (14.5):

$$\begin{aligned} &\frac{dy}{dt} = \frac{d}{dt}y = \frac{d}{dt}\left(-\frac{dx}{dt}\right) = -\frac{d^2x}{dt^2} = x \\ &\frac{d^2x}{dt^2} = -x \text{ with the solution } x = sint \end{aligned} \tag{14.5}$$

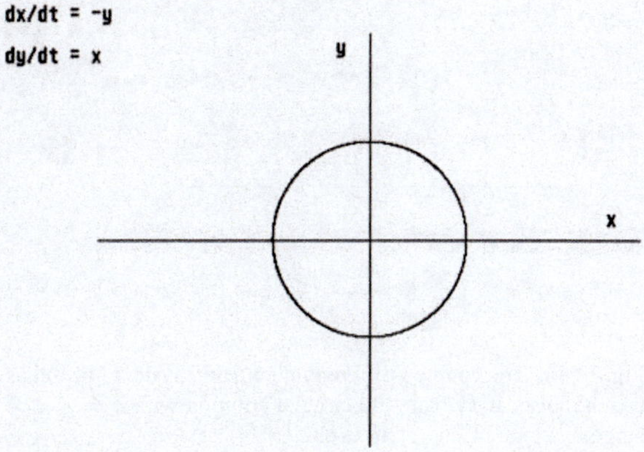

Fig. 14.9 The cyclic movement of the entity's resources and tools in the *x*/*y*-phase space corresponding to Eq. (14.4)

As already indicated, the production process of a Paleolithic society of Homo Sapiens or Neanderthals can possibly be described in a simplified manner using this system of equations.

With the already patriarchally structured society of the Sioux, the prairie Indians of North America, one finds the circle as the decisive symbol in rituals such as the *sun dance*, and they say: 'Everything moves in circles'. This may be an indication that in much older societies these circular structures of production are carried out in the simple way just described by the pendulum equation.

The model could also be refined somewhat by taking into account that a small but largely constant excess g is always produced due to the work, and that the production of new tools y depends on the number of tools available. These results in the differential equation system (14.6):

$$\begin{aligned} \frac{dx}{dt} &= -y + g \\ \frac{dy}{dt} &= x + ay \end{aligned} \tag{14.6}$$

The solution of this system of equations leads to an expanding spiral (Fig. 14.10), which is narrower the smaller the constants a and g are. The system is therefore unstable, which, on the other hand, is also a requirement for the development possibility of a system [38]. If the spiral is narrow enough, i.e., if the additional tool product generated is hardly recognizable, and tool production is practically not dependent on the number of tools available, the spiral will hardly differ from the circle for a long time. Such a society, which develops only very slowly, would appear to its members to be completely stable (circularly stable), although it is unstable.

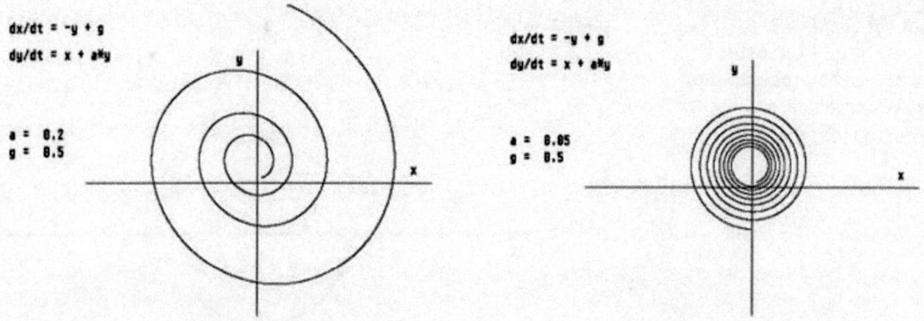

Fig. 14.10 The coupled differential equation system (14.6) has an unstable solution for $\boldsymbol{a}, \boldsymbol{g} > 0$, which moves away from the circular solution with $\boldsymbol{a} = \boldsymbol{g} = 0$ towards infinity in the form of an outgoing spiral. The system explodes

14.9.1 Inherent Information Bounded in the Tools Arise from Their Use

The manufacture of tools in turn requires the tools, which presupposes that information has accumulated and been objectified in the tools and their use. Let us call this bounded information "inherent information" z(for example: handicraft knowledge). In this way, the tools are used to store information that arises in the working process with the tools.

The change of this information $\frac{dz}{dt}$ over time is proportional to the product of the change of the tools $\frac{dy}{dt} = x$ with the available information z that is required for their operation (Eq. 14.7):

$$\frac{dz}{dt} \propto xz \tag{14.7}$$

However, it decreases quadratically with the amount of information. This is due to the fact that wear and tear of the tools during work results in a loss or destruction not only of the tools, but also of the information inherent in them. Due to the aging of the tools and the associated aging of their manufacturing processes, the information embodied in the tools disappears. In other words, even if the tools are not used, the information contained in them is also forgotten and destroyed.

The fact that the destruction process of the inherent information is assumed to be quadratic means that when there is little information, the destruction of the information increases only sub-proportionally and only above a threshold value super-proportionally.

$$\frac{dz}{dt} \propto -z^2 \tag{14.8}$$

In order to keep the inevitable destruction process of inherent information as low as possible, appropriate social structures had to be developed. It was safest if the relevant workshops were located in the holy districts, as in the fairy tale by Frau Holle, or the carriers of the information, toolmakers themselves became demigods or even gods.

A well-known example of this is the story of the limping sacral king and blacksmith Hephaestus [39], who lived in the cave of the "Vulkanon" on Lemnos or Mount Etna in Sicily, according to mythographers, as the "blacksmith god". The three "ring- or one-eyed" Cyclopes, sons of "Mother Earth", are said to have been there his blacksmiths journeymen. The production of general tools with the help of special tools, as operated by the blacksmiths, was in a sense a "divine" affair. Therefore, Hephaestus is thrown out of Olympus by the gods and then brought back into it [40].

The blacksmith Wieland was also such a limping sacred king as Hephaestus [39]. He had learned his blacksmithing from the divine blacksmith Mimir and named his famous "steely" sword "Mimung" after him [41]. He was also famous for his knowledge of runic writing and acted as an advisor to Odin from the Aesir family. The Aesir lost the battle against the Vanes. According to mythographers, Mimir was beheaded as a scourge by the Vanes gods. His head was given back to the Aesir and continued to be venerated by Odin as his counselor.

It is about the highly complex inherent knowledge of the manufacture of tools using suitable tools. This absolutely had to be protected and could not be distributed and thus lost. Such a loss of information would have been too great for the respective society. This is expressed by the quadratic term "$-z^2$". Such a loss of information has certainly taken place in Europe, for example when the blacksmiths of the Early Iron Age who lived and worked in the "castles" left the castles with their new material, "iron".

The historical development in Europe knows further structures of the necessary secrecy of this inherent knowledge. It is the monasteries and later the guilds, and today the confidentiality agreements and patents, which ensure that the inherent knowledge that has been developed at great expense is not lost to the respective social structures.

The example of the patent release for corona vaccines shows how central it is to avoid the loss of inherent knowledge even today. Interestingly, the fact that knowledge about the manufacture of corona vaccines cannot be released is justified with the argument that the complex practical-technical "know-how" of the chemical-biological processes does not allow the patents to be released quickly.

However, beside this loss of inherent knowledge, there will also be a constant influx b of information due to new inventions, so that ultimately the following Eq. (14.9) applies to the change in information over time:

$$\frac{dz}{dt} = xz - cz^2 + b \tag{14.9}$$

The information naturally has an influence on the work process itself, i.e., on the processing of the resources, so that altogether the three-dimensional differential equation system (14.10) results:

$$\frac{dx}{dt} = (-y - z) + g$$
$$\frac{dy}{dt} = x + ay \qquad (14.10)$$
$$\frac{dz}{dt} = xz - cz^2 + b$$

The resulting three-dimensional differential equation system has an unstable limit cycle (see illustration in Fig. 14.11), so that the endeavors of a society that wants to remain "stable" must always be directed towards balancing this unstable state through constant, minimal readjustments.

Thus, the conscious development of the tools beyond their simple use, does not stabilize the social system, but it creates the possibility of skillfully keeping it in the balance of the unstable limit cycle.

However, if the social conditions change too quickly and seriously, for example due to internal developments or external influences such as contact with other cultures, it will be inevitable that the unstable system will lose its balance and either "perish" (trickle) or undergo a stormy, very dangerous development—the system explodes in a spiral. Such a foreign contact could either completely destroy the previous production structure (e.g., reduce c) or enable new, more productive manufacturing processes (e.g., increase c). The decisive factor will therefore be the influence that the new culture has on the dynamics of information processing.

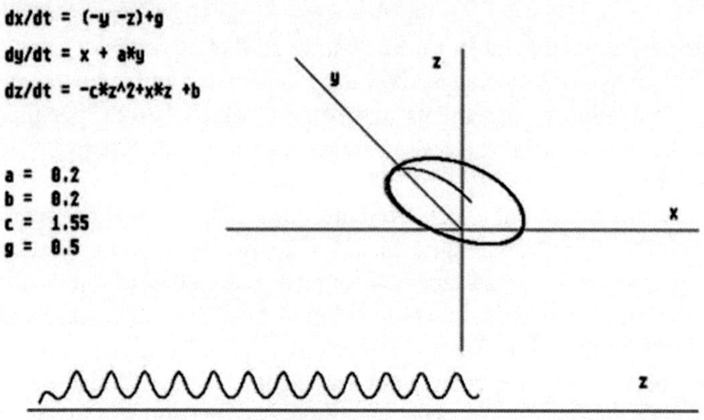

Fig. 14.11 The solution of the differential equation system Eq. (14.10) has an unstable limit cycle from which the system moves away only very slowly when it is in its immediate vicinity. For < 1.55 the system swirls into a fixed point inside the limit cycle (it collapses) (see also Fig. 14.13, Fig. 14.2), for $\boldsymbol{c} > 1.55$ it explodes (see also Fig. 14.12). For $\boldsymbol{c} \cong 1.55$, it can be balanced

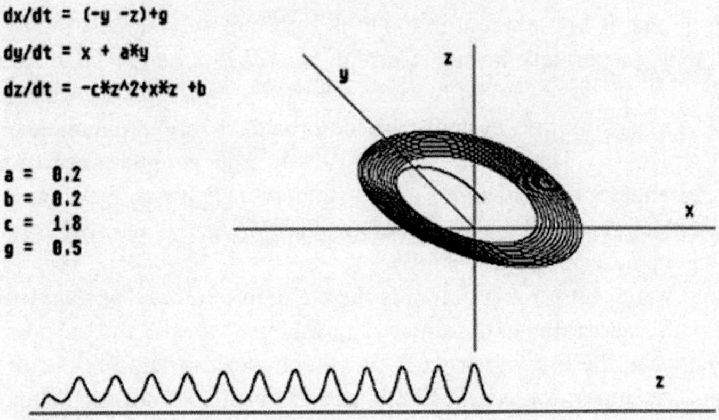

Fig. 14.12 Expiring spiral of the unstable limit cycle for $c = 1.8$; the system (see Eq. 14.10) explodes in a spiral

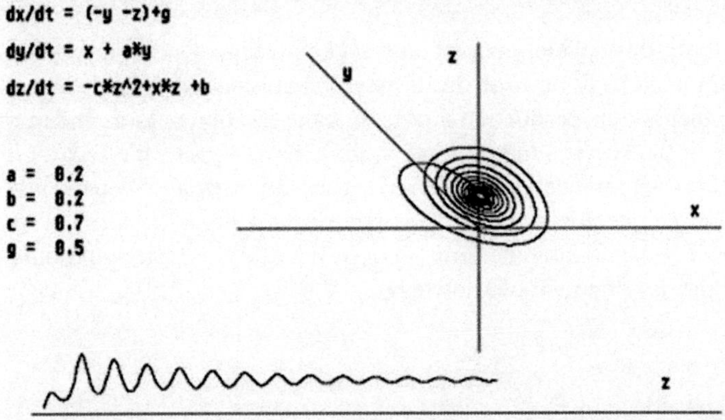

Fig. 14.13 Incoming spiral of the unstable limit cycle for $c = 0.7$; the system (see Eq. 14.10) collapses by approaching a fixed point

14.10 Abstract Knowledge, Freely Available Can Stabilize Unstable Systems Via Transformation Into Chaotic Ones

14.10.1 Paradise is the Birthplace of Science

In the long run, internal and external influences will cause any system, no matter how well balanced, to collapse or explode. Since the problem lies in the developing information z, a structure must be created whose dynamics automatically "balance" the information z without directly influencing its development. On the other hand, it must be similar to the essence of information in the sense that it is a type of "information" that is not inherent in the tool itself and only comes to light through its use, but exists separately from it. Current, but "*abstract knowledge*" w has such a structure.

Usually, socio-economic systems are described by order parameters such as resources x and tools y. Here, however, we are using order parameters of a completely different kind: inherent knowledge z and abstract knowledge w. Nevertheless, these are order parameters, entirely in the sense of the synergetics with which the social dynamics can be described successfully.

The social solution to this problem is the creation of a real "*second world*", the creation of paradise parallel to the already known "first world", and the possibility of switching between these two worlds. This second world—paradise—is the system of conscious creation of the possibility to acquire "abstract knowledge" in the form of academies of priestesses, the garden of Eden, philosophical schools and seminars, monasteries and universities. In short, *paradise is the birthplace of science.*

The originally unstable system becomes chaotic as a result, but it also becomes stable and, in this way, retains great variability in order to be able to react in different ways to new situations over and over again.

Accordingly, this development of abstract knowledge $\frac{dw}{dt}$ is to a very high degree religious or ideological, but, and this is essential, it is not completely detached from the development of the production process and the creation of tools and information. In the form of the two products $xyandzy$, these have a positive effect on the development of abstract knowledge. At the same time, an internal, independent scientific development f is required and a necessary dismantling of traditional knowledge (forgetting), represented by the term $-ew$ in Eq. (14.11), that would only stand in the way of the development of knowledge.

$$\frac{dw}{dt} = dxy + zy - ew + f \tag{14.11}$$

It finally results in the four-dimensional system of non-linear differential Eqs. (14.12):

$$\begin{aligned}
\frac{dx}{dt} &= (-y - z) + g \\
\frac{dy}{dt} &= x + ay \\
\frac{dz}{dt} &= xz - cz^2 + b \\
\frac{dw}{dt} &= dxy + zy - ew + f
\end{aligned} \tag{14.12}$$

However, this system is incomplete because it does not contain any reaction of abstract knowledge w on the production process. As a result, it is unstable and explodes: the inherent information z grows exponentially in an oscillating manner.

The retroactive effect of the abstract knowledge solely on the work process alone according to equation: $\frac{dx}{dt} = (-y - z)w + g$ with the same parameters as in Eq. (14.12), e.g., $a = 0.2$; $b = 0.2$; $c = 3.5$; $g = 0.5$; $d = 0.63$; $e = 0.1$; $f = 1.2$—leads to a fixed-point behavior of the overall system and does not allow any further development. In this sense, the system collapses or congeals respectively.

On the other hand, if the feedback only affects the generation of the tools $\frac{dy}{dt} = (x + ay)w$ with the same parameter set, the highly unstable system will explode momentarily.

Only when the abstract knowledge w acts back on both, $\frac{dx}{dt}$ and $\frac{dy}{dt}$, a stable chaotic system is obtained by two re-injections of the system onto itself:

$$\begin{aligned} \frac{dx}{dt} &= (-y - z)w + g \\ \frac{dy}{dt} &= (x + ay)w \\ \frac{dz}{dt} &= xz - cz^2 + b \\ \frac{dw}{dt} &= dxy + zy - ew + f \end{aligned} \tag{14.13}$$

This is the system of equations (Eq. 14.13, see also Eq. 14.1) that has the desired behavior and that is suitable for interpreting the fairy tale by Frau Holle.

It should be emphasized once again at this point that the mathematical model describes the dynamics of a society, but not that of an individual of its members. In other words, it does not describe what happens to an individual in society. The characters acting in fairy tales are projections of social events onto hypothetical persons and gods or goddesses. However, these hypothetical people do not exist in this way.

It seems strange that the chaotic system has an infinite trajectory that never returns to itself, and that it should describe social life as well as the life of the individual. The two Mary's of the fairy tale only get to Frau Holle's second world once and then come back to the first world again. Each of them therefore only traverses a small part of the infinite trajectory of the chaotic system. Such a section, i.e., an excerpt from the trajectory of the 4D system of equations in the 3D subspace $\{x, y, w\}$, is shown as an example in Fig. 14.5, and only this is described as a typical event in the fairy tale.

It cannot be seen from the figure whether the x/y plane divides the new, four-dimensional "world" into two half-spaces, which was the basic assumption for the interpretation of the fairy tale.

However, a $(n - 1)$-dimensional subspace divides a (n)-dimensional metric space into two (n)-dimensional half-spaces. Thus, the three-dimensional subspace $\{x, y, z\}$ of the real world should be used as the space with which the new four-dimensional world is intersected.

According to the fairy tale, we demand that the description of the real "first world" as the $\{x, y, z\}$-subspace of the new world also allows for "miracles". For this reason,

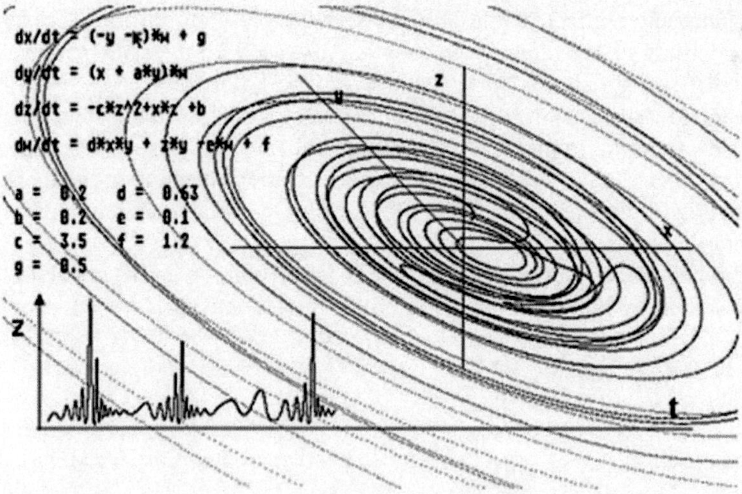

Fig. 14.14 The chaotic movement in the three-dimensional $\{x, y, z\}$-subspace only takes place in the positive half-space $\{x, y, z > 0\}$. The variable z is always positive definite: $z > 0$

it is necessary to take a closer look at the movement in the $\{x, y, z\}$-subspace (see Fig. 14.14).

The movement in the 3D subspace $\{x, y, z\}$ takes place only in the positive half-space $\{x, y, z > 0\}$. With small values of x and y, the direction of rotation of the trajectory is reversed and the incoming spiral becomes the outgoing spiral. Conversely, for "large" values of z, the outgoing spiral becomes an incoming spiral again. We now suspect that these turning points in the directions of rotation of the trajectory are related to their crossing from one to the other four-dimensional half-space $\{x, y, z > 0, w \geq 0\} \leftrightarrow \{x, y, z > 0, w < 0\}$ (see Fig. 14.15).

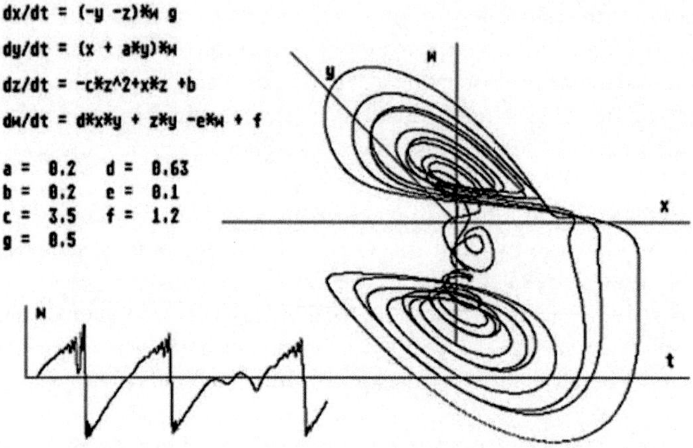

Fig. 14.15 Excerpt from the trajectory of the 4D system of Eq. (14.13) in the 3D subspace $\{x, y, w\}$. The sign of the variable w of the abstract knowledge decides in which of the two subspaces $\{x, y, z > 0, w \geq 0\}or\{x, y, z > 0, w < 0\}$ the system (14.13) is located

The movement of the system in four-dimensional space mostly takes place in the two half-spaces and the transition from the positive to the negative half-space occurs quite suddenly. During the "gate evaluation" (transition from $\{\boldsymbol{x}, \boldsymbol{y}, \boldsymbol{z} > 0, \boldsymbol{w} < 0\}$ to $\{x, y, z > 0, w \geq 0\}$), there may be temporary multiple changes of the partial system's stay in the two half-spaces or the two worlds. In a sense, the system staggers between the two worlds which may be comparable with the autonomous flip-flop of spin orientation in quantum physics.

The change between the two half-spaces is a consequence of the feedback of the abstract knowledge w on the work process, i.e., on the resources $\frac{dx}{dt}$ and on the creation of the tools $\frac{dy}{dt}$. In this way, the entire production process is divided into two sub-processes: the non-sacred area of "normal life" $\{x, y, z > 0, w \geq 0\}$ and the equally real, but sacred area of "paradise"$\{\boldsymbol{x}, \boldsymbol{y}, \boldsymbol{z} > 0, \boldsymbol{w} < 0\}$.

A decoupling of abstract knowledge from the production process leads either to an overall unstable system or to a fix-point behavior, i.e., to a rigid, stagnant system, as shown above.

All well and good, but how can we understand "negative abstract knowledge"? Perhaps, the answer given by bankers and economists will help us here when asked. "Where did the money actually come from?" Then they say: "Debt generates money." However, for the common person, debts are not money, at most negative money.

Does this answer help us further? Well, we might understand negative abstract knowledge as a lack of the knowledge we need to solve problems at hand. For example, we lack the necessary social knowledge to reverse or avoid the climate changes caused by today's lifestyle.

An interesting conclusion emerges from what has been said: the second world of paradise is not a sunken world of meditation, but a highly productive working world in which abstract knowledge is increased and becomes the essential moment of the overall process. Nothing else describes the fairy tale of Frau Holle with the stay of the two girls in the realm of Frau Holle—it is a hard-working life, this service to the goddess.

14.10.2 Chaos, Hyper-Chaos and the Blue-Sky Catastrophe

In his article, "Chaos", Otto E. Rössler [42] describes the creation of a continuous chaos in the three-dimensional space with the help of the *reinjection principle.* It is based on the fact that an unstable system in a two-dimensional space swirls "outwards" and is thrown back onto itself via the third dimension.

> There is an unstable spiral 'downstairs' on a Z-shaped slow manifold. The same spiral, only somewhat displaced in a direction parallel to the edges of the slow manifold, applies 'upstairs'. As a result, a 'reinjection loop' is formed which, depending on the geometry of the arrangement, may have the form shown in the figure. In Fig. 14.10b (Fig. 14.16b in this chapter, the author's) we see a paper model (of the kind considered in the preceding section) that fits right into the flow of Fig. 14.10a (Fig. 14.16a in this chapter, the author's).
>
> The flow of Fig. 14.10a (Fig. 14.16a, the author's) and the paper generalization (Fig. 14.10b (Fig. 14.16b, the author's)) were described ... as an implementation of a 'soft

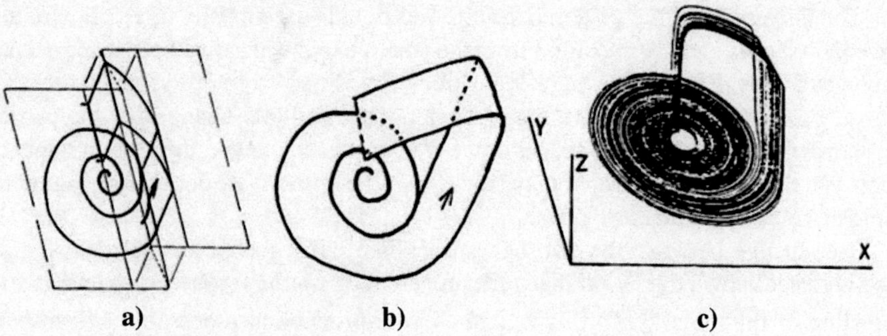

Fig. 14.16 **a** An abstract dynamical system of relaxation type and **b** corresponding paper flow; **c** continuous chaos in Eq. (14.14), compare Fig. 14.16a

watch'. A simple differential equation of the system of Fig. 14.10a (Fig. 14.16a, the author's) is [42]:

$$\begin{aligned} \dot{x} &= -y - z \\ \dot{y} &= x + 0.15y \\ \varepsilon\dot{z} &= \left(1 - z^2\right)\ (x - 1 + z) - \delta z \end{aligned} \tag{14.14}$$

He also successfully applies this principle to four-dimensional space and in this way creates a form of chaos that he describes as 'hyper chaos' [43].

4D Hyperchaos: "In analogy with the construction of Eq. (14.2) Eq. (14.14) in this chapter, the author's), perhaps a, (this time 3-variable) linear system plus switching variable can be found as a straight forward realization of the paper flow. The following equation is more simple (although its cross-section involves a few additional folding's):

$$\begin{aligned} \dot{x} &= -y - z \\ \dot{y} &= x + 0.25y + w \\ \dot{z} &= 2.2 + xz \\ \dot{w} &= -0.5z + 0.05w \end{aligned} \tag{14.15}$$

Eq. (14.4) (Eq. (14.15) in this chapter, the author's) contains only one nonlinear (quadratic) term. A simulation is presented in Fig. 14.20 (Fig. 14.17 in this chapter, the author's). The bottom line shows the excitable ('switching') variable z as a function of t. It is activated only from time to time, namely, whenever the linear 'main flow' (in x,y,w-space, presented stereoscopically on top) exceeds z's threshold (at $x = 0$) from below. During the brief periods of activation of z, a 'reinjection event' takes place in the main flow."

Here, an outward swirling system that is unstable in three-dimensional space is thrown back onto itself via the four-dimensional space.

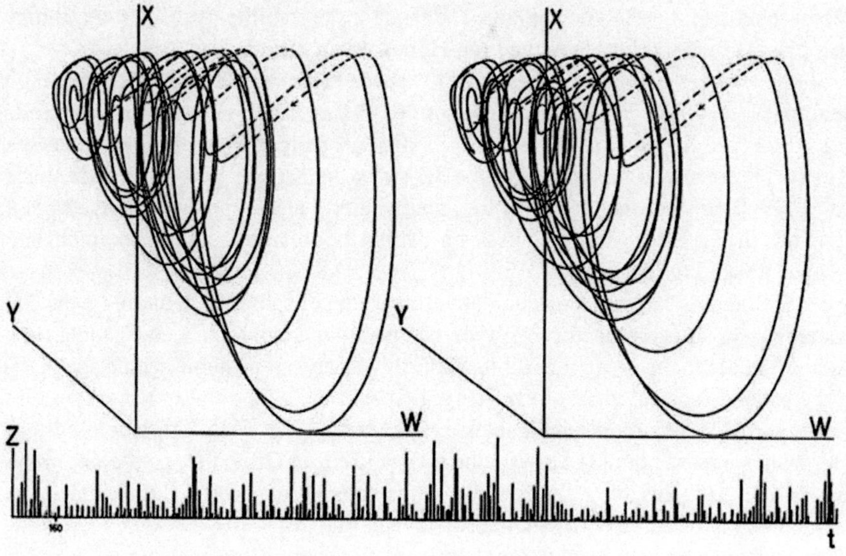

Fig. 14.17 "Continuous 'hyperchaos' in Eq. (14.4) (Eq. 14.15 in this chapter, the author's) … The flow in *x,y,w*-subspace is shown in two different parallel projections (stereoplot) …" [43]

In the case of the "Frau-Holle differential equation system", Otto Rössler's reinjection principle is applied twice, so to speak. For positive values of w, the system swirls "outward" and is thrown back via the four-dimensional space to another, now "inward" swirling sub-system, from which it "escapes"—again over the four-dimensional space—in the direction of the positive w-axis into the original, outward swirling sub-system.

This reversal of the direction of rotation seems to us to be one of the new properties that may characterize chaos in four-dimensional space and that does not appear in three-dimensional chaos. It therefore seems to us to be appropriate to give this special attractor its own name: "the miracle attractor" or the "Frau Holle attractor".

If one only considers the $\{x, y, z\}$-subspace of the generally known craft world, then the trajectory describes a "miracle"—it almost returns to itself at the transition from the $\{x, y, z, w \geq 0\}$-half-space to the complementary half-space $\{x, y, z, w < 0\}$. It is a "miracle" because in this $\{x, y, z\}$-space one cannot observe and understand the reason of turning back—the change of the $\{x, y, z, w\}$-half-spaces.

For a closer examination of the chaotic behavior of the 4-dimensional Frau-Holle equation system (14.13), we should actually examine the correlated 7-dimensional parameter space with the parameters $a, b, c, d, e, f, andg$ of this equation system. However, that would go completely beyond the scope of this article and contradict our intention. In the following, we will largely limit ourselves to an examination of the system against a variation of the parameter c. This parameter c played already a decisive role in the 3-dimensional system of equations. It was responsible for the decrease of the inherent knowledge z and thus for the stability of the system. It may also play a comparable role in the Frau-Holle system of Eqs. (14.10).

If one changes the parameter c in the 4D space $\{x, y, z > 0, w\}$, which in the 3D system (Eq. (14.10)) caused this to collapse ($c < 1.55$, see Fig. 14.13) or to explode ($c > 1.55$, see Fig. 14.12) or else described an unstable limit cycle ($c \approx 1.55$, see Fig. 14.11), then one recognizes in the 3D subspace $\{x, y, z > 0\}$ of the 4D space $\{x, y, z > 0, w\}$ that the trajectory for small values c also approaches a fixed point here, i.e., the system almost collapses, but avoids the collapse at the last moment and escapes it (see Fig. 14.18).

In Shilnikov's language of qualitative theory of dynamic systems [44], the 3D system $\{x, y, w\}$ is a trajectory between two heteroclinically connected saddle foci, a stable saddle focus O_1 in the stable manifold W_1^s and an unstable saddle focus O_2 in the unstable central manifold W_2^c (Fig. 14.19).

For small values of the parameter c, $0.2 <\approx c <\approx 0.45$, the whole system approaches a fixed point F between the two saddle foci O_1 and O_2 (red point in the Fig. 14.20). This lies at the intersection of the heteroclinic orbit Γ_1, the connection between the two saddle foci with the stable manifold W_0^s.

If we change the initial conditions a little bit, for example by choosing $w_0 = -2.5$ as starting value for the variable w, and slightly varying the start values for the other variables, then we can see for otherwise quite similar parameters ($c = 0.45$) that the system still runs towards a stable fixed point in the end, but that this point is additionally surrounded by an unstable cycle L^u (limit cycle) in its immediate vicinity (see Fig. 14.21).

It is well known that a fixed point located on a surface, for example, is always surrounded by an unstable limit cycle. All trajectories that lead to the stable fixed-point start from it. This limit cycle can be located in infinity or it can have a finite

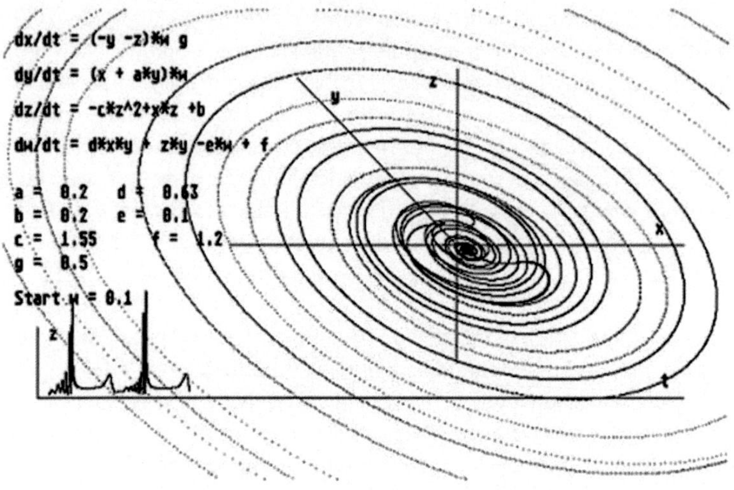

Fig. 14.18 The reversal points ($w < 0$) → ($w \geq 0$) are particularly emphasized by an enlarged representation. The reversal points for ($w \geq 0$) → ($w < 0$) are not shown in the illustration

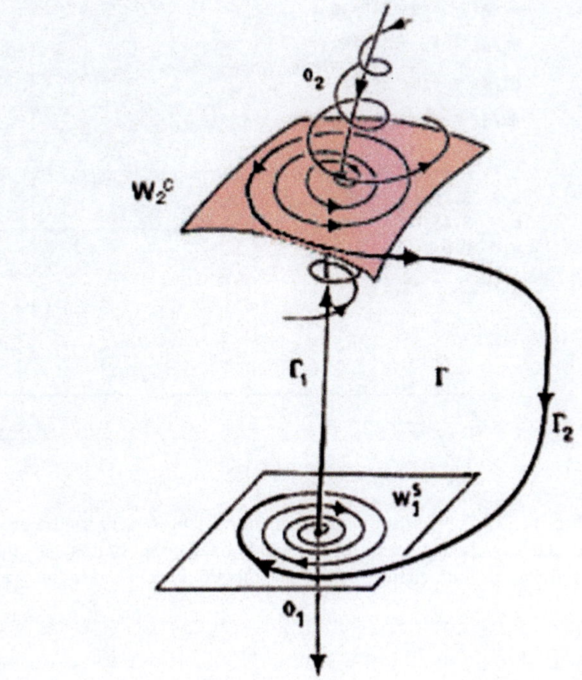

Fig. 14.19 Sketch of the qualitative description of the "Frau Holle-system of equations" for the three-dimensional $\{\mathbf{x}, \mathbf{y}, \mathbf{w}\}$-subspace. The transitions $\boldsymbol{\Gamma} = \{\boldsymbol{\Gamma}_1, \boldsymbol{\Gamma}_2\}$ between the manifolds $\mathbf{W}_1^s$ and $\mathbf{W}_2^c$ with the saddle focus points $\mathbf{O}_1$ and $\mathbf{O}_2$ are shown symbolically

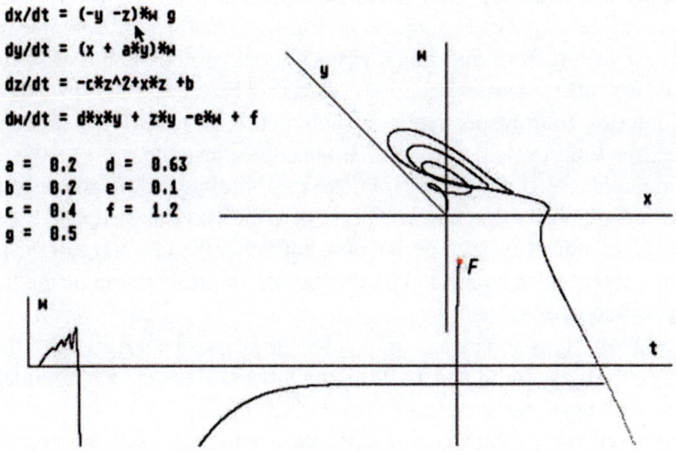

Fig. 14.20 For $c = 0.4$, the system runs onto a stable node F (fixed point, marked red)

radius, as in the present case. If such a fixed point becomes unstable, a stable limit cycle is usually generated in its immediate vicinity.

For $c \approx 0.5$, this fixed point F becomes unstable and turns to a "limit cycle" of period one in the three-dimensional $\{x.y.w\}$-space. However, as the further analysis

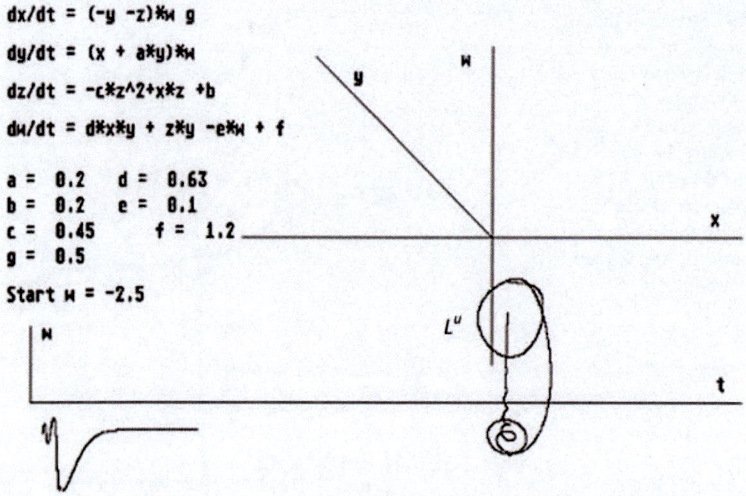

Fig. 14.21 The starting values for the variables $\boldsymbol{x}$, $\boldsymbol{y}$, zund$\boldsymbol{w}$ are set to lie almost exactly on an unstable cycle that surrounds the stable fixed point. In this way, the system circles for a short time in the immediate vicinity of the unstable limit cycle and in this way makes it "visible"

will show, this is not a simple Hopf bifurcation, i.e., fixed point $\rightarrow$ limit cycle. It seems that the stable fixed point, which is a knot, breaks up into a saddle knot and a stable limit cycle in the immediate vicinity (bifurcated), see Fig. 14.22.

This "limit cycle" L itself (Fig. 14.22a) becomes unstable as the c-value increases and in the case of a three-dimensional system a saddle node could arise on it (see Fig. 14.23). Now the system of equations discussed here is a system with four variables or dimension four, respectively. In such a case, a "blue-sky catastrophe" can develop, i.e., the limit cycle L (Fig. 14.23b) becomes instable.

If you now increase slightly the parameter c, a "limit cycle" of the period of two seems to develop. Considering the possibility of a blue-sky catastrophe, this does not seem to be a classic limit cycle of period two, but rather the first stage of the blue-sky catastrophe, which is also suggested by the "fuzzy" representation of the trajectory in the $\{x, y, w\}$-subspace.

Shilnikov gives in his book a nice sketch for the shape of a trajectory of the "blue-sky catastrophe" (Figs. 14.23 and 14.24), which we find again in our model for the value $c = 0.62$ almost one to one as shown in Fig. 14.25.

As an autonomous region, the blue-sky catastrophe is stable for a certain, small range of c-values. As the c-values increase slightly, the blue-sky catastrophe gets bigger and bigger (see Figs. 14.26, 14.27 and 14.28) and reaches upward to the unstable center manifold W_2^c and downward to the stable manifold W_1^s.

If you increase the c-value even further, e.g., $c = 1.55$ or up to $c = 3.7$, the blue-sky catastrophe no longer exists as an autonomous, stable structure in the vicinity of the former fixed point; it has long since become unstable. But even as an unstable, non-autonomous structure, it still has a strong effect on the system. If the trajectory

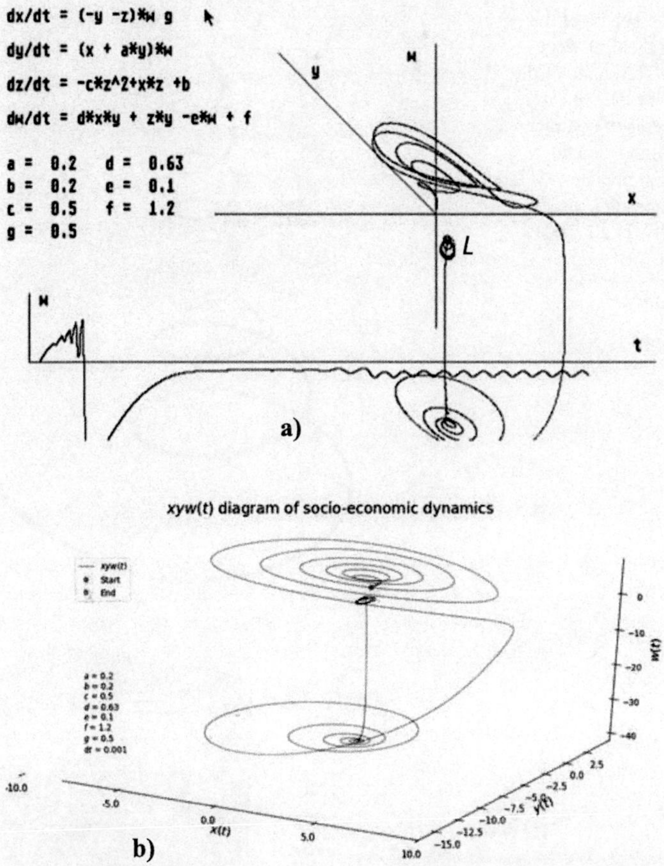

Fig. 14.22 **a** Creation of a "limit cycle" of period one instead of the former fixed point, which has now become unstable or has turned into a saddle node and a stable limit cycle. This stable limit cycle L is still surrounded by the unstable limit cycle, not shown here, the existence of which has already been documented above. **b** Another situation in three-dimensional space: Here the fixed point can only become unstable in one dimension. It becomes the saddle node to which a stable limit cycle is immediately adjacent; this is surrounded by an unstable cycle that includes the saddle knot

Fig. 14.23 Transition from the knot **a** to the saddle knot, **b** with an adjacent limit cycle. (Graphics from. L.P. Shilnikov et al. (2001) p. 543 and p. 811 [45])

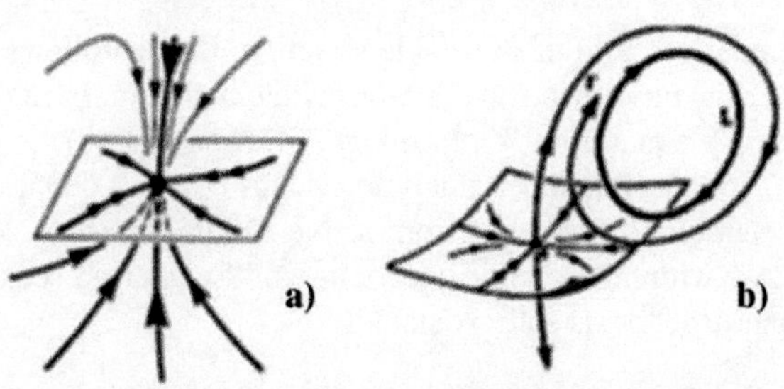

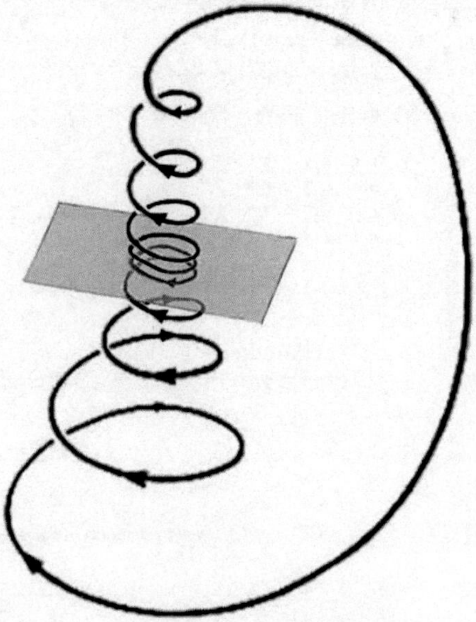

Fig. 14.24 A sketch of the trajectory of the blue-sky catastrophe. The shape of the periodic orbit looks like a helix that seems to contract onto a two-dimensional cycle in the immediate vicinity of the saddle node. (L.P. Shilnikov et al. (2001) p. 810 [46])

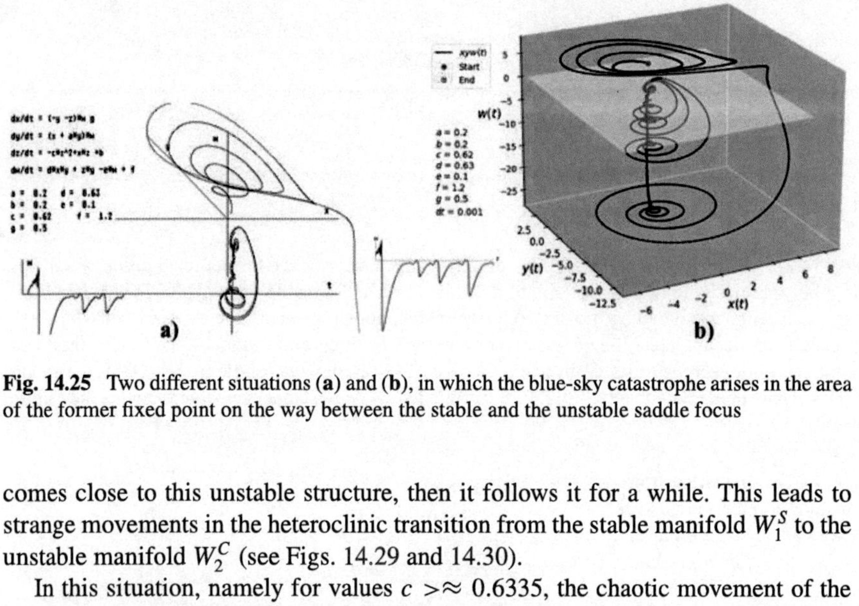

Fig. 14.25 Two different situations (**a**) and (**b**), in which the blue-sky catastrophe arises in the area of the former fixed point on the way between the stable and the unstable saddle focus

comes close to this unstable structure, then it follows it for a while. This leads to strange movements in the heteroclinic transition from the stable manifold W_1^S to the unstable manifold W_2^C (see Figs. 14.29 and 14.30).

In this situation, namely for values $c >\approx 0.6335$, the chaotic movement of the heteroclinic transition from stable saddle focus O_1 to the unstable saddle focus O_2—with the dynamic uncertainty of "staggering" caused by the unstable blue-sky catastrophe—is entirely stable!

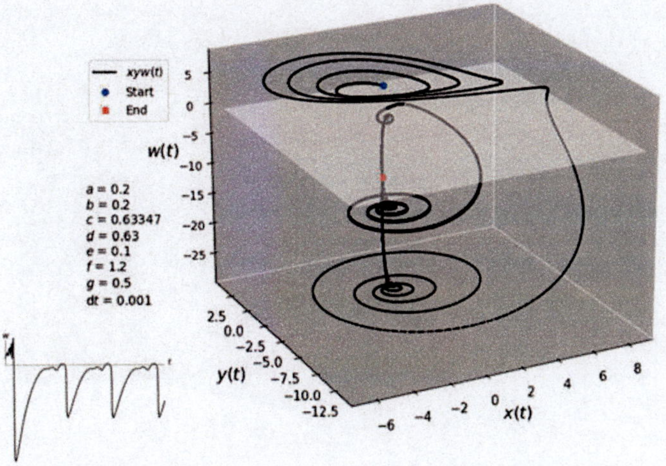

Fig. 14.26 Still autonomous blue-sky catastrophe for $c = 0.63347$. The area of the blue-sky catastrophe almost touches the catchment area of the unstable central manifold $\boldsymbol{W}_2^c$ above and below the stable manifold $\boldsymbol{W}_1^s$

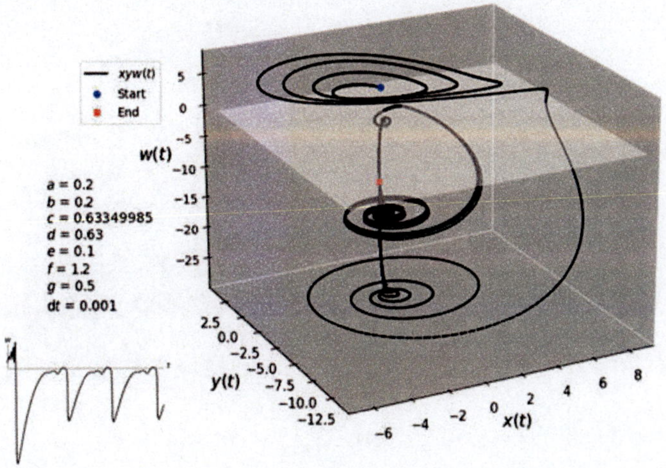

Fig. 14.27 The upper limit for autonomous blue-sky catastrophes is the value $\boldsymbol{c} \approx 0.63349905$. At this $\boldsymbol{c}$-value, the blue-sky catastrophe is barely autonomous in the context of this test

This staggering movement, when the trajectory briefly comes into the catchment area of the unstable blue-sky catastrophe, shortly before it reaches the unstable manifold describes the "indeterminacy of the world change" that an initiate or a test subject of the mystery experiences.

As long as the system ends in a fixed point, a limit cycle or in the stable blue-sky catastrophe (for $c <\approx 0, 6335$), the observable chaotic transition Γ_2 from the upper,

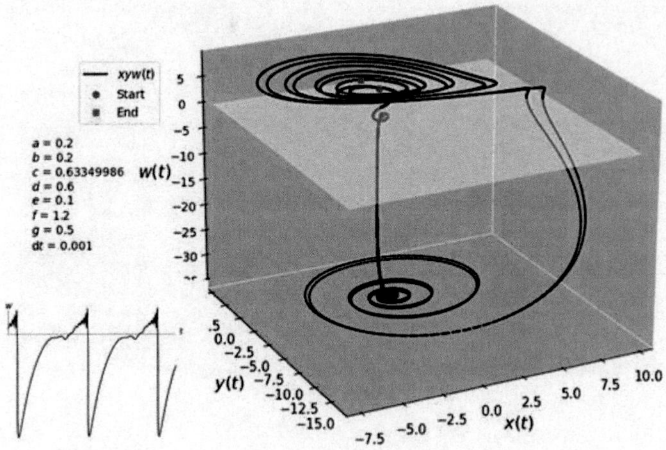

Fig. 14.28 Captured blue sky catastrophe for the value $c = 0.63349906$. With this value, the previously autonomous blue-sky catastrophe is captured by the unstable manifold W_2^C in the fifth period, i.e., in this case it runs through a cycle in the immediate vicinity of this manifold, which can be easily recognized by the positive w-values in the time series $w = w(t)$

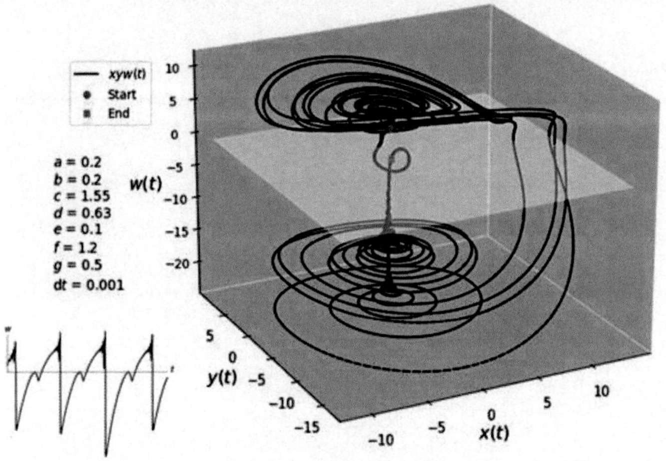

Fig. 14.29 The staggering movement in the $\{x, y, w\}$-sub-space with $c = 1.55$: For the same set of parameters, the inner turning points, which make up the staggering motion, were shown in the $\{x, y, z\}$-subspace above

unstable saddle focus O_2 to the lower, stable saddle focus O_1 was itself unstable as a whole system.

It is a quite astonishingly complex, albeit unstable, world that surrounds these stable structures.

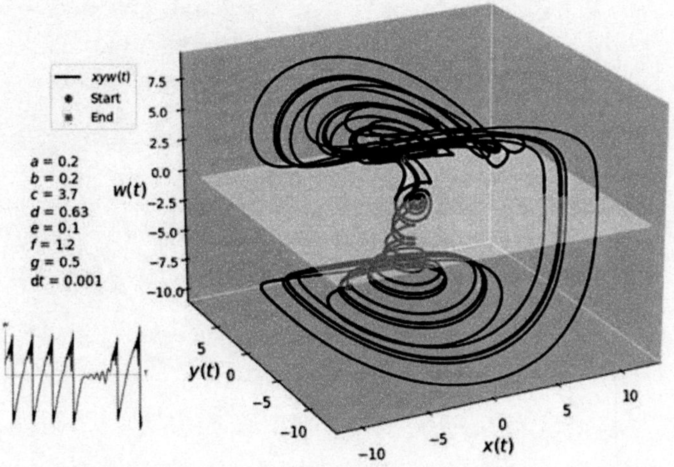

Fig. 14.30 The staggering movement in the $\{x, y, w\}$-sub-space with $c = 3.7$: The trajectory "staggers" and then finally leads with the opposite direction of rotation into the area of the unstable saddle focus O_2 located above. The whole chaotic system is stable as such!

But it is even more astonishing that this previously unstable world changes to stability with further increased complexity if the c-values are increased and the formerly stable, thoroughly complex world, e.g., the blue-sky catastrophe, now becomes unstable in turn. Of course, one can observe similar processes also with variation of the other parameters (see, for example, the variation of parameters d and e in Fig. 14.31) and not only with that of the c-value, where they can admittedly be observed very nicely.

"d" is the parameter that determines the growth of abstract knowledge due to the work with the tools y on the resources x: $\frac{dw}{dt} \propto xy$. The slight reduction in d ($d = 0.63 \rightarrow d = 0.60$) leads to a change in the chaotic behavior with negative values of w, i.e., in the second world. However, it does not significantly change the staggering motion. The whole chaotic system remains stable.

As already mentioned above, the parameter "e" is the one that controls the necessary degradation of the abstract knowledge by "forgetting", $\frac{dw}{dt} \propto -w$, which would prevent the development of new knowledge (compare Chap. 13, "Creativity" of this book). Increasing e from $e = 0.1 \rightarrow e = 0.15$, the unstable blue-sky catastrophe is intensified while the chaotic overall system is more banded.

But what is the c-value all about? What does this parameter stand for? It describes the strength of the information change $\frac{dz}{dt}$ through the loss of information due to the work itself, the wear and tear of the tools and the forgetting of techniques, in short, the lifetime of the tools and processes $\frac{dz}{dt} \propto -z^2$; "$c$" is the corresponding proportionality constant.

This makes it clear how changes in the world of work can have a structure-forming effect on the flow of information in society and its structures, which is of great importance not only for the time of Frau Holle, but also for our time, even if a system of equations for our time will certainly look a little different.

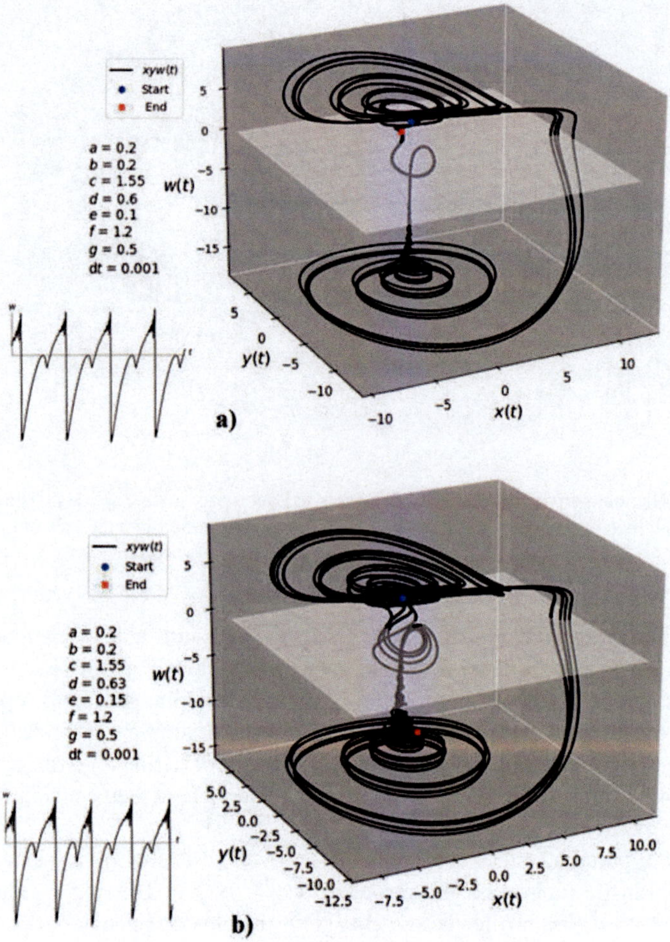

Fig. 14.31 Comparison of the attractors with variation of the parameters d and e with the same critical value of $c = 1.55$; **a** $d = 0.60$, $e = 0.1$ and **b** $d = 0.63$, $e = 0.15$

References

1. Plath, P.J.: Vom Märchen zur Mär oder <Drei mal Drei ist Neune- ihr wißt ja wie ich's meine>—Strukturale Märchenanalysen, pp. 173–218. Logos Verlag, Berlin (2012)
2. Grimm, J., Grimm W.: Deutsche Sagen, Wissenschaftliche Buchgesellschaft Darmstadt (1972), Nachdruck der dritten Auflage von 1891 mit Vorwort zur ersten Auflage von (1816) und (1818)
3. Grimm, J.: Deutsche Mythologie. Akademische Druck- und Verlagsanstalt, Graz, Lizenzausgabe (1992)
4. Grimm, J.: Deutsche Grammatik, vol. 1. Verlag Dieterichsche Buchhandlung, Göttingen (1819)
5. Grimm. https://www.de.wikipedia.org/wiki/Br%C3%BCder_Grimm. Accessed 13 June 2021
6. Ranke-Graves, R. von: Griechische Mythologie—Quellen und Deutung, Vol. I and II, rowohlts deutsche enzyklopädie, rororo 113/114 und rororo 115/116, Rowohlt Taschenbuch Verlag GmbH, Reinbek bei Hamburg (1960)

7. Haken, H.: Synergetik—eine Einführung (translated to German by A. Wunderlin), 3. Edition, Springer Verlag Berlin, Heidelberg (1990) S. 237–238
8. https://www.grimmstories.com/language.php?grimm=024&l=en&r=de. Accessed 16 June 2021
9. Grimm, J.: Deutsche Mythologie, Band I bis III, Drei Lilien Verlag, Wiesbaden (1992); License issue: Reprint (1968) of 4th edition (1875 – 1878) by Akademische Druck-u. Verlagsgesellschaft, Graz; especially Vol I, Cap. XIII, p. 207
10. Grimm, J.: Deutsche Mythologie, vol. I, p. 222. Drei Lilien Verlag, Wiesbaden (1992)
11. Grimm, J.: Deutsche Mythologie, vol. I, p. 223. Drei Lilien Verlag, Wiesbaden (1992)
12. Gimbutas, M.: Die Zivilisation der Göttin, p. 261. Verlag Zweitausendeins, Frankfurt/Main (1996)
13. Gimbutas, M.: Die Zivilisation der Göttin, p. 342. Verlag Zweitausendeins, Frankfurt/Main (1996)
14. Antoni, D.: Malus sylvestris—Oldest crabapple tree in Germany, published under the GNU Free Documentation License (2006). https://www.de.wikipedia.org/wiki/Datei:Stubbendorf_Wildapfelbaum.jpg. Accessed 22 July 2021
15. Gimbutas, M.: Die Zivilisation der Göttin, p. 196. Verlag Zweitausendeins, Frankfurt/Main (1996)
16. Gimbutas, M.: Die Zivilisation der Göttin, Verlag: Zweitausendeins, Frankfurt/Main (1996), S. 195
17. Kuczinski, J.: Die Intelligenz—Zur Soziologie und Geschichte ihrer Grossen, Pahl-Rugenstein Kleine Bibliothek 468, pp. 58–59. Pahl-Rugenstein Verlag GmbH, Köln (1987)
18. Childe, V.G.: Der Mensch schafft sich selbst, Dresden (1959) S. 60
19. Rössler, O.E.: Endophysik—Die Welt des inneren Beobachters, in: Weibel, P. (Hrsg.), Merve Verlag GmbH, Berlin (1992) S. 103; Originalversion: O.E. Rössler, Explicite Observers. In: Plath, P.J. (ed.) Optimal Structures in Heterogeneous Reaction Systems, Springe Series in Synergetics. Springer Verlag Berlin, vol. 48, pp. 131–132. Heidelberg, New York, London, Paris, Tokyo, Hong Kong (1989)
20. Ranke-Graves, R. von: Die Weiße Göttin—Sprache des Mythos, rowohlts enzyklopädie, Rowohlt Taschenbuchverlag, Reinbeck bei Hamburg, 6th edition (1999) S. 335–336; and Graves, R., The White Goddess: A Historical Grammar of Poetic Myths, Grevel Lindop (Ed.), Farrar, Straus & Giroux, New York (2013) pp. 274–275
21. von Ranke-Graves, R.: Griechische Mythologie—Quellen und Deutung, vol. I, p. 22. Rowohlt Taschenbuch Verlag GmbH, Reinbeck bei Hamburg, Rowohlts deutsche Enzyklopädie (1960)
22. von Ranke-Graves, R.: Griechische Mythologie—Quellen und Deutung, vol. I, p. 27. Rowohlt Taschenbuch Verlag GmbH, Reinbeck bei Hamburg, Rowohlts deutsche Enzyklopädie (1960)
23. Rössler, O.E.: Chaos. In: Güttinger, W., H. Eikemeier, (Eds.), Structural Stability in Physics, Springer Series in Synergetics, Vol. 9, Springer-Verlag, Berlin, Heidelberg, New York (1979) p. 290
24. Li, T.-Y., Yorke, J.A.: Period three implies chaos. Am. Math. Month. **82**, 985–992 (1975)
25. Ranke-Graves, R. von: Griechische Mythologie—Quellen und Deutung, Bd. I, rowohlts deutsche enzyklopädie, rororo 113/114, Rowohlt Taschenbuchverlag GmbH, Reinbek bei Hamburg (1960) S. 27–28
26. Spitzing, T. (Director): Das Delphi Syndikat—Die geheime Macht des Orakels, Ifage Filmproduktion GmbH, Deutsche Erstausstrahlung (German Premiere Broadcast): Sat 2003–08–30, arte (2003)
27. Plath, P.J.: Vom Märchen zur Mär oder <Drei mal Drei ist Neune—ihr wißt ja wie ich's meine>—Strukturale Märchenanalysen, pp. 3–11. Logos Verlag, Berlin (2012)
28. Kant, I.: Schriften zur Metaphysik und Logik, Bd 1, Werkausgabe Bd. V, Weischedel, W. (Hrsg.), Suhrkamp Taschenbuch Wissenschaft, stw 188 (1988) S. 55–57
29. Plath, P.J.: Jenseits des Moleküls—Raum und Zeit in der Chemie, Vieweg Reihe: Facetten—Interdisziplinäre Wissenschaft (Series Editor: H. Schuster), Friedr. Vieweg & Sohn Verlagsgesellschaft mbH, Braunschweig/Wiesbaden (1997) S. 67 ff.
30. Plath, P.J.: Jenseits des Moleküls—Raum und Zeit in der Chemie, Vieweg Reihe: Facetten—Interdisziplinäre Wissenschaft (Series Editor: H. Schuster), Friedr. Vieweg & Sohn Verlagsgesellschaft mbH, Braunschweig/Wiesbaden (1997) S. 64–68

31. Ebeling, W., Peschel, M., Weidlich, W. (Eds.): Models of Selforganization in Complex Systems. Mathematical Research, vol. 64. Akademie Verlag, Berlin (1991)
32. Parthey, H.: Das Neue—Seine Entstehung und Aufnahme in Natur und Gesellschaft. Akademie-Verlag, Berlin (1990)
33. Haken, H.: Synergetik—Eine Einführung (translated by A. Wunderlin), Springer-Verlag Berlin, Heidelberg, New York, London, Paris, Tokyo, Hong Kong, 3rd Edition (1990)
34. Haken, H., Plath, P.J., Ebeling, W., Romanovsky, Y.M.: Beiträge zur Geschichte der Synergetik—Allgemeine Prinzipien der Selbstorganisation in Natur und Gesellschaft, Springer Spektrum—Research, Springer Fachmedien Wiesbaden (2016)
35. Haken, H., Wunderlin, A.: Die Selbststrukturierung der Materie—Synergetik in der unbelebten Welt, Reihe Facetten. Friedr. Vieweg & Sohn Verlagsgesellschaft mbH, Braunschweig/Wiesbaden (1991)
36. Wildgen, W.: Mythos und Religion—Semiotik des Transzendenten. Verlag Könighausen & Neumann, Würzburg (2021)
37. Witzke, Th.: Die neolithischen Feuerstein-Gruben von Rijkholt bei Maastricht, Niederlande, GAG—Grubenarchäologische Gesellschaft, https://www.untertage.com/publikationen/19-europa/62-die-neolithischen-feuerstein-gruben-von-rijkholt-bei-maastricht-niederlande.html. Accessed 05 May 2021
38. Plath, P.J.: Komplexität, Chemie und Leben. In: Huber, L. (Ed.) Wie das Neue in die Welt kommt—Phasenübergänge in Natur und Kultur, pp. 71–94. WUV Universitätsverlag, Wien (2000)
39. Ranke-Graves, R. von: Die Weiße Göttin—Sprache des Mythos, rowohlts enzyklopädie, Rowohlt Taschenbuchverlag, Reinbeck bei Hamburg, 6th edition (1999) S. 393
40. von Ranke-Graves, R.: Griechische Mythologie—Quellen und Deutung, vol. I, pp. 75–77. Rowohlt Taschenbuch Verlag GmbH, Reinbeck bei Hamburg, Rowohlts deutsche Enzyklopädie (1960)
41. Mime (Schmied), Wikipedia. https://www.de.wikipedia.org/wiki/Mime_(Schmied). Accessed 16 May 2021
42. Rössler, O.E.: Chaos, in: Güttinger, W., and H. Eikemeier, H. (Eds.), Structural Stability in Physics, Springer Series in Synergetics, vol. 9, pp. 290–309. Springer-Verlag, Berlin, Heidelberg, New York (1979)
43. Rössler, O.E.: Chaos. In: Güttinger, W., and H. Eikemeier (Eds.), Structural Stability in Physics, Springer Series in Synergetics, vol. 9, p. 302. Springer-Verlag, Berlin, Heidelberg, New York (1979)
44. Shilnikov, L.P., Shilnikov, A.L., Turaev, D.V., Chua, L.O.: Methods of qualitative theory in nonlinear dynamics, Part II, World Scientific Series on Nonlinear Science, Series A, vol. 5 (Series Editor: Leon L. Chua), World Scientific New Jersey, London, Singapore, Hong Kong, p. 447 and pp. 470–471 (2001)
45. Shilnikov, L.P., Shilnikov, A.L., Turaev, D.V., Chua, L.O.: Methods of qualitative theory in nonlinear dynamics, Part II, World Scientific Series on Nonlinear Science, Series A, vol. 5 (Series Editor: Leon L. Chua), World Scientific New Jersey, London, Singapore, Hong Kong, p. 543 and p. 811 (2001)
46. Shilnikov, L.P., Shilnikov, A.L., Turaev, D.V., Chua, L.O.: Methods of qualitative theory in nonlinear dynamics, Part II, World Scientific Series on Nonlinear Science, Series A, Vol 5 (Series Editor: Leon L. Chua), World Scientific New Jersey, London, Singapore, Hong Kong, p. 810 (2001)

Part V
Kaleidoscope

Peter J. Plath
Ernst-Christoph Haß

Introductory Remarks

“Kaleidoscope”, could also be called “Research Splinter” because it contains short research papers on synergetics in chemistry and biology and many new suggestions that require further intensive research. All the more, all this work testifies to the enthusiasm that filled the young or older researchers when they successfully dealt with these questions. It was wonderful to experience how positive the actors are about their almost forgotten works when we told them that we shall write about it in our book.

We would like to particularly emphasize the contribution by Ms. Monika Waltersdorf to the Chap. 15 on the blue wonder. Years later, in her long forgotten and hidden documents, she found photos of the experiments that she had taken during her research internship with Prof. Plath at the University of Bremen. These photos were important for our reworking of the pattern formation in the methylene blue catalyzed oxidation of sulfides in polyacryle amide gel (PA-MBO-S system).

While we were working on the contribution to the fractal structure of the aggregating Dictyostelium discoideum cells (Chap. 16), we managed to find the former students Inga Hense and Olaf Kniemeyer. They were both very pleased that we were unearthing their previous work, but they had unfortunately no further documentation on it.

Special mention should also be made of Ms. Gesa Patzelt, whose data and tables provided to us as part of her diploma thesis served as the basis for our calculations of the kinetics of foam decay.

Working through the old documents in this way, we were faced with many new questions that we had to answer and we felt like presenting new problems ourselves in a short and concise manner.

This applies above all to the question of the description of the decay of beer foam using kinetics (Chap. 17) that correspond to a consecutive reaction with feedback. On the other hand, by chance we became aware of the fascinating structures in the granite kitchen top of the Plath family, which in our opinion represent very, very old crystallized Liesegang structures in the cold magma (Chap. 18). With the help of the stonemason Olaf Borwig, who made this slab from the Parakiwi granite Karelia red 20 years ago, we were able to clarify a number of questions about the origin and crystallinity of the structures in these granites, for which we thank him very much.

Chapter 15
The Blue Wonder

The Methylene Blue-Oscillator (MBO-Sytem)

Peter J. Plath

Fig. 15.1 Structure formation in the blue miracle in a Petri dish after 2 min. The glucose gel was produced from 20 ml glucose 2 M to 10 ml NaOH 2 M and then covered with 20 ml methylene blue 10^{-3} M. The resulting layer height was approx. 9 mm. (Photo: S. Huth, N. Tobias and P.J. Plath)

P. J. Plath et al., *Imagery Synergetics*, Understanding Complex Systems,
https://doi.org/10.1007/978-3-030-95607-3_15

15.1 Introduction

This striking title "The Blue Wonder" (Fig. 15.1) (also named as Blue Bottle experiment) [1] is not an advertising slogan for a new product on the market, although it is not infrequently found there, but rather it describes a well-known experiment from the inorganic beginners' lecture, which H.W. Roesky describes in his "Chemische Kabinettstücke" as follows [2]:

Set up: 1,000 ml Shaking cylinder, pipette.

Chemicals: Glucose, Sodium hydroxide pills NaOH, Methylene blue

Preparations: Prepare the following solutions:

A: 0.4 g of methylene blue in 200 ml of water

B: 10 g sodium hydroxide and 80 g glucose in 800 ml water

Procedure: Solution B and 10 ml of solution A are added to the 1 L shaking cylinder. The shaking cylinder is then closed. After a short time, the solution is colorless. If the colorless solution is shaken vigorously, it will turn blue again. This color will disappear again after about a minute. When shaken again, the color reappears. This process can be repeated until the oxygen in the cylinder is used up. If you ventilate, then the game can begin again.

Explanation: The discoloration of the methylene blue by reduction is based on the conversion into the leucomethylene blue, the aldehyde group of the grape sugar is oxidized to the carboxyl group.

$$R - \mathrm{CHO} + \mathrm{H_2O} + \text{Methylene blue} \rightarrow \mathrm{R\text{ - }COOH} + \text{Leukomethylene blue} \tag{15.1}$$

$$\text{Leukomethylene blue} + {}^{1}\!/_{2}\mathrm{O_2} \rightarrow \text{Methylene blue} + \mathrm{H_2O} \tag{15.2}$$

Schneider and Münster describe this reaction in their book "Nonlinear Structures in Chemistry" [3] in the chapter "Turing structures: chemical morphogenesis" a little less spectacular than the "PA-MBO system", better called PA-MBO-S System. They replaced the gel-forming glucose with sodium sulfide and instead introduced acrylamides as a gel-forming agent, which is polymerized into polyacrylamide (PA). In this way they could better control the gelation process.

In this PA-MBO-S reaction, the sulfide ions (S) of the sodium sulfide are oxidized by the oxygen present in the solution with the aid of the catalyst methylene blue MB, whereby the colorless leuco-methylene blue MBH is formed by reduction (Eqs. 15.3 and 15.4). This is a radical reaction mechanism.

It was very likely that the response of the Blue Wonder would also lead to chemical oscillations. In 1984 Maria Burger and Richard Field published a new oscillating chemical reaction that contained no metal ions catalysts [4]: "The chemical oscillator is composed of sulfide ion, sulphite ion, methylene blue (which serves as catalyst)

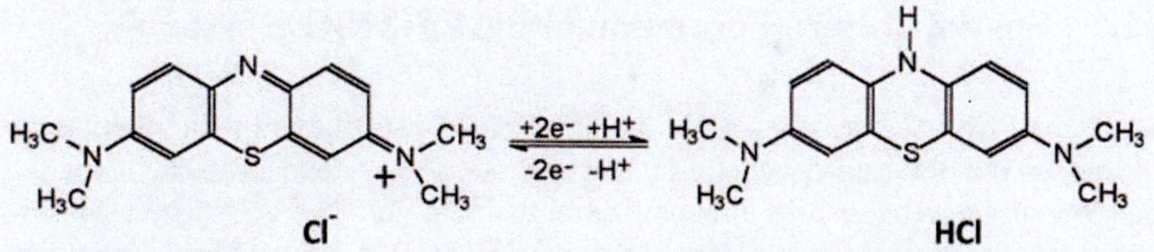

Fig. 15.2 One of the mesomeric structures of methylene blue cation MB^+ (left formula) with absorption maxima at 668 nm which is the transformed into colorless leuko-methylene blue MBH (right formula)

and dissolved oxygen. During the reaction, the methylene blue catalyst oscillates between its oxidized, coloured form (MB^+) and its reduced, leuko form (MBH)."

$$2SH^- + O_2 \rightarrow 2S + 2HO^- \tag{15.3}$$

"This new oscillator was intentionally designed by the expedient of coupling an apparent autocatalytic reaction with a return mechanism. Reaction (Eq. 15.4), the reduction of MB^+ by HS^-, is the apparent autocatalytic reaction" (Burger and Field [4])

$$HS^- + MB^+ \rightarrow MBH + S \tag{15.4}$$

The reduction of the methylene blue MB^+ to leuko-methylene blue MBH is understood as a pH-dependent process (Eq. 15.5) (Fig. 15.2).

$$\begin{aligned} MB^+ + e^- &\rightleftharpoons MB \\ MB + e^- + H^+ &\rightleftharpoons MBH \end{aligned} \tag{15.5}$$

Although the oscillations in this system (oxidation of sulfide by oxygen using the catalyst methylene blue in alkaline solution) only occur in a fairly small range of flow rates through the reactor, this discovery of a new oscillating system by Burger and Field led to numerous further systematic investigations in the continuous stirred tank reactor (CSTR) [5]. Based on these investigations, a rather complicated proposal for a mechanism of the Methylene-Blue-HS^--O_2 oscillator was developed based on corresponding investigations in the Belousov–Zhabotinsky reaction (BZ reaction) [6].

This system was then examined in detail in the groups of Field (Montana, USA) and Schneider (Würzburg, Germany), regarding the development of a mechanism for an oscillating reaction without autocatalysis [7] and the effect of fluctuations on the oscillating system [8], respectively.

15.2 Spatial Pattern Formation in the PA-MBO-S system

Since the MBO-S system is an oscillating reaction, it is not surprising that, as in the case of the Belousov–Zhabotinsky reaction, attempts were made to create spatial patterns in the Petri dish by immobilization using a gel. The gel was created by polymerizing acrylamide (c.f. Figs. 15.4 and 15.12). Watzl and Münster from the Würzburg Institute were quite successful in creating transient Turing-like spatial structures [9]. Since ionic components are involved in the reaction (Eqs. 15.3 and 15.4), the reaction should also be able to be influenced by electrical fields. This leads to controllable Turing-like structures [10, 11].

Two groups of students [12, 13] took up the production of Turing-like structures (Fig. 15.3) in the polyacrylamide-methylene blue-sulfide-oxygen system in the winter semesters 1996/1997 and 1997/1998. This work was part of my practical courses on chemical synergetics and discrete chemistry.

The group of Thorsten Kaese and Carsten Knapp used the following procedure for their investigations:

Solution 1: 20% solution of Acrylamide (20 g in 100 mL aqueous solution)

Solution 22% solution of N.N-Methylene-bisacrylamide (0.5 g in 25 mL water)

Solution 3 20% solution of Ammonium peroxodisulfate (2 g in 10 mL water)

Solution 4 30% solution of Tris-ethanolamine (3 g in 10 mL water)

Solution 5Methylene blue chloride solution MBCl: 0.006–0.03 mol/L

Solution 6Sodium sulfide (Na_2S*9 H_2O) solution: 0.03–0.5 mol/L

Solution 7 Sodium sulfite solution: 0.01 mol/L (0.03 g in 25 mL water)

The solutions are mixed as follows:

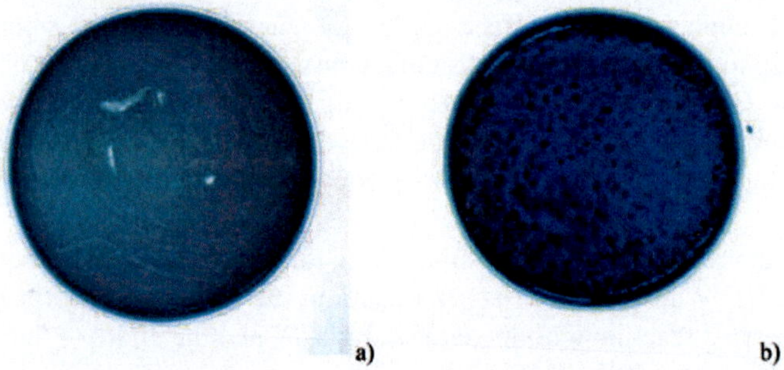

Fig. 15.3 Turing-like patterns in the polyacrylamide-methylene blue-sulfide-oxygen system. **a** Irregular stripes, MBCl 0.012 mol/L, Na_2S 0.09 mol/L, 0.90 ml solution 3; **b** irregular dark dots MBCl 0.03 mol/L, Na_2S, 0.036 mol/l, 0.20 ml solution 3: (photos: T. Kaese, C. Knapp, and P.J. Plath)

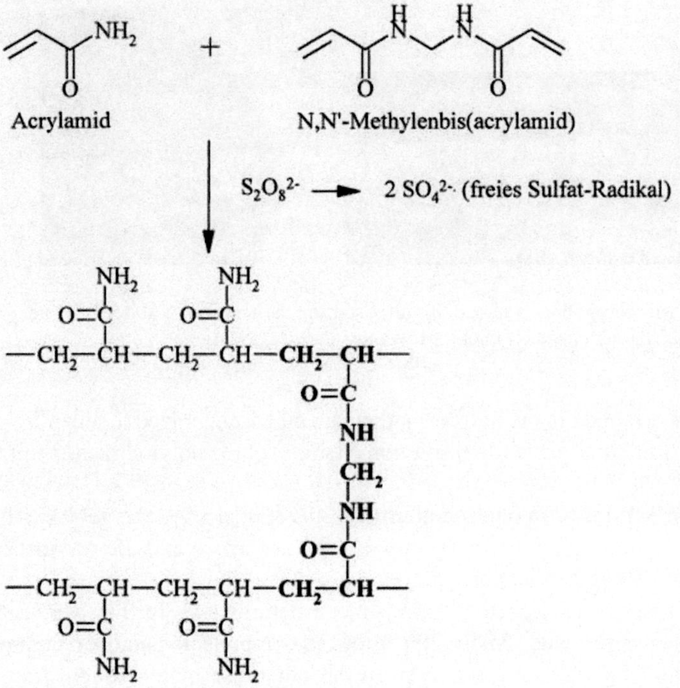

Fig. 15.4 Sketch: mechanism of polymerization of acrylamide and cross-linking of their polymerization chains [14]

Mixture a: 5.5 mL sol. 1 + 0.47 mL sol. 2 + 0.43 mL sol. 4

Mixture b:0.5 mL sol. 5 + 4.0 mL sol. 6 + 1.5 mL sol. 7

The mixtures a and b are placed in a stand-alone Petri-dish. The polymerization is then started with 0.18 mL sol. 3. After approx. 10 min, the gel is covered with a small amount of Methylene blue chloride solution. After a further five minutes, the gel is carefully lifted off the floor.

The group of Monika Waltersdorf and Jan Kaiser took a slightly different approach by increasing drastically the concentration of the polymerization starter Ammonium peroxodisulfate (sol. 3) in the final solution [12]. For their experiments reported here, they used 0.38 mL (experiment 17) and 1.0 mL (experiment 18) (sol. 3). For sodium sulfide, they used the concentration 0.5 M.

For their experiments, they used a Plexiglas Petri dish (diameter 8.5 cm) with a lid. The petri dish was placed over a light table. A thick Plexiglas pane between the light table and the Petri dish prevented the solution from heating up from below in order to prevent the formation of Bènard structures.

The components (except the Triethanolamine and Sodium sulfate) were arranged side by side in the Petri dish using disposable syringes and pipettes. This should prevent the individual components from reacting prematurely with one another. If all

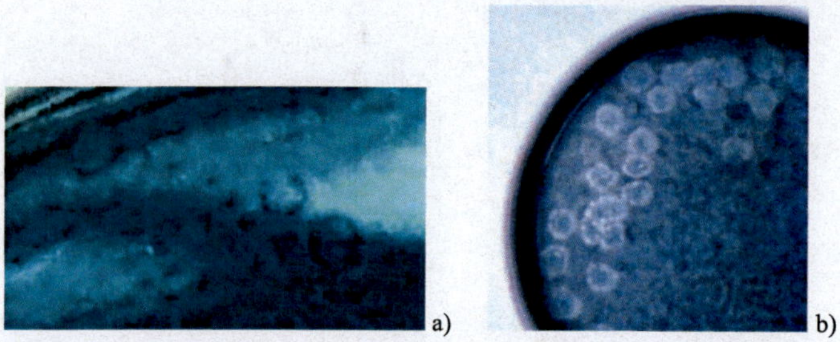

Fig. 15.5 "Mini-eruptions" **a** occur 20 s after starting the reaction; **b** about 10 s later (size of the Petri-dish 8.5 cm in diameter) (Photo: M. Waltersdorf, P.J. Plath)

species are arranged in the Petri dish, they are all homogenized together by carefully shaking. After homogenization, solution 3 (sol. 3) of the polymerization initiator was added.

"Mini-eruptions" can be seen after just 20 s (Fig. 15.5). The gel has a thickness of approx. 2 mm. After about 10 min, a gel has formed and after a further hour a Turing-like pattern can be observed.

In addition to the expected Turing-like structures, two further, very interesting observations were made: About 20 s after the components had been homogenized, bright spots were suddenly observed in the otherwise blue solution. Larger spots showed a dark spot in the middle. All spots occur spontaneously and locally and usually have one round to oval shape. (Fig. 15.6) We call them "*mini eruptions*". It is also very nice to see that an area develops around the "center of the eruption" that remains colorless for a longer period of time. It could be that the light spots in Fig. 15.3a are comparable to the "micro eruptions" described here.

This indicates a spontaneous reduction of the methylene blue to MBH, which is prevented from re-oxidation to MB^+. A blue border surrounds the area, which is darker than the rest of the solution. In the further course, a labyrinth-like pattern formation takes place. After a while, the pattern becomes blurred.

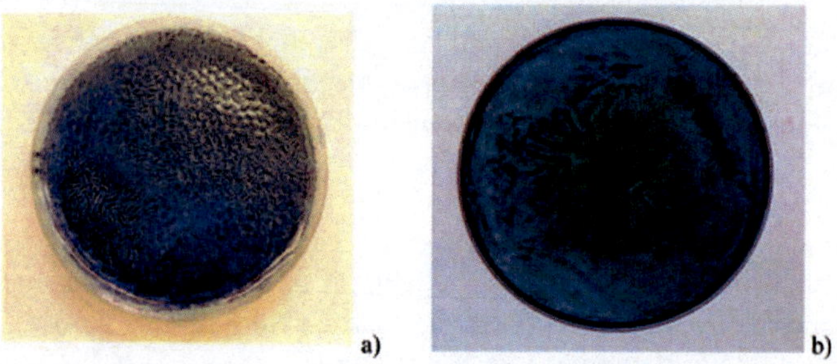

Fig. 15.6 Various Turing-like structures of the PA-MBO-S system, such as **a** hexagonal patterns and stripes and **b** tractor tire tracks. (diameter of petri-dish 8.5 cm) (Photo: M. Waltersdorf, P.J. Plath)

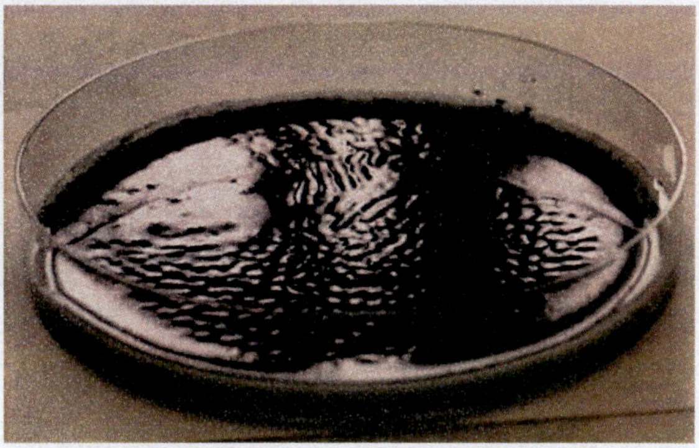

Fig. 15.7 Wavy surface of the polyacrylamide gel. The turing-like structures of the PA-MBO-S system, such as hexagonal patterns and stripes, can also be clearly seen on the surface structure of the gel. (Photo: M. Waltersdorf, P.J. Plath)

Figure 15.6a shows one of the well-known zigzag patterns made with the PA-MBO sulfide system. The second interesting observation we were able to make, is the formation of a pattern that looks like tractor tire marks. The formation of this pattern is probably due to an overlay of dot patterns (Fig. 15.6b). The surface structure of a hexagonal dot pattern can be seen in Fig. 15.7. The reflection of the chemical processes within the gel in the wavy surface structure of the gel indicates that the differences in the local surface tension are caused by the different concentrations of the substances involved in the reaction.

15.3 The Blue Wonder (MBO-G-System) and the Catalytic Memory [15]

Stefan Braun, Jörg Mühlisch and Torsten Ueckert examined the "Blue Wonder" presented by Roesky [2] (1984) in their research internship in 1992 and 1993 [16]. They investigated the kinetics of the methylene blue-catalyzed oxidation of glucose by oxygen in an alkaline solution in the continuously stirred tank reactor (CSTR).

If you carry out the classic "Blue Wonder" experiment of Roesky by adding the components methylene blue, sodium hydroxide and glucose to the Petri dish and stirring them with a glass rod, you immediately get a homogeneously looking blue solution (Fig. 15.8a).

Immediately after the mixture has been homogenized by stirring, the solution begins to discolor, as the methylene blue is reduced to leuko-methylene blue by the glucose in the solution. However, this discoloration does not take place uniformly. In some places, the solution remains a strong blue color. These blue lines represent the old flow pattern of stirring the solution, which has long since ceased to flow in this way, but is completely stationary now (Fig. 15.8b).

Along these blue lines, the leuko-methylene blue that occurs everywhere else is oxidized again to methylene blue by the supply of oxygen from the adjacent air. In the viscous glucose solution, when stirring, oxygen-rich areas were created where layers

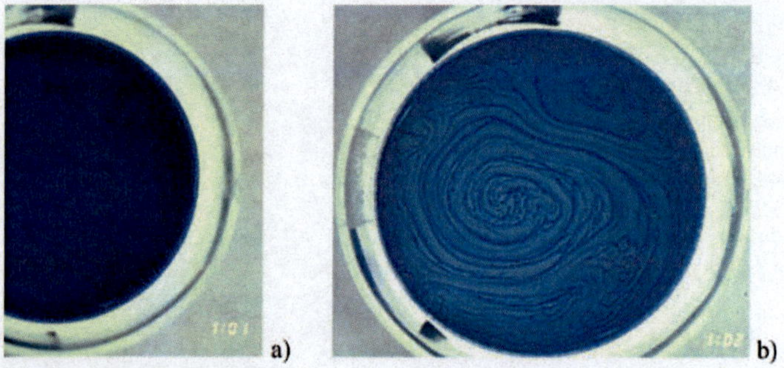

Fig. 15.8 After stirring the mixture of methylene blue, glucose and sodium hydroxide. **a** 1 min after combining the various solutions by stirring; the solution appears completely homogeneous; **b** Another minute afterwards, the patterns of stirring of the viscose solution dominate the picture, while the methylene blue in the rest of the solution has already been considerably reduced. (Photos: M. Schröder and P.J. Plath)

of liquid of different viscosity and speed flowed past one another. At these former borderlines, there is also an increased transport of oxygen into the now completely "solidified" gel layer for a long time. In this way, due to the catalytic oxidation of the glucose and the leuko methylene blue taking place on these lines, the former flow pattern remains recognizable. It seems as if the gelled solution has retained a "catalytic memory" of the former flow pattern for a while.

After about two hours, however, all the methylene blue that was still present is now reduced to leuko methylene blue by the glucose (Fig. 15.9a).

The surface of the solution is continually exposed to atmospheric oxygen and the glucose gel "ages". In this way, oxygen slowly enters the gel again. However, this does not happen in a homogeneous way; instead a very fine-grained convection structure is formed (Fig. 15.9b). Like a string of pearls, small punctiform centers emerge from which often several fine blue lines emanate. The oxygen reaches all these points in the aging gel and oxidizes the leuko-methylene blue back to methylene blue.

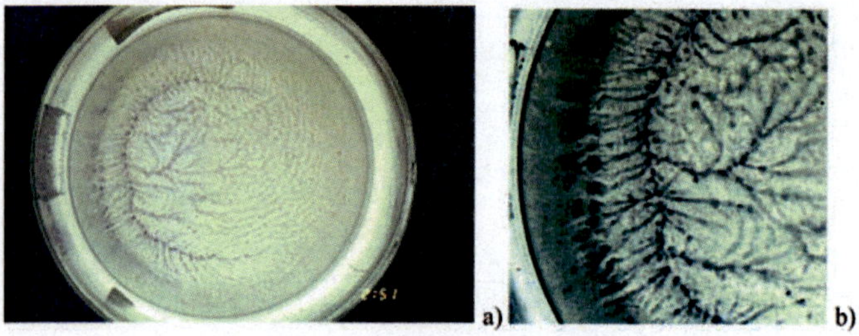

Fig. 15.9 One hour and 51 min after the start of the experiment. **a** New structures emerge in the aging gel. **b** Contrast-enhanced enlargement of (**a**) for a better visualization of the newly occurring reaction convection patterns in the aging gel. (Photos: M. Schröder and P.J. Plath)

The pattern formation described here is very complex. The gradient of the chemical potential, e.g. of the oxygen penetrating the layer, could be a driving force in this structure formation in the very complex reaction process. Unfortunately, the available data do not allow a decision about the reason for this structure formation at the moment.

Based on the attempts by M. Schröder [15] (1990) and the work of Stefan Braun, Jörg Mühlisch und Torsten Ueckert [16] in 1992/93, the student group Stefan Huth and Niels Tobias carried out again the MBO-Glucose system (MBO-G system) in the Petri dish in the winter-semester 1997/1998 [17].

The viscosity of the glucose solution depends on the pH-value of the solution. Moreover, the reduction of methylene blue to leuko methylene blue requires an alkaline solution. That is why Stefan Huth and Niels Tobias have changed both the concentration of the methylene blue and the glucose content. Morever, they varied the alkaline pH-values drastically in their attempts to perform the blue miracle in the Petri dish.

They observed interesting pattern formation in the blue miracle in a Petri dish after 2 min (Fig. 15.1). The glucose gel was produced from 20 ml glucose 2 M to 10 ml NaOH 2 M and then covered with 20 ml methylene blue 10^{-3} M. The resulting layer height was approx. 2.8 mm.

Even 30 s after starting the reaction, the homogeneously blue solution began partially structuring (see Fig. 15.10a), but a few minutes later this structure disintegrated again (see Fig. 15.10b). S. Huth and N. Tobias related their observations to the formation of rolling cells in the Bénard experiment, which is schematically represented in Fig. 15.11.

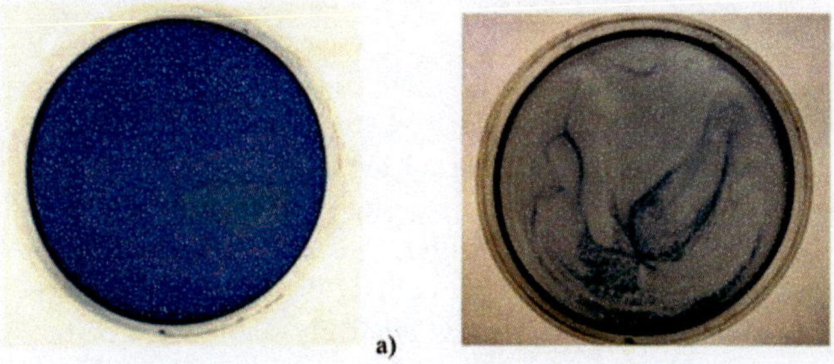

a) b)

Fig. 15.10 Structure formation in the blue miracle in a petri dish of 15 cm in diameter; **a** after 0.5 min and **b** after 4 min. The glucose gel was produced from 20 ml glucose 2 M to 10 ml NaOH 2 M and then covered with 20 ml methylene blue 10^{-3} M. The resulting layer height was approx. 2.8 mm. (Photo: S. Huth, N. Tobias and P.J. Plath)

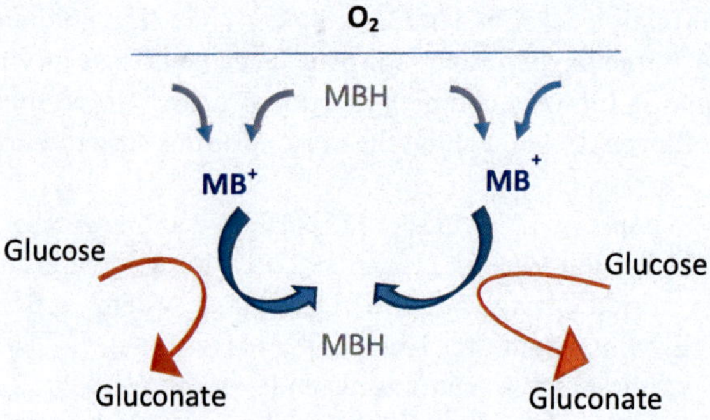

Fig. 15.11 Sketch of the transport and reaction processes in the glucose gel layer the height of which is almost 2.8 mm, which is in the order of magnitude of the cells between the limiting reaction lines from colorless leuco-methylene blue to methylene blue (see Figs. 15.9b and 15.10b)

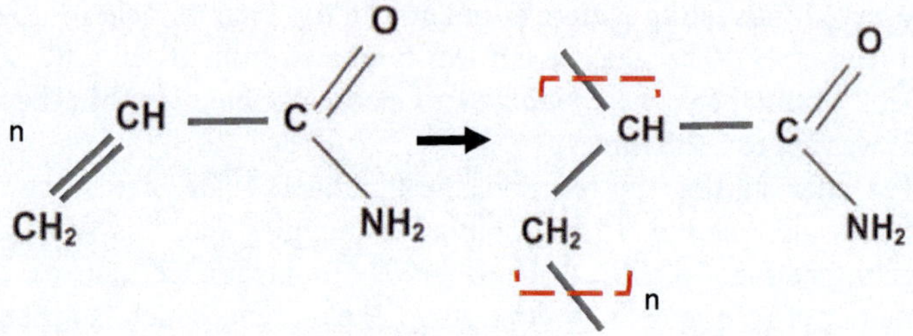

Fig. 15.12 Symbolic description of the polymerization of Acrylamide to Polyacrylamide with *n* subunits

15.4 Some Final Remarks

First of all, it is absolutely clear that for all the experiments reported here which are carried out in Petri dishes or, as in Roesky's case, in a shaking funnel, it is true that they represent closed systems with regard to the substances to be oxidized, be it sulfides or glucose. This means that none of the structures observed here can be stable. At most, they can be viewed as relatively stable for a certain period of time. The long-term tests carried out over almost two hours prove this very clearly. However, this does not mean that such transient structures should not be regarded as scientifically interesting. After all, all living structures only exist as temporary structures.

Gel formation through the polymerization of acrylamide or of glucose in alkaline solution is crucial for what is happening and the structures that can be observed. The respective gel not only serves to suppress undesired convection but is itself an essential reaction partner.

The polymerization of acrylamide in aqueous solution is known to be a radical chain reaction (see Fig. 15.12). Ammonium peroxodisulfate $(NH_4)_2S_2O_8$ serves as a radical starter [18] (see also Fig. 15.4). On the other hand, the oxidation of the sulfide (Eq. 15.3) catalyzed by the methylene blue is also a thoroughly radical reaction, as Resch, Field, Schneider and Burger already show in their reaction mechanism for the MBO-S system [6] published in 1989. Fechner, Strasser Eiswirth, Schneider and Münster (1999) also emphasize this in their *simplified chemical mechanism* for the same MBO-S system if acrylamide is also taken into account explicitly [19]. The observed very local “mini-eruptions” seem to be a typical structure which is closely connected to the radical character of both the reactions together. Something similar applies to the coupling between the surface tension and the chemical reaction, in particular the polymerization of the acrylamides.

In the articles from 1989 and 1991, Field et al. assume that the methylene blue is essentially involved in the oxidation of the sulfide ions via a radical reaction mechanism, for example via Eq. 15.6:

$$MB^{+} + HS^{-} \rightarrow MB^{\circ} + HS^{\circ} \tag{15.6}$$

We assume that this is also the case with the oxidation of glucose in a similar way. The essential radical reaction mechanism is closely related to any gel formation. The gel is part of the overall reaction process! However, this makes it difficult to understand the process of the blue miracle on the basis of classical chemical kinetics, because if the gel is involved, essential requirements for its use are omitted: the space is no longer homogeneous, or the space plays a decisive role in the reaction.

References

1. https://de.wikipedia.org/wiki/Blue-Bottle-Experiment (read 2021-07-15), and https://av.tib.eu/search?f=CREATOR%3Bhttp://d-nb.info/gnd/134081366 and https://av.tib.eu/search?f=CREATOR%3Bhttp://d-nb.info/gnd/134081366
2. Roesky, H.W.: Chemische kabinettstücke (Teil 2). In: Kontakte (Merck-Darmstadt), vol. 2, S, pp. 44–45 (1984)
3. Schneider, F.W., Münster, A.F.: Nichtlineare Dynamik in der Chemie, pp. 189–191. Spektrum akademischer Verlag GmbH, Heidelberg, Berlin, Oxford (1996)
4. Burger, M., Field, R.J.: A new chemical oscillator containing neither metal nor oxyhalogen ions. Nature **307**, 720–721 (1984)
5. Resch, P., Field, R.J., Schneider, F.W.: The methylene-blue-HS–O_2 oscillator: mechanistic proposal and periodic perturbation. J. Phys. Chem. **93**, 2783–2791 (1989)
6. Resch, P., Field, R.J., Schneider, F.W., Burger, M.: Reduction of methylene blue by sulfide ion in the presence and absence of oxygen: simulation of the methylene Blue-O_2-HS^--CSTR oscillation. J. Phys. Chem. **93**, 8181–8186 (1989)
7. Zhang, Y.-X., Field, R.J., Simplification of the methylene-blue-HS^--O_2 CSTR oscillator: a homogeneous oscillatory mechanism with nonlinearities but no autocatalysis. J. Phys. Chem. **95**, 723–727 (1991)
8. Resch, P., Münster, A.F., Schneider, R.W.: A subcritical hopf-bifurcation in the mehylene blue oscillator: effects of imposed fluctuations. J. Phys. Chem. **95**, 6270–6275 (1991)
9 Watzl, M., Münster, A.F.: Turing-like spatial patterns in a polyacrylamide-methlene-blue-sulfide-oxygen system. Chem. Phys. Lett. **242**(3), 273–278 (1995)

10. Watzl, M., Münster, A.F.: Control of mosaic and turing patterns by light and electric field in the methylene blue–sulfide–oxygen system. Phys. Chem. A **102**(15), 2540–2546 (1998)
11. Schneider, F.W., Münster, A.F.: Nichtlineare dynamik in der chemie. Spektrum akademischer Verlag GmbH, pp. 189–192 and 216–217. Oxford, Heidelberg, Berlin (1996)
12. Kaiser, J., Waltersdorf, M.: Herstellung turing-ähnlicher muster im polyacryl-methylenblau-sulfid-sauerstoff-system. In: Plath, P.J., Veranstalter (eds.) Protokoll zum Praktikum Diskrete Chemie. WS (1996/1997), unpublished results
13. Kaese, T., Knapp, C.: Erzeugung turing-ähnlicher muster im polyacryl-methylenblau-sulfid-sauerstoff-system. In: Plath, P.J., Veranstalter (eds.) Protokoll zum Praktikum Chemische Synergetik, (Plath, P.J., Veranstalter) WS (1997/1998), unpublished results
14. Kaiser, J., Waltersdorf, M.: Herstellung turing-ähnlicher muster im polyacrylamid-methylenblau-sulfid-sauerstoff-system. In: Praktikum zur Veranstaltung, Diskrete Chemie' WS (1996/1997), p. 5. Universität Bremen Veranstalter P.J. Plath (1997)
15. Plath, P.J.: Jenseits des Moleküls—Raum und Zeit in der Chemie. In: Schuster, H. (ed.) Reihe Facetten—Interdisziplinäre Wissenschaft, pp. 22–25, 77–82, Tables 6 and 7. Friedrich Vieweg & Sohn Verlagsgesellschaft mbH Braunschweig, Wiesbaden (1997)
16. Braun, S., Mühlisch, J., Ueckert, T.: Kinetische Untersuchungen am Reaktionssystem Methylenblau/Sauerstoff/Glucose in alkalischer Lösung. Bericht zum Forschungspraktikum (SS 1992 bis SS1993) (supervisor: P.J. Plath)
17. Huth, S., Tobis, N.: Protokoll zum Versuch Methylenblau—das blaue Wunder (PA–MBO-G System). In: von Plath, P.J. (ed.) Praktikum zur Veranstaltung, Chemische Synergetik, WS 1997/98. Universität Bremen (1998) (unpublished results)
18. https://de.wikipedia.org/wiki/Polyacrylamid (read 2021-07-23)
19. Fechner, F., Strasser, P., Eiswirth, M., Schneider, F.W., Münster, A.F.: Spatial entrainment of patterns during the polymerization of acrylamide in the presence of the methylene blue-sulfide chemical oscillator. Chem. Phys. Lett. **313**, 205–210 (1999)

Chapter 16
Fractal Aggregation of *Dictyostelium discoideum*

Peter J. Plath

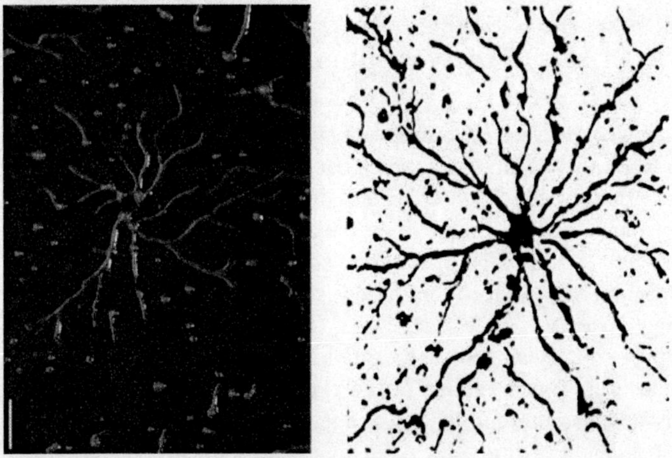

Fig. 16.1 Aggregation cluster of NC4 Dictyostelium discoideum at the beginning of the aggregation, which marks the starting point of the measurements. **Left** Photo of the aggregation cluster (R. Herbst, O. Kniemeyer (2021) Leibniz-HKI, with friendly permission); **right** black and white transformation of a similar photo for better graphical evaluation

"The aggregation of Dictyostelium represents a very interesting phase in the life of this organism, as it shows the transition from single cells to a multicellular organism" [1].

This way, the biology students Inga Hense, Olaf Kniemeyer and Astrid Mielke formulated their motivation in the project "Self-organization in chemical and biological systems" supervised by Michel Vicker and the author in 1994/1995 to deal with the "determination of the fractal dimension during the aggregation of Dictyostelium discoideum". For mathematical support, they worked together with the computer science student Jan K. Plath.

P. J. Plath et al., *Imagery Synergetics*, Understanding Complex Systems,
https://doi.org/10.1007/978-3-030-95607-3_16

The transition from single cells to multicellular organism is still very topical today [2, 3].

At that time, the leading groups studying Dictyostelium discoideum were interested in the issues of information transfer between cells that lead to aggregation. Particularly of interest was the formation of spiral waves that started from the aggregation moud that formed in the center of the aggregation.

However, if one looks at the structure of the "rivers" of cells (see Fig. 16.1) that are formed during the aggregation of cells, then these images have a great similarity with the structures of corresponding source systems of rivers all over the world. There are also certain undeniable similarities with the images of electrolytic zinc deposition discussed in Chap. 5—"Fractal Zinc tree" of this book.

Therefore, the question we asked ourselves was: "Is it possible to describe the temporal change of the structures during the aggregation of the Dictyostelium cells in a similar way?" Are these also fractal structures?

16.1 Way of Life of Dictyostelium

As part of our work, we used two strains of *Dictyostelium discoideum*: the wild strain NC4 and a laboratory strain AX2, which can grow in a liquid nutrient medium.

The wild strain lives in the litter layer of the forest floor as a single, freely moving amoeba. As long as bacteria in the soil serve as food for the dictyostelium cells, they multiply through cell division. If the food source is exhausted, the individual cells gather to form strands, move radially to a self-organized center and thereby form so-called aggregates.

The initially solitary amoeba aggregate in initially random centers when there is a lack of nutrition. The formation of the aggregates is mediated by a chemotactic signal. This happens through the messenger substance cAMP, a universal second messenger within eukaryotic cells. After about five hours of "starvation", some cells release cAMP every 6–8 min [4].

The cells detect the cAMP and migrate in the direction of the signal for a few minutes, after which they themselves release cAMP into the environment and remain insensitive to a new signal for a few minutes. They are in the refractory state. The refractory period means that the signal can only migrate away from the center; the neighboring cells, which are closer to the center, are still refractory when the cell they are stimulating releases its cAMP.

Since we carried out these investigations, many of the questions that we asked ourselves at that time and those that we could not even ask have been comprehensively answered for example by Michael Vicker [5–8], Stefan C. Müller [9, 10], Martin Falcke [11], Cornelius J. Weijer [12, 13] and many other groups. In 2019, Cornelius J. Weijer described more or less the current status in detail in a very readable review article [14]. I will not go into this any further, but will instead focus on the aspects that were of particular interest to us at the that time.

16.2 The Fractal Dimension of the Aggregation of *Dictyostelium discoideum*

The aggregation of Dictyostelium represents a very interesting phase of the life of this organism, as it shows the transition of the single cells to a multicellular, differentiated organism. If our assumption is correct that we can describe the aggregation patterns with the methods of fractal geometry [15], then the temporal change in these structures should be able to be described as the temporal variation of the fractal dimension during the aggregation phase of the cells.

16.2.1 Preparation of the Samples

A SM5 agar was used as the nutrient medium for the NC4 strain, which was then diluted in a ratio of 1:5 with KK_2 buffer. This dilution is necessary because otherwise the bacteria would grow too quickly and thus displace *Dictyostelium discoideum* in the phase that is not yet exponential. First, an *E-coli* bacteria suspension was spread onto the plates. This was followed by inoculation with a few spores of the NC4 strain. These plates were then stored at room temperature in the dark for 48 h.

The liquid medium HL5 was used as a food source for the cells of the AX2 strain. These cells nourish themselves by absorbing the nutrients from this medium.

In order to achieve aggregation in both the NC4 and AX2 strains, the cells were carefully "washed": The cells were rinsed with KK_2 buffer with the agar into a centrifuge tube and centrifuged for two minutes. The supernatant was poured away and the pellet was redissolved with KK_2 buffer.

This is necessary to remove all bacteria from the medium. The cells only aggregate after a certain period of starvation. The washed cells can also be better observed and documented. After washing three times, the cells were placed on plates thinly streaked with KK_2 agar. After 4–6 h, aggregation began in the AX2 strain. In the case of the NC4 strain, the incubation period was 48 h on a SM5 agar with *E. coli* bacteria.

The following cell number concentrations were used for the aggregation:

1. NC4 378 cells/mm^2

   Due to the poor yield after centrifugation, the aggregation could only be carried out with this number of cells. In spite of this low cell concentration, aggregation occurred.
2. AX211,000 cells/mm^2 (compare Hall et al. 1988) [16]

16.2.2 Observation of the Aggregation

The aggregations of the wild strain NC4 and the cultivated strain AX2 were recorded directly from the microscope with the aid of a video camera (CCD-500). Individual structures were then selected from the video film and photographed with a reflex camera. The time interval between the recordings was adapted to the rate of aggregation. These photographs were enlarged to photographs measuring 10 × 15 cm. The photos of the aggregates were then copied onto an overhead transparency that could be attached directly to the computer screen. With the help of the computer mouse, the individual structures were circumnavigated and transferred to the computer for numerical evaluation.

We used the box-counting method to determine the fractal character of the aggregations [17].

Let $N(s)$ be the number of squares with side length s, which are necessary to fill a square area F in parquet, then it is known that the following applies.

$$N(s)s^2 = F \tag{16.1}$$

This can easily be transformed into the double-logarithmic expression:

$$ld(N(s)) = -2ld(s) + ldF \tag{16.2}$$

or more general:

$$log(N(s)) = -Dlogs + logf \tag{16.3}$$

where D is the dimension of the space to be measured. But D does not have to be a natural number. D can also be a positive rational number. Then the structure examined is a fractal. This is the case when, as in the case of the aggregation images, the area is not uniformly filled and, for example, only the squares in which there are Dictyostelium cells are counted.

In order to determine the fractal dimension of the aggregation structure, square grids of various mesh sizes with $s = 4, 5, 6, 8, 10, 12, 15, 20, 40$and 60 pixels are placed over the structures. $N(s)$ is now the number of boxes with the side length of s pixels, in which cells are located. This box is then called a black box and according Peitgen et. al. the fractal dimension D becomes now the special designation box-counting dimension D_b.

For the two *Dictyostelium discoideum* cell strains NC4 and AX2 in Figs. 16.2 and 16.3, each show only one aggregation structure and the associated calculations of its fractal box-counting dimension D_b.

Table 16.1 shows an overview of the time course of the obtained fractal dimensions of the aggregation structures in the two cell strains. The fractal dimension D_b of NC4 is in the range of approx. 1.46–1.41, but for AX2 it is in the range of 1.71–1.59. This suggests that the fractal dimension could be characteristic of the aggregation ability of the respective cell strain. This is also suggested by the visual impression of the aggregation pattern.

Moreover, a monotonous decrease in the box-counting dimension could be expected in the course of continued aggregation. A decrease in dimension occurs

Fig. 16.2 **a** Aggregation cluster of NC4 *Dictyostelium discoideum* 30 min after the first photo (see Fig. 16.1). **b** Estimation of the fractal dimension of the aggregation cluster shown in (**a**) concerning Eq. (16.3); $D = 1.414$

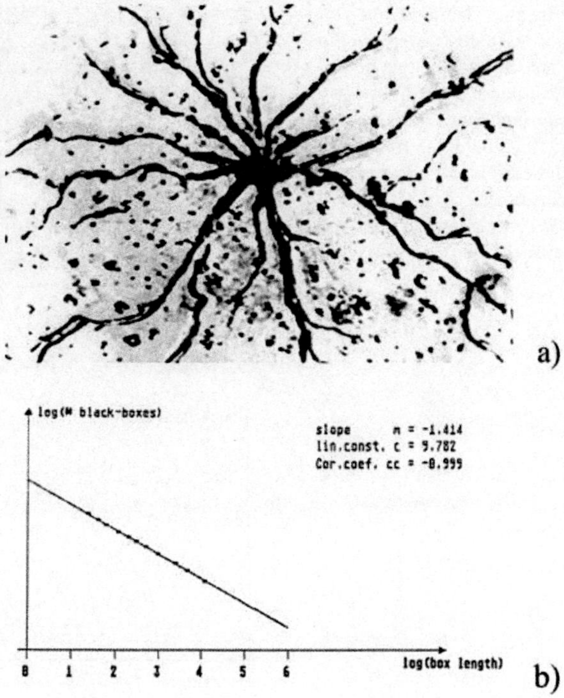

over the entire observed period of aggregation in both cell lines, but it is not monotonic. However, the correlation coefficients “cc” are in all cases so good with 0.999 or 0.998 that they cannot explain the fluctuations that occur over time. The question of whether this is due to a lack of professionalism in the handling of these systems or the complexity of the experimental system itself can best be answered by looking at the relevant recent literature.

Using the example of STO-MEF cells, Bitter [18] shows in his doctoral thesis what influence cell density has on aggregation and thus on the experimentally determined fractal dimension. Strong variations of the fractal dimensions are observable within time especially for lower cell concentrations *P*. Bitter claims that the mean particle size of the aggregates are also responsible for the estimated fractal dimensions. For sizes lower than 20 μm he got dimension of about 1.7 which are close to the dimension of the Witten & Sander model of diffusion limited aggregation [19].

Although P. Bitter investigated a different cell system and a different aggregation pattern, his results given here can be understood as an explanation for our own results regarding the Dictyostelium aggregation.

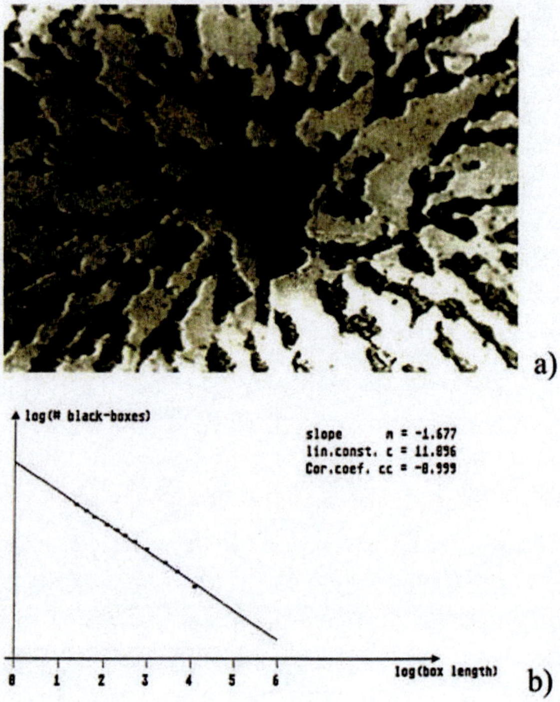

Fig. 16.3 **a** Aggregation cluster of AX2 *Dictyostelium discoideum*; (the first photo, i.e. arbitrary setting of the starting point $t = 0$). **b** Estimation of the fractal dimension of the aggregation cluster shown in (**a**) concerning Eq. (16.3); $D = 1.677$

Table 16.1 Comparison of the fractal dimensions D_b of the aggregation structures of NC4 and AX2 *Dictyostelium discoideum* cell strains

Aggregation phases	Time (min)	NC4 D_b	Time (min)	AX2 D_b
1	0	1.457	0	1.677
2	30	1.414	12	1.69
3	60	1.44	32	1.707
4	90	1.408	78	1.587

References

1. Hense, I., Kniemeyer, O., Mielke, A.: Bestimmung der fraktalen Dimension während der Aggregation von *Dictyostelium discoideum.* Bericht zum Projekt. Selbstorganisation in chemischen und biologischen Systemen. SS 1994–WS 1994/95 (Supervisors: P.J. Plath and M. Vicker) (July 1995)
2. Yuan Wang S., Pollina, E.A.,I-Hao Wang, Pino L.K., Bushnell, H.L., Ken Takashima, Fritsche C., Sabin, G., Garcia, B.A., Lieberman Greer, P., Eric Lieberman Greer, E.: Role of epigenetics in unicellular to multicellular transition in Dictyostelium. Genome Biol. **22**(1), 134 (4 May 2021). https://doi.org/10.1186/s13059-021-02360-9
3. Koryu Kin, Schaap, P.: Evolution of multicellular complexity in the dictyostelid social amoebas. Genes **12**, 487 (2021). https://pubmed.ncbi.nlm.nih.gov/33801615/; PMID: **33801615,**

PMCID: PMC8067170. https://doi.org/10.3390/genes12040487

4. Bozarro. S.: Dictyostelium: from unicellularity to multicellularity. In: Russo, V.E.A., et al. (eds.) Development—the molecular genetic approach. Springer, Berlin (1992)
5. Killich, T., Plath, P.J., Xiang Wei, Bultmann, H., Rensing, L., Vicker, M.G.: The locomotion, shape and pseudopodial dynamics of unstimulated dictyostelium cells are not random. J. Cell Sci. **106**, 1005–1013 (1993)
6. Killich, T., Plath, P.J., Haß, E.-C., Wei Xiang, Bultmann, H., Rensing, L., Vicker, M.G.: Cell movement and shape are non-random and determined by intracellular, oscillatory rotating waves in *Dictyostelium* amoebae. BioSystems **33**, 75–87 (1994)
7. Vicker, M.G., Wei Xiang, Plath, P.J., Wosniok, W.: Pseudopodium extension and amoeboid locomotion in *Dictyostelium discoideum*: Possible autowave behaviour of F-actin. Phys. D: Nonlinear Phenom. **101**(3–4), pp. 317–332 (1 March 1997)
8. Brosteanu, O., Plath P.J., Vicker M.G.: Mathematical analysis of cell shape. In: Alt, W., Deutsch, A., Dunn, G.A. (eds.) Dynamics of Cell and Tissue Motion. Mathematics and Biosciences in Interaction. Birkhäuser, Basel (1997). https://doi.org/10.1007/978-3-0348-8916-2_4
9. Müller, S.C., Mair, T., Steinbock. O.: Traveling waves in yeast extract and in cultures of *Dictyostelium discoideum*. Biophys. Chem. **72**(1–2) 37–47 (1998). https://doi.org/10.1016/s0301-4622(98)00121-5. Affiliations PMID: 9652085
10. Foerster, P., Müller, S.C.: Geometric parameters of aggregation waves in *Dictyostelium Discoideum*. In: Holden, V., Markus, M., Othmer, H.G. (eds.) Nonlinear wave processes in excitable media, part of the NATO ASI Series, Series B: Physics, vol. 244, pp 277–279. Plenum Press New York, London (1991)
11. Falcke, M., Levine, H.: Pattern selection by gene expression in *Dictyostelium Discoideum*. Phys. Rev. Lett. **80**, 3875–3878 (1998)
12. Vasiev, B., Siegert, F., Weijer, C.J.: A Hydrodynamic model for *Dictyostelium discoideum* mound formation. J. Theor. Biol. **184**(4), 441–450 (1997)
13. Singer, G., Araki, T., Weijer, C.J.: Oscillatory cAMP cell-cell signalling persists during multicellular *Dictyostelium* development. Commun. Biol. **2**, 139 (2019). https://doi.org/10.1038/s42003-019-0371-0
14. Bretschneider, T., Othmer, H.G., Weijer, C.J.: Progress and perspectives in signal transduction, actin dynamics, and movement at the cell and tissue level: lessons from *Dictyostelium*. Interface Focus **2016**, 20160047 (2016). https://doi.org/10.1098/rsfs.2016.0047
15. Mandelbrot, B.B.: The Fractal Geometry of Nature. W.H. Freeman and Company, New York (1983)
16. Hall, A.L., Schlein, A., Condeelis, J.: Relationship of pseudopod extension to chemotactic hormone-induced actin polymerization in amoeboid cells. J. Cell Biochem. **37**(3), 285–299 (1988). https://doi.org/10.1002/jcb.240370304
17. Peitgen, H.O., Jürgens, H., Saupe, D.: Bausteine des Chaos – Fraktale, pp. 256–265. Springer-Verlag, Berlin, Heidelberg, New York, Klett-Cotta Stuttgart (1992)
18. Bitter, P.: Dynamics and structure of cellular aggregation. Doctor thesis 17.12. 2014, pp. 73–80. Fachbereich Physik, Philipps-Universität Marburg (2015). http://archiv.ub.uni-marburg.de/diss/z2015/0211/pdf/dpb.pdf (Accessed 2021–07–04)
19. Witten, T.A., Sander, L.M.: Phys. Rev. B **27**(9), 5686–5697 (1983)

Chapter 17
Segregation and Growth—Consecutive Kinetics of Beer Foam Decay

Ernst-Christoph Haß, Peter J. Plath, and Gesa J. Patzelt

Fig. 17.1 Beer foam in a glass: Does a good beer take seven minutes to froth? (Photo: P.J. Plath)

P. J. Plath et al., *Imagery Synergetics*, Understanding Complex Systems,
https://doi.org/10.1007/978-3-030-95607-3_17

17.1 Introduction

The decay of beer foam is observed millions of times a day (Fig. 17.1), but hardly ever studied scientifically. An exception to this is the work of Leike [1] as well as of Dale, Walker and Lyddiatt [2]. Sauerbrei [3, 4] was the first to present careful physicochemical studies on the course of beer foam decomposition by standardizing the start-up conditions using ultrasonic foaming and paying careful attention to temperature constancy, defined foam volume, etc. She could show that the time evolution of the foam volume during the foam decay can be described by a formula of the type given in Eq. (17.1):

$$V(t) = V_0 e^{-at} e^{bt^{-2.5}} \text{ or, equivalently,} \\ \ln(V(t)) = \ln(V_0) - at - bt^{2.5} \quad (17.1)$$

V_0 represents the foam volume at $t = 0$, i.e., immediately after foaming up; a is the coefficient at the beginning of the exponential foam decay which is mainly caused by drainage of the liquid beer, and b corresponds to diffusion due to rearrangement of the bubbles in the further course of the foam decay.

Furthermore, she observed the formation of Apollonian structures in the reorganization phase and determined for the first time the evolution of the bubble size distribution functions during the decay of the foam. The main focus of this work was to investigate the development of these distribution functions. However, in order to formulate kinetics of the evolution of the individual bubble size classes, the amount of available data was too small.

In her student research project (Forschungspraktikum) [5] and her diploma thesis, Patzelt [6] continued this work to investigate the kinetics of individual bubble size classes based on a sufficient amount of data. In doing so, she emphasized on determining the differences in the temporal development of foam decay at uncoated and coated glass surfaces, i.e., in addition to the hydrophilic surface, the influence of a hydrophobic surface is also investigated. The latter serves as a model for detergent residues on a beer glass.

17.2 Experimental Results

The measurements were carried out in the rectangular glass vessel of 92 mm × 51 mm × 150 mm (length × width × height) shown in Fig. 17.2, which was coated with a hydrophobic silane layer on the right side (see red arrow). This coated glass vessel was manufactured by the company Krüss GmbH and given to the working group of Prof. Plath for these experiments.

The measuring glass was filled with 140 ml beer (Haake Beck Pils) at a temperature of 24 ± 1 °C. In order to produce as little foam as possible and to prevent the formation of a monolayer of beer foam, the vessel was tilted and the neck of the beer bottle was

Fig. 17.2 Side view of the glass vessel provided by Krüss GmbH with silane coating on the side indicated by the red arrow. (Photo: P. J. Plath)

held against the edge of the vessel. After filling, the beer in the vessel was foamed up in an ultrasonic bath to achieve an approximately completed foaming. After foaming, the vessel was placed on a laboratory jack as quickly as possible and image acquisition was started on the computer using the G&View® program. To record the temporal development of foam decay, several image series were taken on both, the uncoated and the coated, sides of the vessel using a CCD-camera horizontally pointing to each side. Each image series consisted of 30 photos taken at a time interval of 10 s from each other.

To demonstrate the foam decay, Fig. 17.3a, b show exemplarily two snapshots at the beginning and after 100 s taken on the uncoated glass wall. In these images, one can already see the increase in bubble sizes and the decrease in the number of bubbles

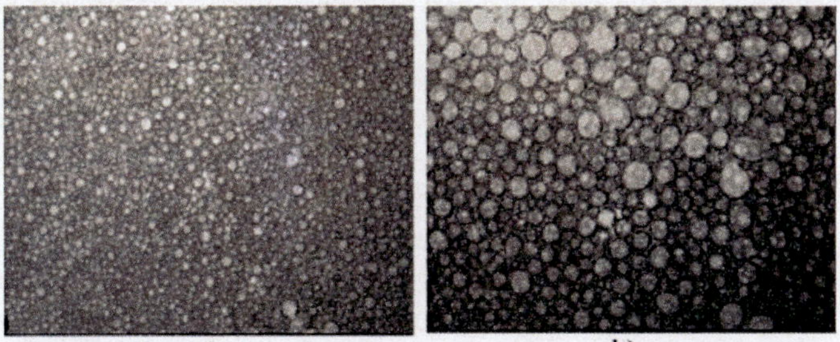

a) b)

Fig. 17.3 Snapshots during beer foam decay taken at the uncoated wall of the glass; **a** immediately after frothing up, **b** after 100 s. (Photo: G. Patzelt, P. J. Plath)

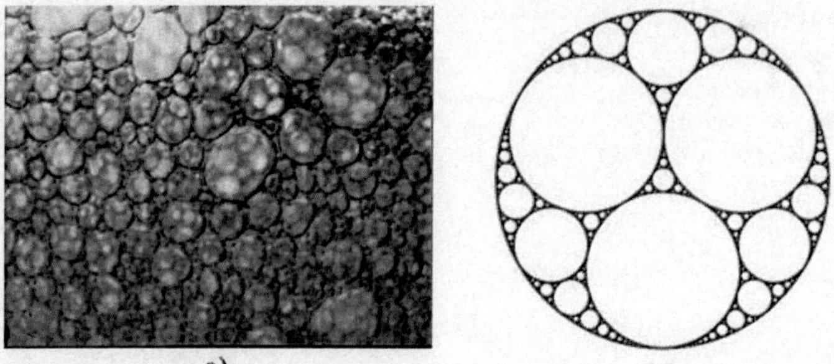

a) b)

Fig. 17.4 **a** Snapshot during beer foam decay taken at the uncoated wall of the glass after 230 s; (Photo: G. Patzelt, P.J. Plath) **b** schematic representation of an Apollonian gasket (published by Mathematica Stack Exchange [9])

during the first time-intervals. When the foam decay is progressing, rearrangement becomes dominant and many smaller bubbles are replaced by few large bubbles.

At the end of the beer foam decay, several small bubbles arrange themselves in the spaces between large bubbles (see Fig. 17.4a), which roughly corresponds to an Apollonian arrangement. (The Apollonian packing of bubbles was shown even more pronounced in earlier publications [4, 7].) In the Apollonian construction of surface coverage by circular packings, circles are embedded close together by an infinite process, with another circle drawn in each space between three touching circles [8]. A finite set of Apollonian or Leibniz packings constructed in this way is called 'Apollonian gasket' (see Fig. 17.4b).

17.3 Temporal Development of Individual Bubble Size Classes

Using the foam analysis program *FoamStar* [10], developed from the MIRChem GmbH in the group of Prof. Plath, the first 26 images of each series were evaluated with respect to the bubble size distribution within the individual bubble size classes. For this purpose, the bubble sizes were divided into 11 size classes of 50 μ each, i.e., the size class with the smallest bubbles covered the size range 0–50 μ, the next one >50–100 μ etc., and that with the largest bubbles contained all ones with a diameter larger than 500 μ.

Automatic foam analysis as well as semi-automatic identification by drawing bubble diameters into the bubble picture was used to detect the individual bubbles, determine their size and then classify them into the respective size class. The resulting absolute bubble size distribution in the case of an uncoated glass wall is shown exemplarily in Fig. 17.5 using 3-D images in two different views.

If one denotes the number of counted bubbles in the k-th size class at time t by $N_k(t)$ and the number of all counted bubbles at time t by $N(t)$, the relative frequency

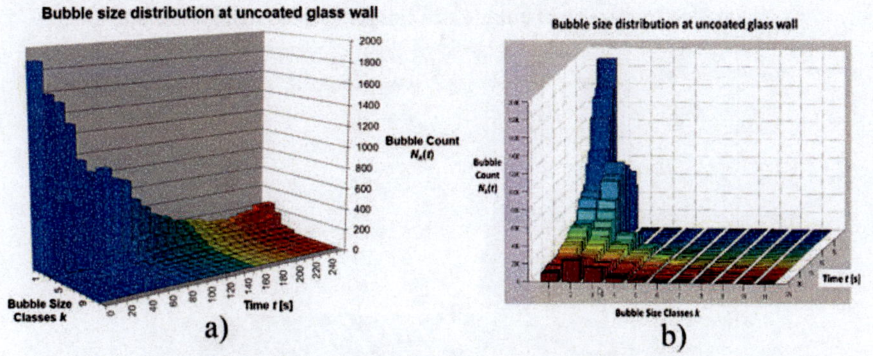

Fig. 17.5 3D-diagram of the bubble size distribution $N_k(t)$ of bubble sizes as a function of bubble size classes k and time t, obtained from images of beer foam decay at uncoated side of the glass vessel; **a** view from time axis and **b** view from axis of bubble size classes k

$p_k(t)$ at time t of bubbles in this interval is given by Eq. (17.2):

$$p_k(t) = \frac{N_k(t)}{N(t)}, k = 1, \cdots, 11 \tag{17.2}$$

The bubble size distribution $p_k(t)$ as a two-dimensional function of bubble size classes k and time t is shown in Fig. 17.6a, b for the time series of beer foam decay photographed on uncoated and coated walls, respectively.

From Fig. 17.6a, b, it can roughly be seen that the number of bubbles $N_1(t)$ in the first size class $k = 1$, which contains all bubbles immediately after foaming, is more or less monotonically decreasing within time. In the size classes with the next higher bubbles, the number of bubbles initially increases and then decreases again to some extent, while in the size classes with the largest bubbles, the number of bubbles is comparatively low over time. In particular, in the case of time series of photographs taken at the uncoated glass wall, the number of bubbles is slightly increasing again in the end of foam decay.

To get a better idea of the reaction kinetics of foam decay, the function curves of the relative frequencies $p_k(t)$ as a function of time are presented in multi-line diagrams for selected bubble size classes k, see Fig. 17.7a, b, which reproduces the time behavior described above.

17.4 Kinetic Modelling By A Multi-Step Consecutive Reaction

The time dependence of the curves shown in Fig. 17.7 can be compared with that of the concentration curves of the reaction kinetics of a multi-step consecutive reaction

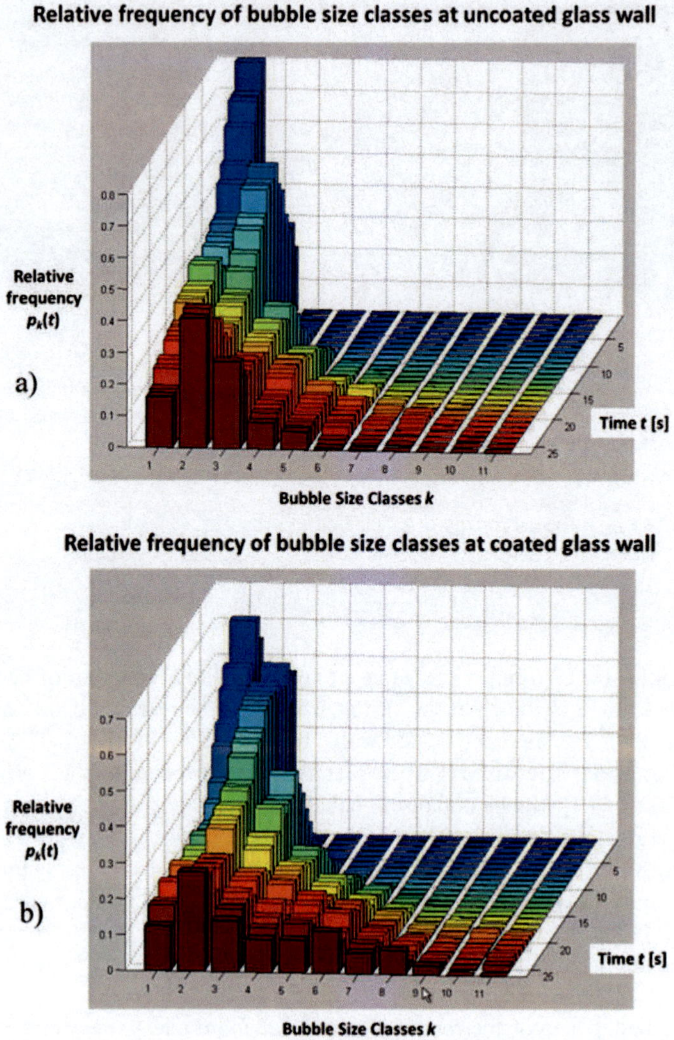

Fig. 17.6 3D-diagram of the relative frequency $p_k(t)$ of bubble sizes as a function of bubble size classes k and time t, obtained from images of beer foam decay, **a** at uncoated side of the glass vessel and **b** at coated side

$A \rightarrow B \rightarrow C \rightarrow D \rightarrow \ldots$ with no fundamental difference between the curves for foam decomposition on the uncoated and the coated vessel side.

The reaction rates of a simple consecutive reaction of first order $A \stackrel{k_1}{\rightarrow} B \stackrel{k_2}{\rightarrow} C$ with the starting substance A, the intermediate B and the end product C as well as the reaction constants k_1 and k_2, where the sum of concentrations $c_A + c_B + c_C$ is constant, are given by the well-known differential equation system (17.3),

a)

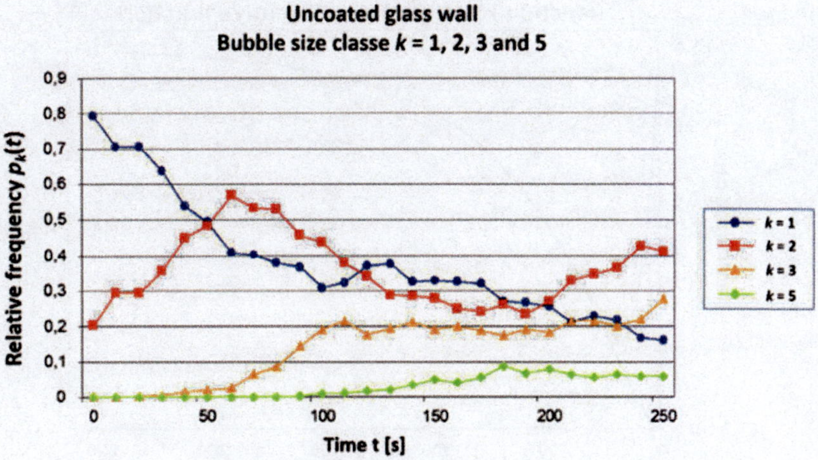

b)

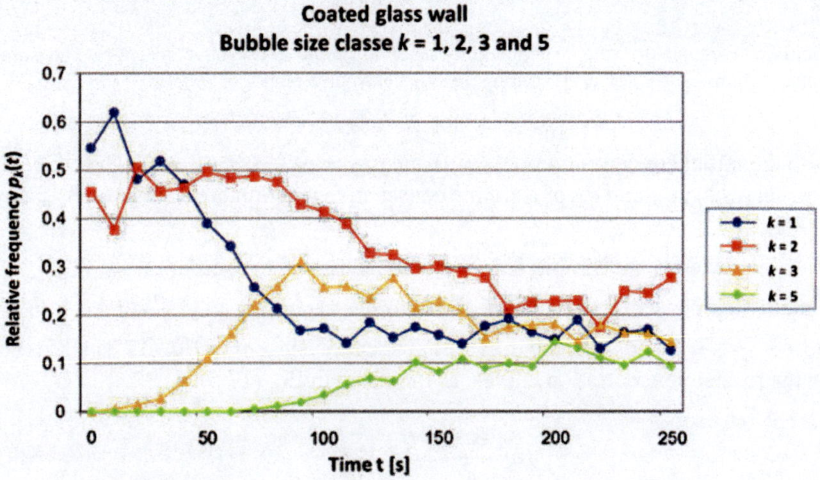

Fig. 17.7 Temporal development of the relative frequency $p_k(t)$ of bubble sizes for selected size classes k, **a** in the case of uncoated side of the glass vessel and **b** in the case of coated side

$$\begin{aligned} \frac{dc_A}{dt} &= -k_1 c_A \\ \frac{dc_B}{dt} &= k_1 c_A - k_2 c_B \\ \frac{dc_C}{dt} &= k_2 c_B \end{aligned} \tag{17.3}$$

The solution of this equation system is shown in Eq. (17.4):

$$\begin{aligned} c_A &= c_{A0} e^{-k_1 t} \\ c_B &= c_{A0} \frac{k_1}{k_2 - k_1}\left(e^{-k_1 t} - e^{-k_2 t}\right) \\ c_C &= c_{A0}\left(1 - \frac{k_2 e^{-k_1 t} - k_1 e^{-k_2 t}}{k_2 - k_1}\right) \end{aligned} \tag{17.4}$$

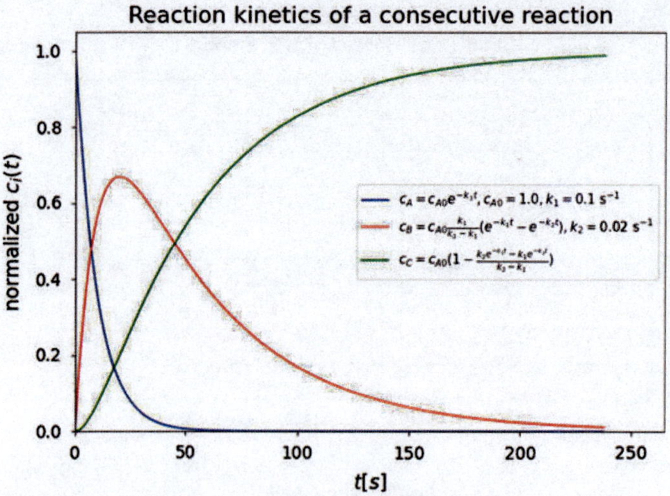

Fig. 17.8 Example for the reaction kinetics of a simple consecutive reaction. The sum of concentrations $c_A + c_B + c_C$ is constant (= 1) at each time point t

with the initial concentrations $c_{A0} > 0$ and $c_{B0} = c_{C0} = 0$ at $t = 0$. The concentration curves c_A, c_B and c_C of a simple consecutive reaction are shown as an example in Fig. 17.8.

If one extends the simple consecutive reaction of first order $A \overset{k_1}{\to} B \overset{k_2}{\to} C$ to a multi-step consecutive reaction of first order according to $A \overset{k_1}{\to} B1 \overset{k_2}{\to} B2 \overset{k_3}{\to} B3 \overset{k_4}{\to} \cdots \overset{k_n}{\to} Bn$, $n \geq 2$, the differential equation system of Eq. (17.3) is extended by the further concentration curves, as indicated in Eq. (17.5):

$$
\begin{aligned}
&\text{(a)} \quad \frac{dc_A}{dt} = - \ k_1 c_A \\
&\text{(b)} \quad \frac{dc_{B1}}{dt} = k_1 c_A - k_2 c_{B1} \\
&\text{(c)} \quad \frac{dc_{B2}}{dt} = k_2 c_{B1} - k_3 c_{B2} \\
&\text{(d)} \quad \frac{dc_{B3}}{dt} = k_3 c_{B2} - k_4 c_{B3} \\
&\quad \ldots \\
&\text{(e)} \quad \frac{dc_{Bn}}{dt} = k_n c_{Bn-1}, n \geq 2
\end{aligned}
\tag{17.5}
$$

This corresponds formally to the reaction kinetics of the consecutive intermediates $B1$, $B2$ and $B3$ and the final product Bn. Here, again the sum of all concentrations is constant and the concentration of the final product approaches this constant at the

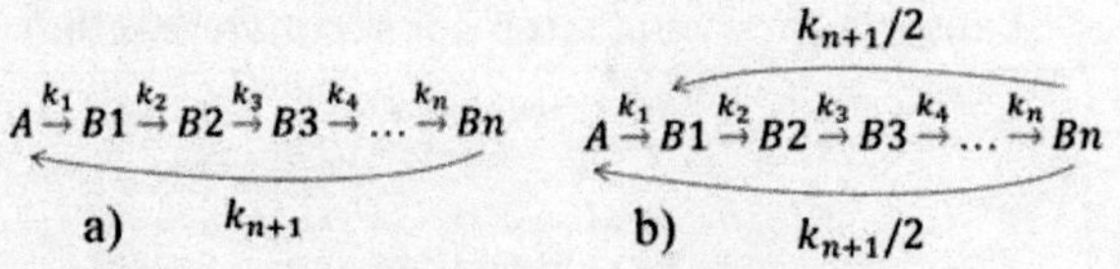

Fig. 17.9 Consecutive kinetics with feedback; **a** to initial bubble size class $k = 1$ and **b** to bubble size classes $k = 1, 2$. The sum of all concentrations is constant

end of the reaction for large t-values, whereas those of the other substances converge to zero (compare Fig. 17.8.)

On the other hand, the time developments of the relative frequencies $p_k(t)$ of the bubble size classes k tend to constant values greater than zero at the end of foam decay and, especially in the case of the uncoated wall, increase slightly again for small k. To reproduce this time behavior with consecutive kinetics, we introduce feedback of the concentration of the final bubble size class to that of the initial classes (see e.g., Fig. 17.9).

The rows (a) and (e) or (a), (b) and (e) of Eq. (17.5) then change according to Eqs. (17.6) and (17.7), respectively.

$$
\begin{aligned}
\text{(a)}\quad \frac{dc_A}{dt} &= -\ k_1 c_A + k_{n+1} c_{Bn} \\
&\dots \\
\text{(e)}\quad \frac{dc_{Bn}}{dt} &= k_n c_{Bn-1} - k_{n+1} c_{Bn}
\end{aligned}
\tag{17.6}
$$

$$
\begin{aligned}
\text{(a)}\quad \frac{dc_A}{dt} &= -\ k_1 c_A + {}^{k_{n+1}}/_{2} c_{Bn} \\
\text{(b)}\quad \frac{dc_{B1}}{dt} &= k_1 c_A - k_2 c_{B1} + {}^{k_{n+1}}/!_{2} c_{Bn} \\
&\dots \\
\text{(e)}\quad \frac{dc_{Bn}}{dt} &= k_n c_{Bn-1} - k_{n+1} c_{Bn}
\end{aligned}
\tag{17.7}
$$

Examples for both, Eqs. (17.6) and (17.7), are shown in Figs. 17.10 and 17.11, where all reaction constants k_i are set to 1.0; the sum of all concentrations is constant equal $c_A(t = 0)$.

Figures 17.10 and 17.11 show that the temporal developments of the relative frequencies $p_k(t)$ of bubble size classes k can in principle be described by consecutive kinetics with feedback. By varying the reaction constants k_i, a more detailed fit to the reaction patterns is possible as demonstrated in Fig. 17.12.

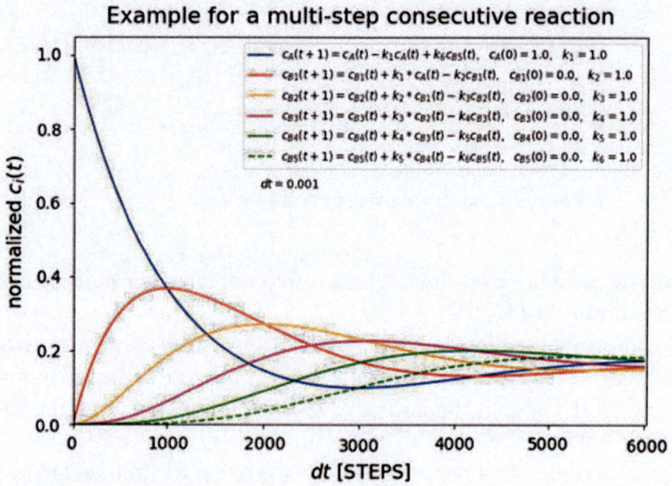

Fig. 17.10 Example for the consecutive reaction kinetics with feedback of bubble size class $k = 6$ to bubble size class $k = 1$ and all reaction constants $k_i = 1.0$. The sum of all concentrations is 1.0 for all STEPS dt

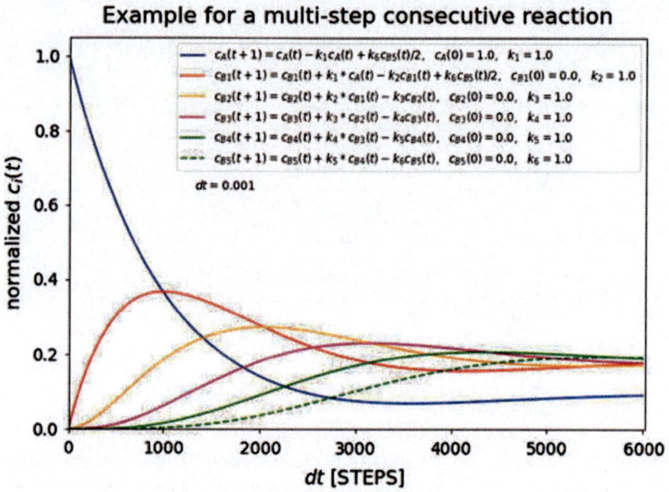

Fig. 17.11 Example for the consecutive reaction kinetics with feedback of bubble size class $k = 6$ to bubble size classes $k = 1$ and $k = 2$ and all reaction constants $k_i = 1.0$. The sum of all concentrations is 1.0 for all STEPS dt

17.5 Segregation and Agglomeration of Bubbles

We would like to emphasize at this point that the temporal developments of foam decay shown in Figures 17.6 and 17.7 are not due to concentrations of chemical

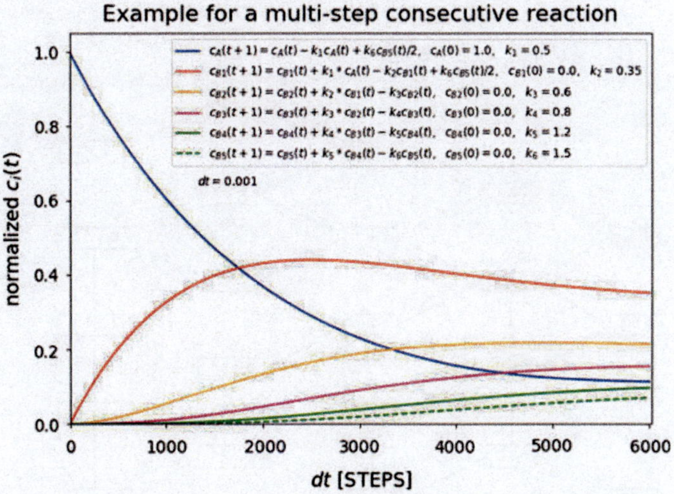

Fig. 17.12 Example for the consecutive reaction kinetics with feedback of bubble size class $k = 6$ to bubble size class $k = 2$ with varying reaction constants k_i. The sum of all concentrations is 1.0 for all STEPS dt

substances, but rather to time dependencies of the relative frequencies $p_k(t)$ of the bubble size distribution in the bubble size classes k.

Another method of visualizing the evolution of such distributions is applying diagram or partition lattices according to Ruch [11] which we have already used extensively in our previous work [4, 12–14]. Figure 17.13 shows exemplarily the 11-dimensional partition lattice with the time sequence of the rounded bubble size distributions determined from the relative frequencies $p_k(t)$ and normalized to $k = 11$ at the individual time points t during foam decay on the uncoated glass walls.

From both, the kinetics and the course in the partition lattice, one can see that the numbers of small bubbles decrease in the beginning of foam decay, but increase again at the end. Furthermore, as the bubble sizes grow in the initial time intervals, they are more uniformly distributed than at the end of the foam decay where segregation and agglomeration of bubbles takes place. These effects are particularly pronounced when decaying on uncoated glass walls.

In our previous publication, we explained this behavior by drainage in the beginning of foam decay and reorganization of bubbles to Apollonian arrangements at the end. But what could be the reason that bubbles grow to a certain size or agglomerate and in the spaces in between many smaller bubbles are distributed (see Fig. 17.14)?

We assume that during foam decay, especially on uncoated glass walls with a surface that is not smooth at all, a few large bubbles will burst at the defects of the glass surface, i.e., they will simply be broken. The bursting of these large bubbles then leads to the formation of many small bubbles again. It should be emphasized that this only happens at the end of the foam decay, making the difference between uncoated and coated glass walls. In our kinetic model of a multi-step consecutive

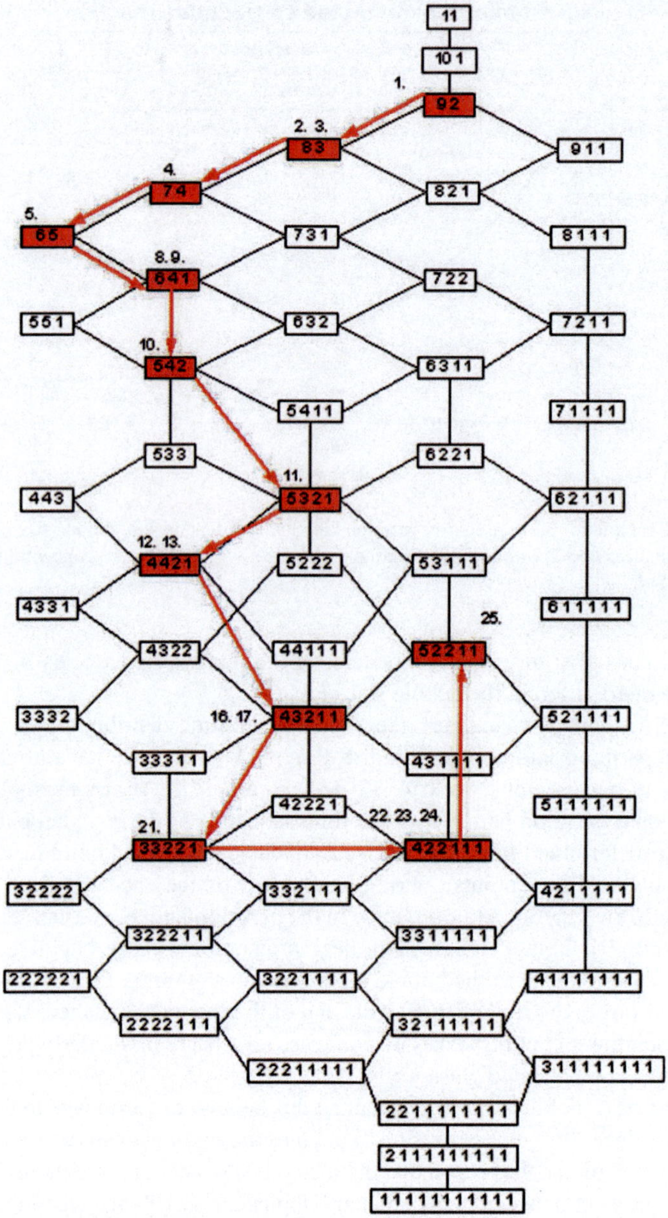

Fig. 17.13 11-Dimensional partition lattice with the time sequence of the rounded bubble size distributions at each time points t during beer foam decay on the uncoated glass walls. The partitions of bubble size distributions are shown in the boxes with red background; the numbers of the time steps t are indicated above or to the right of the boxes. The red arrows denote the transition between the different time steps

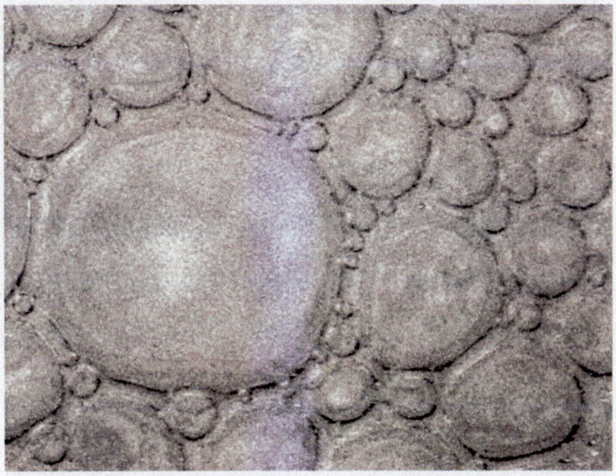

Fig. 17.14 Detailed representation of an Apollonian bubble size arrangement at the end of foam decay. (Photo processed by P. J. Plath)

reaction, this effect of segregation is simulated by the feedback of bubble size classes with larger k-values to those of smaller k-values. In the case of the partition lattice, this corresponds to transitions to higher ordered distributions.

Furthermore, we suppose that this decay of single large bubbles into many small bubbles occurs spontaneously and is a fractal event. Referring to a work by Schwietering on fractal sounds [15], we would like to compare this process with an ice cube thrown into a liquid producing a strange sequence of cracking sounds, or with a drop of water falling into a hot pan where it explodes into many smaller drops, then these drops fall back into the oil and explode again, etc., until everything is vaporized.

The beer foam decay of the liquid enriched with CO_2 bubbles, which is characterized by increasing bubble sizes at its beginning and spontaneous decay of individual large bubbles at its end, can also be interpreted as *a liquid/gas system that segregates itself*. In the following Chap. 18 of this book, we discuss another self-segregating system in which alkali feldspar and plagioclase are separated in the liquid granitic magma under suitable pressure/temperature conditions.

Well then, cheers!

References

1. Leike, A.: Demonstration of the exponential decay law using beer froth. Eur. J. Phys. **23**, 21–26 (2002)
2. Dale, C.J., Walker, S.G., Lyddiatt, A.: Dynamic changes in the composition and physical behaviour of dispensed beer foam. J. Inst. Brewing **99**, 461–466 (1993)
3. Sauerbrei, S.: Die Dynamik schnell zerfallender Schäume—Untersuchungen an Bierschaum. Diplomarbeit an der Universität Bremen (2004)
4. Sauerbrei, S., Haß, E.C., Plath, P.J.: The apollonian decay of beer foam. bubble size distribution and the lattices of young diagrams and their correlated mixing functions. Discret. Dyn. Nat. Soc. Article ID 79717, 1–35 (2006)

5. Patzelt, G.: Forschungspraktikum SS 2006, Betreuer P.J. Plath. Universität Bremen (2006) (unpublished results)
6. Patzelt, G.: Zerfall von Bierschaum – Dynamik auf Partitionsdiagrammen. Diplomarbeit, Universität Bremen (2007)
7. Haß, E.C., Ottensmeyer, R. und Wittstock, A.: Schaumcharakterisierung mit optischen Messtechniken und PC-gestützter Bildverarbeitung. tm Technisches Messen **71**(11), 603–612 (2004)
8. Mandelbrot, B.B.: The Fractal Geometry of Nature, pp. 186–188. W.H. Freeman and Company, San Francisco (1977)
9. https://mathematica.stackexchange.com/questions/79239/minimalistic-code-challenge-on-apollonian-gaskets (Accessed 28 Aug 2021)
10. Haß, E.C. und Wittstock, A.: Bedienungsanleitung *FoamStar* 1.5. MIR-Chem GmbH, 10 Feb 2021
11. Ruch, E.: The diagram lattice as structural principle. Theoret. Chim. Acta (Berl.) **38**, 167–183 (1975)
12. Sauerbrei, S., Sydow, U., Plath, P.J.: On the characterization of foam decay with diagram lattices and majorization. Z. Naturforsch. **61a,** 153–165 (2006)
13. Sauerbrei, S., Plath, P.J.: Diffusion without constraints. J. Math. Chem. **42**, 153–175 (2007). https://doi.org/10.1007/s10910-006-9085-x
14. Sauerbrei, S., Knicker, K., Hass, E.C., Plath, P.J.: Weak majorization as an approach to non-equilibrium foam decay. Appl. Math. Sci. **1**(11), 527–550 (2007)
15. Schwietering, J.: Fractal sounds. In: Encarnação, J.L., Peitgen, H.-O., Sakas, G., Eglert, G. (eds.) Fractal Geometry and Computer Graphics, pp. 64–73. Springer Verlag Berlin, Heidelberg, New York (1992)

Chapter 18
Rapakiwi Granite—An Ancient Fossilized Liesegang Experiment?

Peter J. Plath

Fig. 18.1 Karelian red—a Rapakivi granite in our kitchen, with the inner ellipse diameters 5.3 and 6.2 cm (*Photo* P.J. Plath)

When we bought our old house, which stands on large granite stones as is customary in the country, we had a kitchen buffet made of red granite built into it (Fig. 18.1). It matched the style of the house and we liked it. I hadn't thought about the pattern of this granite. However, when I was writing the chapter on Liesegang structures, one day I noticed the special structures in our kitchen buffet: are this also Liesegang patterns? So, first I asked our stonemason what kind of granite he had sold us at the time. "That's Karelian Red" he told me. This not exactly comprehensive information

P. J. Plath et al., *Imagery Synergetics*, Understanding Complex Systems,
https://doi.org/10.1007/978-3-030-95607-3_18

still helped me. The granite "Karelia red" occurs on the border between Finland and Russia and is a popular granite for representative buildings.

The Karelia red granite belongs to the group of Rapakivi granites [1] (rapakiwi, Finnish: bad or rotten stone) which are named this way because of its light weathering. It contains oval inclusions made from orthoclase feldspars, which are often surrounded by a plagioclase fringe. They originated in the Young Algonkian at the end of the Precambrian [2].

However, Karelian Red, also called Baltic Red [3], in our kitchen is a frost-resistant and polishable natural stone which is distributed worldwide. Baltic Red was created 1.2 billion years ago in the Precambrian. The large, rounded potash feldspar deposits are randomly distributed in the granite and usually contain biotite crystals. Around the large ovoide, there is a border of red potassium feldspars, quartz and biotite [4].

There is a rather extensive geological literature on the formation of Rapakivi granite, especially of the Wyborg type (Wyborgite), which also includes the Baltic Red Granite [5]. In particular, the question of how the round alkali feldspar crystals (ovoid) come about, which are quite often surrounded by plagioclase and quartz crystals, is discussed. The other question deals with the formation of rings, e.g. from quartz crystal, in the feldspar ovoides.

It is assumed that "the cores of the investigated feldspars and quartz crystals were formed under pressures of 5–6 kbar (at a depth of around 20 km) and temperatures of around 680–720 °C. The outer parts of the same crystals only formed at pressures of 1–2.5 kbar, i.e. in the uppermost crust of the earth and at temperatures of 650–750 °C [5, 6]."

In order to explain such a sequence of rings in the ovoid as shown in Fig. 18.2 in this way, an extremely complicated sequence of geological events is required, which, however, seems to me to be quite unlikely.

"To explain such formations, the alkali feldspars must have been repeatedly exposed to changing environmental conditions. Presumably these crystals got into different hot or chemically differently composed areas through convection movements within the magma. An intense movement of the crystals in the magma until shortly before solidification would also make the juxtaposition of ovoides with and without fringes and the appearance of ovoides in the midst of idiomorphic crystals more understandable [5]."

In my kitchen countertop, there are a large number of very different, large and small ovoides with none or several rings, and they are close together. Looking on our own experiments on Liesegang ring formation (compare Chap. 3 in this book), I come to the assumption that there has to be a much simpler explanation for the rings in the feldspar ovoides!

I just see my granite countertop as my very, very old experimental laboratory. The Wiborg Parakivi granite, i.e. my Karelia red granite slab, is approx. $1.45–1.65 \cdot 10^9$ years old [5]. I can't change anything about the experiments that were carried out in the ascending magma at that time. However, I can compare them with experiments that could be carried out today.

Fig. 18.2 Karelian Red ovoide with several internal plagioclase rings and grey plagioclase areas in the outer part (top left) and dark brown red quartz crystals around the ovoide. The central alkali feldspar ovoide is by no means a monolithic single crystal but consists of many crystals. (*Photo* P. J. Plath)

18.1 Segregation of a Fluid and Subsequent Ostwald Ripening

When I see how densely and irregularly the many different sized feldspar ovoides lie in front of me on the kitchen worktop, an image immediately comes to my mind: *a liquid that segregates itself.*

The liquid granitic magma will segregate at a suitable pressure and/or temperature if the conditions overstep a critical point. If this process runs slowly enough (very carefully slowing down the parameters, otherwise segregation would not appear to such an extent), according to the process of Ostwald ripening, the many very small droplets that were created first will dissolve again in favor of fewer larger droplets, which will grow more and more. At the end of this process, only the more or less large feldspar ovoides will remain.

18.2 Transformation to a Liesegang System

In this way, the feldspar ovoides are created as large droplets of liquid with a high surface tension compared to the surrounding silicate liquid. It is easy to imagine that an amorphous, very tough feldspar gel is created from the liquid feldspar in the ovoides when the pressure and temperature conditions change slightly.

In the areas between the feldspar ovaries transformed into a gel, i.e. their surroundings, there is still the other, may be still liquid, silicate part of the now separated granitic magma. That would be an almost classic Liesegang arrangement, as we know it from test tube experiments. In the case of the corresponding 2D Petri dish experiments, the electrolyte outside the gel system is usually in the center. This electrolyte can definitely be present in solid form as a salt, as the original experiments by Liesegang show. In our case, the "outer electrolyte" is located in the spaces between the feldspar gel ovaries. This is an *inverse Liesegang arrangement*. The "outer electrolyte" is really outside the gel body.

In the classic Liesegang experiment, ions of the "outer electrolyte" diffuse into the gel and react there with the ions of the "inner electrolyte" present here, causing a precipitation reaction. These reaction diffusion systems, in which locally fixed pattern formation of the precipitates occurs, usually has ring disks in the test tube or a sequence of rings around the center of the external electrolyte in the 2D case of the Petri dish experiments or spherical shells around the center of the sphere in suitable 3D experiments. (See Chap. 3 in this book).

If in case of inverse Liesegang arrangement, ions migrate from the surface of the ovoid into the center, which lead to precipitation reactions, then this should also lead to the formation of ovoid surfaces or spherical surfaces in the body of the ovoid as well. Of course, these do not have to be closed areas. They can also have a fractal structure, so that local accumulation of precipitation products can occur within the respective areas.

Now the gel itself is a thermodynamically unstable arrangement. The gel will crystallize if it only has enough time for it, e.g. if it does not dry out in the meantime. But the precipitated products will also change from the amorphous to the crystalline state for the same reason because of their instability. This would happen primarily in the places where the larger accumulations in the fractal spherical surfaces were previously.

What kind of "electrolyte" could it be in this case? The alkali feldspars found in the basic body of the ovoid can be described by the general formula: $Na_xK_{1-x}Al\ Si_3O_8$, while the plagioclase feldspars, which especially form the edges of the ovides, are classified by the formula: $Na_xCa_{1-x}Al_{2-x}Si_{2+x}O_8$. If the potassium ions of the alkali feldspar are exchanged by calcium ions, then the alkali feldspars cum grano salis are converted into plagioclase feldspars. Figure 18.3 shows that this is not a simple process, but is accompanied by a variety of conversions. Even more, there can be miscibility gaps and associated segregations! In addition, the alkaline feldspars and the plagioclase feldspars differ in their crystal dress [7, 8].

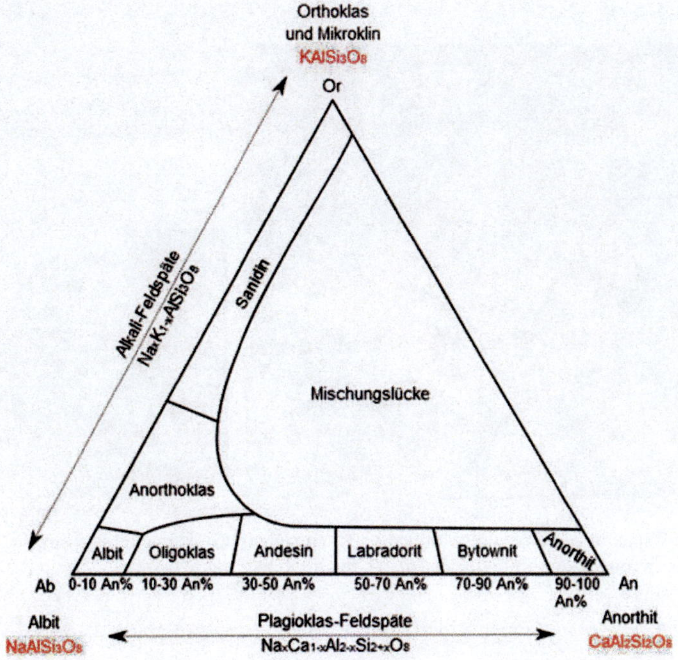

Fig. 18.3 Ternary phase diagram of the feldspars (at 900 °C) modified from N.N. Greenwood, A. Earnshaw—Chemistry of Elements (1998) - S. 357, Pergamont Press (1990)-S. 414 [9]

The phase diagram in Fig. 18.3 shows the situation in the rising and cooling magma only in a very simplified manner, because it only takes into account the situation at 900 °C and does not take the pressure into account. If the two physical quantities pressure and temperature are explicitly taken into account, further miscibility gaps occur.

The Parakiwi Karelia red, however, shows a lot of well-developed biotite crystals in the edge of the ovoides. These are also known under the names magnesium iron mica or dark mica with the general formulae: $K(Mg,Fe^{2+},Mn^{2+})_3[(OH,F)_2|(Al,Fe^{3+},Ti^{3+})Si_3O_{10}]$ (Fig. 18.4).

Biotite crystallizes from rising igneous rock melt just before feldspar and quartz when the magma cools to around 800–700 °C (phase of main crystallization). It forms parageneses with potassium feldspar and plagioclase. So it is geologically often associated with these minerals [11]. The biotite belongs to the leaf or sheet silicates where one silicon atom of the underlying Si_4O_{10} unit is replaced by an aluminum atom [12]. Even if the biotites according to Strunz are no longer regarded as independent minerals but as mixed crystals of the phlogopite $KMg_3[(F,OH)_2|AlSi_3O_{10}]$ [13], this does not change our argumentation.

Now I have almost all the necessary ingredients together to explain the idea of the self-organized Liesegang structures of the Rapakiwi Karelian Red on my kitchen table in a comprehensible way. The leafy aluminosilicate biotite contains, among

Fig. 18.4 **a** Rapakiwi Karelian red; biotite crystals surround the alkaline feldspar ovoid. The inner grey Liesegang ring made of plagioclase also contains some biotite crystals. (*Photo* P. J. Plath); **b** Biotite crystal [10]

other things, very easily mobile divalent ions that are required to balance the charge for the aluminum atoms built into the layered silicate. They represent the source for the diffusion of calcium ions or magnesium ions, which are responsible for the formation of the pagioclase rings in the feldspar gel or in the tough, just liquid alkali feldspar.

The precipitation of compounds insoluble in aqueous gels, which is so typical for the Liesegang structures, is replaced in the Parakiwi by the precipitation or crystallization of plagioclase in the still liquid or gel-like alkali feldspar. Under these conditions, there is a strong miscibility gap between alkali feldspar and plagioclase, which corresponds to the insolubility of the precipitation products in the classic Liesegang experiment.

18.3 Special Plagioclase Liesegang Patterns

It is plausible to assume that when the solutions of alkali feldspar and plagioclase are separated, the previously crystallized biotite mica is localized in the plagioclase solution.

The segregation of the liquid igneous rock can be compared well with the formation of the beer foam (see Chap. 17). If the beer bottle is opened, the beer/CO_2 solution separates and the CO_2 gas bubbles are created, which are surrounded by the liquid lamellae of the beer. This is the beer foam. It is well known that foams of this type are very well suited to separating wettable particles from non-wettable ones, which is used in the many flotation processes in engineering and in the household. An example of this is ore enrichment through flotation, e.g. of low-grade copper ores, where the valuable copper ore in the lamellae of the foam is separated from the lighter gangue, which sinks to the bottom [14].

In our case of the separating igneous liquid, the still liquid alkali feldspar ovaries are comparable to the gas bubbles and the still liquid plagioclase fringes correspond to the liquid lamellae of the foam. The previously crystallized biotite mica crystals are held in them (see Fig. 18.5).

Like the plagioclase lamellae, these biotite mica crystals now represent the reservoir for the diffusion of divalent cations such as Ca^{2+} and Mg^{2+}, Fe^{2+} into the alkali feldspar.

A first precipitation or crystallization of plagioclase in the interior of the alkali feldspar gel occurs when the "solubility product" is exceeded, i.e. when the alkali cations Na^+ and K^+ present there are exchanged by the divalent alkaline earth cations in a sufficient number of alkali-feldspar units. Since these compounds are not miscible with the alkali feldspar, they must precipitate as separate very small plagioclase and biotite crystals.

If the process proposed here, which is very similar to the Liesegang precipitation process, describes the situation largely correct, the corresponding pattern formation phenomena as in the classic Liesegang experiments should also be observed. This includes in particular the formation of several consecutive rings. In Figs. 18.1 and 18.2, this form of pattern formation can be seen quite well and thus fully confirms our assumptions.

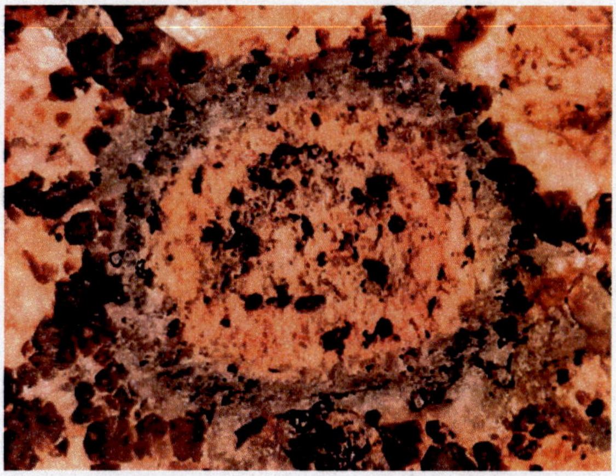

Fig. 18.5 Karelian red; the gray plagioclase hem of the alkali feldspar ovoides (size of the axes of the inner ellipse are 1.5 cm and 2.0 cm) can be seen as the "zeroth Liesegang ring". This ring is surrounded on the outside by biotite crystals. The "first" plagioclase Liesegang ring inside the ovoid is also clearly recognizable, especially since it is associated with the biotite mica. Several neighboured alkali feldspar ovoides separated by plagioclase lamellae containing biotite crystals mark the picture. (*Photo* P. J. Plath)

18.4 Fixation of the Pattern by Crystallization

It is regarded as an unmistakable sign of the presence of a single crystal in granite when the accessible surface reflects light as a compact surface [5]. Figure 18.6 shows, using the example of an unworked or unpolished surface of Karelian Red Granite, that it only reflects light at a certain angle, while otherwise it appears to the observer as a rough, red surface. In Fig. 18.6a the hexagonal structure of the single crystal of alkali feldspar is clearly recognizable. However, there are no Liesegang pattern observable.

The formation of such large single crystals is possible if the time for the transformation of the gel or the viscous liquid is long enough and if no other crystal formation, such as e.g. by the formation of plagioclase or biotite, the formation of the feldspar crystals disturbs. An alkali feldspar single crystal that fills the entire ovoid and at the same time contains one or more plagioclase rings with biotite crystals is a contradiction in terms. Of course, three-dimensional lattice defects can also occur in single crystals, but these would not take on macroscopic dimensions.

"Inclusions are solid foreign phases that belong to this type of three-dimensional lattice defects in the crystal. Excretions (precipitates) are special cases of inclusion, in which the foreign phase from the crystal itself is formed. This is the case, for example, when the minority component in the interior of the crystal forms its own phase when a solid solution cools. Since volume defects distort the surrounding

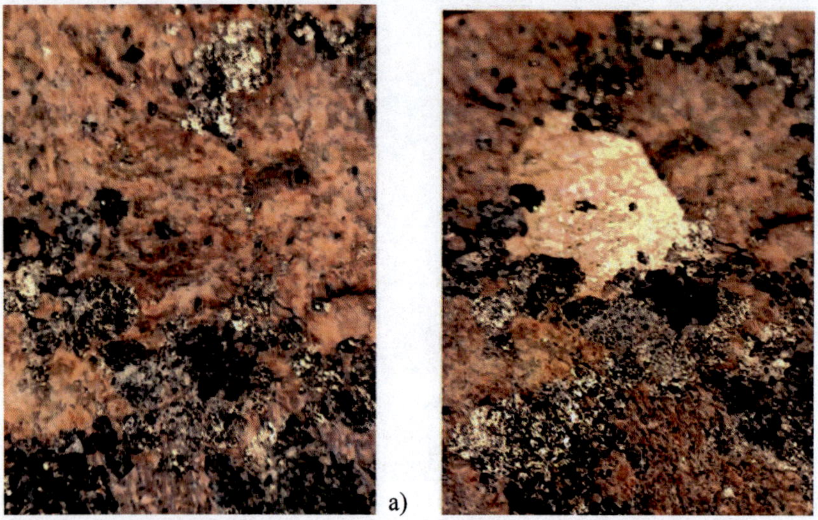

a) b)

Fig. 18.6 Just the same unpolished Karelian red granite slab: **a** two larger, separate areas of alkali feldspar can be seen, only the biotite mica flakes glitter in the light; **b** by varying the angle of incidence of light, one of the two alkali feldspar areas can be recognized as a single crystal through the reflection. The length of the reflecting single crystal is about 2.0 cm (*Photo* P. J. Plath) crystal, they are surrounded by a zone of higher concentration of low-dimensional lattice defects [15]."

However, if these three-dimensional lattice defects reach macroscopic sizes, this should no longer be understood as a disruption of the original crystal lattice, whose lattice should have the translational symmetry on the length scale of the unit cells of the crystal. In the macroscopic case of the disturbance, however, this translation symmetry is no longer given, not even piece by piece. In the case of Fig. 18.7a, the

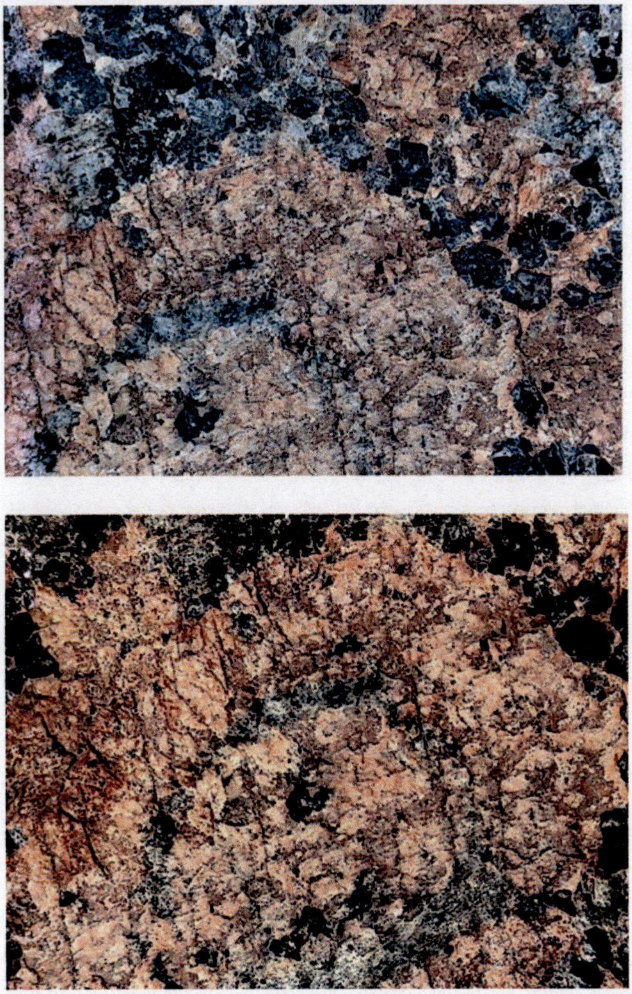

a)

b)

Fig. 18.7 Just the same Alkali feldspar ovoide of unpolished Karelian Red with a Liesegang ring in the ovoid; **a** glimpse of reflections on the different crystals; **b** no reflection at all; However, one can detect all the different single crystals of alkali feldspars and plagioclase and biotite slight degree of reflection shows that the orientation of the crystal planes is nevertheless retained in individual sub regions. Nevertheless, due to the "free" orientation of a large number of small crystals in the ovoid to a certain extent, these can be filled in practically any shape without the need for a previous melting of these areas, as is sometimes claimed [5].

These considerations show, however, that the Liesegang pattern formation precedes the crystallization and cannot only take place in its aftermath. Liesegang's crystallization of plagioclase in the alkali feldspar gel is therefore also the reason for the three-dimensional lattice disorder and, as a result, the poly-crystallinity. The formation of the biotite crystals fits into this picture of the crystalline Liesegang pattern formation, because the "easy" displaceability of the crystal planes in the mica enabled them to adapt to the occurring shear forces particularly well. Biotite or phlogopite is still stable even at a pressure of 70 kbar, which corresponds to a depth in the earth of over 200 km.

"Biotite or phlogopite is still stable even at a pressure of 70 kbar, which corresponds to a depth in the earth of over 200 km" [13]. "Biotite and other phyllosilicates tend to recrystallize laterally when exposed to hot pressure, while quartz and feldspars tend to remain granular. The cleavage areas correspond to the direction of the maximum tectonic shear forces [11]."

18.5 Three Dimensional Plagioclase Hems of the Alkali Feldspar Ovoides

Three dimensional plagioclase hems of the alkali feldspar ovoides.

There is a discussion about why the plagioclase fringes of the alkali feldspars are present one time and completely absent another time, even in the same stone [5].

It seems that the disputants are completely overlooking that the feldspar ovaries are obviously three-dimensional, not two-dimensional structures, and that the plagioclase hems, whether thin or thick, have a certain width. If you now carry out corresponding flat cuts, spherical sections can arise that sometimes show very thick plagioclase hems (cf. Fig. 18.8) while others hardly show such rings. The plagioclase hems should not be imagined as closed spherical shells, because the liquid plagioclase lamella ruptures when it crystallizes.

Furthermore, it can happen that the cuts are so close below the plagioclase border that no Liesegang rings are affected, but this does not mean that there was no Liesegang structure formation here (cf. Fig. 18.8a).

Fortunately, when cutting my 2.8 cm thick Rapakiwi Karelian red kitchen worktop, a few larger alkali feldspar ovaries were cut through in such a way that the three-dimensional shape of the Liesegang structures was clearly preserved (see Fig. 18.9).

It seems to be certain that the structure formation in the alkali feldspar ovides of the Rapakiwi granites of the Karelian Red type but also in the Karelian Brown type can be understood as a Liesegang pattern formation. However, much remains to be

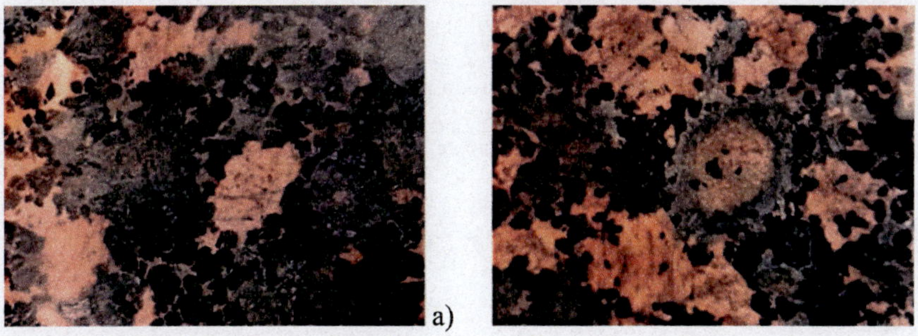

a) b)

Fig. 18.8 Plagioclase hems of the alkali feldspar ovoides with sections in which the stonemason accidentally ground off only the outer shell of the alkali feldspar ovoides located in the center of the figures. **a** Visible part of the ovoide, length 1.5 cm, width 0.7 cm; **b** visible part of the ovoide, length 1.5 cm, width 1.2 cm

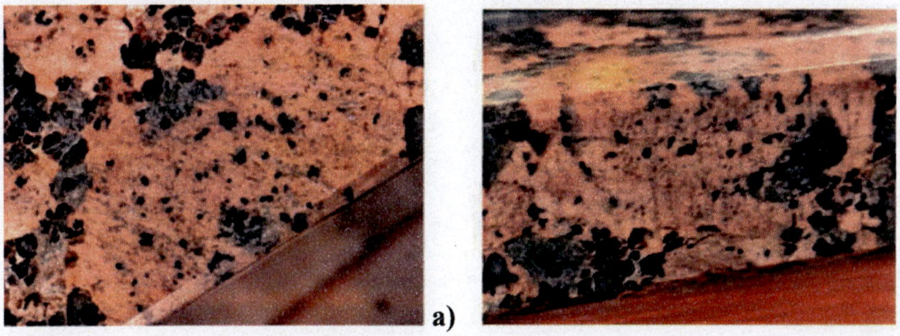

a) **b)**

Fig. 18.9 "Pseudo three dimensional picture" of Rapakiwi granite Karelian Red in my kitchen with an alkali Feldspar ovoid with up to five Liesegang shells of plagioclase and Biotite; the granite slab was photographed (Size of the ovoid along the edge; 4.5 cm; the thickness of the worktop is 2.8 cm); **a** from above and **b** from the side

researched, e.g. the acidity has not been discussed here, nor has the complex temperature, pressure and mixing behavior of the many components been comprehensively discussed. This open ending is perfectly in the spirit of this book.

References

1. https://de.wikipedia.org/wiki/Rapakiwi (Read: 06 Aug 2021)
2. Bülow, von K., Pietzsch, K.: Präkambrium. In: Die Entwicklungsgeschichte der Erde—Brockhaus—Taschenbuch der Geologie, pp. 22–23. VEB F.A. Brockhaus Verlag, Leipzig (1955)
3. https://de.wikipedia.org/wiki/Baltik_Rot (Read: 06 Aug 2021)
4. https://www.kristallin.de/rapakiwis/rapeinlei.htm (Read: 06 Aug 2021)
5. https://www.kristallin.de/rapakiwis/rapeinlei2.htm#Anker1 (Read: 07 Aug 2021)

6. Eklund, O., Shebanov, A.D.: The origin of rapakivi texture by sub isothermal decompression. Precambr. Res. **95**, 129–146 (1999)
7. Särchinger, H.: Geologie und Gesteinskunde, p. 16. Volk und Wissen Volkseigener Verlag, Berlin, (1952)
8. Kleber, W.: Einführung in die Kristallographie, pp. 62, 64. VEB Verlag Technik, Berlin (1961)
9. https://de.wikipedia.org/wiki/Feldspat (Read: 15 Aug 2021)
10. https://nds.wikipedia.org/wiki/Biotit
11. https://de.wikipedia.org/wiki/Biotit (Read: 16 Aug 2021)
12. Hollemann, A.F., Wiberg, E.: Lehrbuch der Anorganischen Chemie, pp. 328–330, and 378. Walter de Gruyter & Co., Berlin (1960)
13. https://de.wikipedia.org/wiki/Phlogopit (Read: 16 Aug 2021)
14. Hollemann, A.F., Wiberg, E.: Lehrbuch der Anorganischen Chemie, p. 450. Walter de Gruyter & Co., Berlin (1960)
15. https://de.wikipedia.org/wiki/Gitterfehler (Read: 18 Aug 2021) (Engl. lattice defects)